12/22

Numbers

Numbers

A CULTURAL HISTORY

Robert Kiely

2230

An Imprint of ABC-CLIO, LLC
Santa Barbara, California • Denver, Colorado

Copyright © 2022 by ABC-CLIO, LLC

All rights reserved. No part of this publication may be reproduced, stored in a retrieval system, or transmitted, in any form or by any means, electronic, mechanical, photocopying, recording, or otherwise, except for the inclusion of brief quotations in a review, without prior permission in writing from the publisher.

Every reasonable effort has been made to trace the owners of copyright materials in this book, but in some instances this has proven impossible. The editors and publishers will be glad to receive information leading to more complete acknowledgments in subsequent printings of the book and in the meantime extend their apologies for any omissions.

Library of Congress Cataloging-in-Publication Data

Names: Kiely, Robert, 1962– author.
Title: Numbers : a cultural history / Robert Kiely.
Description: Santa Barbara, California : ABC-CLIO, [2022] |
　　Includes bibliographical references and index.
Identifiers: LCCN 2022000456 (print) | LCCN 2022000457 (ebook) |
　　ISBN 9781440869334 (hardcover) | ISBN 9781440869341 (ebk)
Subjects: LCSH: Numeration—History. | Number theory—History. |
　　BISAC: HISTORY / Ancient / General | HISTORY / Modern / General
Classification: LCC QA141.2 .K54 2022 (print) | LCC QA141.2 (ebook) |
　　DDC 513.509—dc23/eng20220521
LC record available at https://lccn.loc.gov/2022000456
LC ebook record available at https://lccn.loc.gov/2022000457

ISBN: 978-1-4408-6933-4 (print)
　　　978-1-4408-6934-1 (ebook)

26 25 24 23 22　　1 2 3 4 5

This book is also available as an eBook.

ABC-CLIO
An Imprint of ABC-CLIO, LLC

ABC-CLIO, LLC
147 Castilian Drive
Santa Barbara, California 93117
www.abc-clio.com

This book is printed on acid-free paper ∞

United States of America

513.5
KIELY

Contents

Preface vii

Introduction ix

CHAPTER 1
Numbers in Ancient Mesopotamia *1*

CHAPTER 2
Numbers in Ancient Egypt *25*

CHAPTER 3
Numbers in Ancient Greece *47*

CHAPTER 4
Numbers in the Hellenistic Mediterranean *75*

CHAPTER 5
Numbers in Traditional China *101*

CHAPTER 6
Numbers and the Classical Maya *127*

CHAPTER 7
Numbers in Ancient and Medieval India *145*

CHAPTER 8
Numbers in the Medieval Arabic World *167*

CHAPTER 9
Numbers in Medieval Europe 193

CHAPTER 10
Numbers in Early Modern Europe 221

CHAPTER 11
Numbers in 18th- and 19th-Century Europe 253

CHAPTER 12
Women and Numbers 285

CHAPTER 13
Numbers in the 20th Century 307

Appendix: A Brief Look at Navigation 325

Bibliography 329

Index 335

Preface

Many treatments of the history of mathematics have focused on the development of mathematical operations and concepts themselves. These accounts provide important treatments of the growth of mathematical knowledge over the span of world history. Yet they do not always place such knowledge in intellectual and cultural context, and they are often inaccessible to readers who have not yet acquired advanced mathematical skills or who have lost these skills through disuse. This book seeks to explore mathematics, and the concept of number itself, as a component of cultural and intellectual history; it emphasizes the relationship of mathematics with other forms of thought: mysticism, religion, philosophy, and science. Particular attention is paid to the linkage between mathematics and the understanding of nature, a connection that has played a crucial role in many different cultures.

The movement of ideas serves as another important theme. Ideas, especially in written form, travel easily with people and merchandise. In our own time, they can flash around the globe at the touch of a button or a smart phone. Since antiquity, mathematical concepts have been transmitted from place to place as a result of migration, conquest, or cultural contact. Modern mathematics proceeded from the progressive mixing of traditions from the Middle East, Europe, North Africa, and Asia. The ebb and flow of empires and the voyages of merchants had a profound effect on its complex development.

Numbers affect so many aspects of human society and culture: money and commerce, administration and taxation, timekeeping and calendars, measurement and engineering, mapmaking and navigation—more than can be treated in a single volume. In this book, each chapter begins with a commentary examining important issues in the numerical or mathematical thought of a given culture or time period; subsequent entries explore selected topics or figures in more detail. The commentaries focus on particular themes: the control of nature in ancient Mesopotamia; the importance of measurement in ancient Egypt; the link between geometry and philosophy in Classical Greece; the role of royal patronage in Hellenistic Alexandria; the ancient Chinese notion of order and its link to number; the concept of cyclical time among the medieval Maya; the tradition of Sanskrit poetry and astronomy in India; the Arabic community of thought and criticism in the Middle Ages; the transmission of knowledge from the Arabic world to medieval Europe; the idea of mathematical natural law in early modern Europe; the rise of a culture of mathematical precision in the 19th century; the mathematical achievements of women despite their exclusion from much of the scholarly world until the last few decades. These themes provide a sense of unity and direction to a very broad and diverse subject matter. The book ends with a short series of essays on the role of mathematics in 20th-century history and culture. These are intended to illustrate the tremendous impact of mathematics on modern life through the discussion of selected examples.

Introduction

In 1973 NASA launched the *Voyager 1* and *2* spacecraft, the probes that would visit most of the outer planets of the solar system in succeeding years. As each craft finished its mission, it would exit the solar system and travel off into deep space. Because there was a possibility, however remote, that at some point in the future extraterrestrial beings would encounter these probes, NASA equipped each with a gold-plated copper disc with images and sounds of the planet earth recorded on it. The disc was encased in a gold cover engraved with directions for its use and other information intended to identify its source. The engraving included mathematical information presented in binary code.

The inclusion of such instructions reveals a fundamental assumption on the part of NASA: namely, that any beings intelligent enough to travel in space and encounter *Voyager 1* or *2* would be mathematically sophisticated enough to decipher the message. This assumption was reasonable, for it would be impossible to develop the technology necessary for space travel without a knowledge of mathematics. However, the inclusion of such information also suggests the idea that mathematics serves as a common frame of intellectual reference, shared by all thinking beings.

This leads one to consider the relationship of mathematics, or the concept of number itself, to human culture. If the capacity for mathematical reasoning does indeed link all human beings, that capacity must be expressed in a given cultural framework. Human reason does not exist in a vacuum; it is affected by other components of consciousness and thought, including sensation, belief, and creativity. Thus, it is useful to explore the interaction of mathematics with other forms of intellectual activity: religion, art, philosophy, or science, for example. The notion of number also has a practical aspect. Humans are material beings, and they live in a material world. Number allows for an enhanced understanding of the necessities and luxuries associated with material existence. With number, one may quantify, one may count: so many fruits, so many sacks of grain, so many gemstones. Number and mathematics thus have a close association with the realities of wealth, production, and commerce, all important components of culture.

A study of the history of number and mathematics in the context of different world cultures sheds light on these complex relationships. Such a study reveals interesting patterns. For example, people count things, and this leads them to develop numbers—systems of numeration—in oral or written form. But people have ten fingers and ten toes, and they probably counted with those long before they developed more complex representations of number. It is no accident that many number systems around the world are decimal, based on the number 10; this is true of ancient Egypt, China, and India. However, many of the peoples of Western Africa based their systems on the numbers 5 and 20. The Maya based theirs on 20, while the ancient Sumerians used 60, with 10 as a secondary base. All of these approaches, including the use of 60, can be explained in terms of finger and toe counting. It is clear that local customs—the way one counts—had a profound effect on more complex ideas—the base of a system of numeration. These concepts then became embodied in oral and written language, the words and characters used to express numbers.

Astronomy provides a more elaborate example of the link between mathematical thought and cultural context. The regular movements of the sun, moon, stars, and planets lend themselves to mathematical analysis and interpretation. However, while astronomical phenomena remain largely constant, the nature of such human interpretation varied from culture to culture as people saw the heavens from different intellectual perspectives. The Babylonians believed that the movements of the stars and planets were a source of divine portents and omens, and they compiled a large body of astronomical observations and measurements in the process of interpreting those omens. The Maya associated their gods with the maintenance of time itself and viewed the cyclical movements of the heavens as an expression of the cyclical nature of time. They developed an intricate mathematical understanding of the timing of celestial cycles linked to religious ritual. The Greeks believed that the circle was the most perfect of geometrical forms and created detailed cosmological models based on circular geometry to account for the observed positions of the heavenly bodies. Medieval Indian astronomers also emphasized circular geometry and combined it with their own forms of calculation and with the tradition of Sanskrit poetry. In each of these cases, the mathematical description of the heavens was influenced by other elements of culture: religious beliefs, philosophical assumptions, literary traditions.

The interaction of cultures and the movement of ideas have also been major factors in the history of mathematics. Modern humanity lives in a digital world where information moves almost instantaneously. Yet for most of human history, ideas moved like merchandise, as people traveled from place to place. The spread of mathematical ideas is a fascinating by-product of cultural contact. Thus, Mesopotamian mathematics and Egyptian practical geometry had an influence on Greek geometry and astronomy. Greek astronomy probably made its way to India. The Arabs came into contact with Greek geometry when they conquered the Eastern Mediterranean in the Early Middle Ages; at the same time, they adopted the Indian system of numbers, based on decimal place values (the ancestor of modern numeration). Arab and Persian scholars combined these Greek and Indian elements and added important new concepts. Arabic mathematics, in turn, spread across Northern Africa from the 8th to the 11th centuries CE and was transmitted to Western Europe in the 12th century, where it inspired the growth of European mathematical thought during the Renaissance. In the modern period, Western mathematics moved around the world as a result of European cultural influence and imperialism.

One often thinks of mathematics as an abstract discipline, an exercise in pure thought. However, the application of mathematics to real phenomena has played a crucial role in the development of many cultures. This is true of such practical endeavors as commerce and engineering, and it is equally true with respect to the mathematical description of nature inherent in different astronomical traditions and in modern scientific thinking. The act of measuring is essential for such activities. One measures the volume of wine traded, the height of a wall being built, the angle of a star above the horizon, the velocity of a falling body. Measurement, however, is itself an aspect of human culture and is affected by human judgement and subjectivity.

In the *Measure of All Things* (2002), Ken Alder describes the work that led to the establishment of the meter as the standard unit of length in the metric system. The revolutionary regime of France in the early 1790s wished to do away with the varied measures traditionally used in France and to replace them with a uniform system of measurement. Scholars associated with the French government sought a new scheme based on nature itself; in other words, the units of measurement would reflect actual values found in the natural world (Alder 2002, 84–87). In addition, the system would have a decimal character, based on powers of 10. The Celsius scale of temperature, created by Anders Celsius in 1742, serves as an example of these parameters: the scale is based on the freezing and boiling points of water, and the range of temperature between these values is divided into 100 degrees.

For the measuring of distance, the scholars of the French Revolutionary regime chose to base their fundamental unit on the size of the earth itself.

The meter would be defined as one ten-millionth of the distance from the north pole to the Equator, along a meridian of longitude. The government of France dispatched two mathematicians to measure a substantial portion of this distance in precise terms, a process that would yield an exact value for the meter. The two men, Jean-Baptiste Joseph Delambre and Pierre François André Méchain, worked on their task for more than five years. Delambre covered the northern part of France, while Méchain labored in Spain and southern France. They employed the method of triangulation to determine exact distances, a technique that proceeds from the fact that a triangle may be constructed if all three of its angles and one of its sides are known. This allowed them to create a continuous chain of triangles over land, based on angle measurements between established reference points (Alder 2002, 22).

Delambre successfully completed his part of the enterprise, but Méchain made crucial errors in his measurements and calculations as he worked in the region of Barcelona (Alder 2002, 121). As he proceeded with his work, he sought to hide his error. The length of the meter, as established in 1798, reflected Méchain's mistake. Delambre discovered the discrepancy after Méchain's death and brought the issue to public attention. As Ken Alder points out, the story reveals a great deal about the character of the human drive for quantitative precision and accuracy. While scholars may endeavor to base measurements upon nature itself, the act of measuring is nevertheless a thoroughly human enterprise. Units of measurement are human inventions, intended to serve human purposes, and subject to human error. The notion of "natural" measures adopted by the French Revolutionary authorities was therefore illusory from the start. As the discipline of modern statistics grew in the 20th century, techniques were developed to address the inevitability of error in measurement (Alder 2002, 308). These techniques testify to the limitations of the act of measuring itself, an act that is, in the end, a form of cultural expression.

Many practical aspects of human culture relate to mathematics and number: money and commerce, administration and taxation, engineering and architecture, navigation and mapmaking. While this book will touch upon these topics, its principal focus will be on the development of mathematical ideas; the movement of such ideas as a result of cultural contact; and their relationship to religion, philosophy, views of nature, and scientific thought. The exploration of these issues will illustrate the complex history of mathematics as a component of world cultures.

1

Numbers in Ancient Mesopotamia

Chronology of Dominant Groups
(some dates are approximate)

Sumerian	4000–2335 BCE
Akkadian	2335–2193 BCE
Sumerian Renaissance	2047–1705 BCE
Old Babylonian (Amorite)	1894–1595 BCE
Kassite	1595–1150 BCE
Assyrian	1365–609 BCE
Neo-Babylonian (Chaldean)	626–539 BCE
Persian	559–331 BCE
Seleucid	312–64 BCE

CULTURAL ICON: THE YALE TABLET

The Yale tablet (Figure 1.1) serves as striking evidence of the significant mathematical capabilities of the residents of ancient Mesopotamia. The tablet is a piece of hardened clay small enough to fit in the palm of one hand and probably dates from 1800 to 1600 BCE, a period when much of Mesopotamia was under the control of the Babylonian Empire (perhaps best known for the laws of Hammurabi). The Babylonians adopted many of the cultural innovations of the Sumerians, the group that dominated Mesopotamia until the late third millennium BCE (around 2200). Indeed, Sumerian culture served as a major foundation of the region throughout much of antiquity. The Yale tablet is written in the cuneiform characters developed by the Sumerians and employed by subsequent peoples in the region; the characters represent a mathematical calculation using the sexagesimal, or base 60 numeration system (again, a Sumerian invention—see later discussion). It appears to be a school exercise, a problem in mathematics.

Figure 1.1 The Yale tablet dates from 1800 to 1600 BCE and demonstrates the sophistication of ancient Mesopotamian mathematics. The figure illustrates the length of the diagonal of a square. (agefotostock/Alamy Stock Photo)

The tablet contains a drawing of a square, with two diagonals dividing it into four isosceles triangles. Several sexagesimal numerical expressions in cuneiform accompany the diagram. Along one side of the square is the cuneiform symbol for the number 30, and along one of the diagonals is an approximation of the square root of 2 (taken to five decimal places in modern mathematical terms). Below the square root appears a number that apparently represents the product of the other two terms, 30 times the square root of 2: approximately 42.43. Because Sumerian and Babylonian numbers are written without decimal points, or more properly, sexagesimal points, other interpretations of the tablet are possible. However, the most accepted view of the tablet portrays it as a student's calculation of the length of the diagonal of a square, based on the length of a side.

The figures on the tablet demonstrate that the ancient Babylonians were familiar with the Pythagorean theorem, the relationship between the hypotenuse of a right triangle and the other two sides, as early as 1800 BCE (Pythagoras, for whom the theorem is named, was born around 570 BCE). The theorem is usually expressed in modern terms as $a^2 + b^2 = c^2$, where c represents the length of the hypotenuse and a and b the lengths of the other sides. In this case, the relationship has been applied to the problem of the length of the diagonal of a square. Such a diagonal forms the hypotenuse of an isosceles triangle including the two adjacent sides.

If one assigns the value of 1 to the side of such a triangle, then by the Pythagorean theorem, the length of the hypotenuse must be the square root of 2 (Figure 1.2). Consequentially, the length of the diagonal of any square must be the length of one side multiplied by that square root. The Yale tablet depicts this calculation quite clearly, in this case, with a side value of 30. Further, the square root of 2 is an irrational number; that is, a number that cannot be represented as a fraction of integers. Thus, the solution to this problem requires an approximation of this value, and the tablet includes such an approximation. The resulting value for the length of the hypotenuse reveals considerable precision.

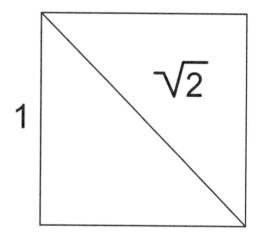

Figure 1.2 Diagonal of a unit square.

Taken as a whole, the Yale tablet reveals a great deal about the mathematical abilities of the Babylonians in the early second millennium BCE, and of Mesopotamian cultures in general. Along with the basic ability to represent numbers and to perform mathematical operations, the command of the Pythagorean theorem shows substantial geometrical expertise. In addition, the routine employment of the square root of 2 indicates the capacity to calculate such values or to approximate them with a high degree of accuracy. Those operations are not represented on the tablet; scholars suggest that the Babylonians constructed useful mathematical tables (multiplication, reciprocals, square roots, cube roots, Pythagorean triples) that could be used quickly for ordinary figuring. In this case, a student or scribe apparently recognized the essential relationship, drew upon a standard value for the square root of 2, did the necessary multiplication, and produced the answer.

Finally, the problem in question, its solution, and the tablet itself suggest something about the nature of Mesopotamian (in this case, Babylonian) mathematics. While the entire exercise could certainly be viewed as an example of pure thought, it seems to have a practical character as well. The use of an approximate figure for the square root of 2 suggests that the problem requires a specific answer, one that may be employed in the real world. Geometry and trigonometry have many concrete applications, from the management of land use to the engineering necessary for construction and irrigation projects. The Yale tablet may give us a glimpse into the training of the literate and mathematically skilled scribes and scholars essential for the complex public works that provided safety and stability in the face of hostile enemies and a hostile nature—perhaps essential for civilization itself.

CULTURAL COMMENTARY: ORDER AND CONTROL

The fertile lands between the Tigris and Euphrates rivers in southwestern Asia fostered one of the great early civilizations in human history. While cities of significant size emerged in Mesopotamia in the late fourth millennium BCE, the region had supported substantial cultivation long before that. The availability of water, combined with the renewal of the soil (as a result of river flooding) made the region extremely productive, a source of real agricultural wealth. This wealth formed the economic foundation of the population centers—the city-states of Sumer—that grew after 3500 BCE.

The ancient Sumerians produced a thriving urban culture that lasted for more than 1,000 years. Supported by the produce of the surrounding farmland, Sumerian city-states developed political systems, religious institutions, and trading networks, along with systems of writing, numeration, and mathematics capable of maintaining the sophisticated records essential for a complex society. Early pictographic systems of writing and numeration evolved into more agile cuneiform characters by the late third millennium. Such written characters, pressed into clay tablets with a stylus, are perhaps the most important legacy of the Sumerians to later peoples of the region. Indeed, over the centuries from 2200 to 500 BCE, other ethnic groups came to dominate Mesopotamia: Akkadians, Babylonians, Kassites, Assyrians, and Neo-Babylonians, to name a few. However, Sumerian culture, adopted and adapted by these peoples, continued to play a dominant role. For the cultures of Mesopotamia, the institutions of society—agriculture, religion, government—represented order and security, a respite from the chaos and danger of an uncertain world and an untamed nature. Writing and mathematics formed a crucial part of this sense of security, for such powerful intellectual tools represented the capacity of men and women to plan, to record, to administer—in short, to have some degree of control over their surroundings.

Perhaps the most famous work of literature from the region, the *Epic of Gilgamesh*, serves as a potent representation of Mesopotamian culture. Found in an Assyrian library of the 7th century BCE, it undoubtedly has roots in the far older civilizations of Sumer and Babylon; like the cuneiform characters of which it is composed, it represents the cultural continuity of the region over three millennia. The story treats a number of themes, most famously, the human quest for immortality and the ultimate failure of that quest. However, it also deals with issues of human agency and control—the ability to be master of one's fate. Indeed, one might argue that the desire for immortality is merely one aspect of this larger concern.

Building and agriculture, the hallmarks of early civilization, embody the desire for control. Agriculture allows humanity to command its own food supply, and this sets humans apart from the rest of nature. It is often argued that early civilizations grew from the need to cooperate on such large-scale agricultural projects as the construction of irrigation systems. In the *Epic of Gilgamesh*, the hero's future companion Enkidu emerges from the forest where the gods created him; he is described as wild, untamed, and animal:

> He knew not the cultivated land ...

Here, agriculture and cultivation are identified with civilization itself, with security and control. The wilderness, the first home of Enkidu, is the opposite of all this, a place where humanity is at the mercy of disordered nature.

As ruler and priest-king, Gilgamesh serves as the ultimate embodiment of order and control. In the opening of the epic, he is described as beautiful and powerful, but he is also credited with building the walls of Uruk—the outer walls that guard the city and the walls of temples that serve as centers of worship:

> In Uruk he built walls, a great rampart, and the temple of blessed Eanna for the god of the firmament Anu, and for Ishtar the goddess of love. Look at it still today: the outer wall where the cornice runs, it shines with the brilliance of copper; and the inner wall, it has no equal. Touch the threshold, it is ancient. Approach Eanna the dwelling of Ishtar, our lady of love and war, the like of which no latter-day king, no man alive can equal. Climb upon the wall of Uruk; walk along it, I say; regard the foundation terrace and examine the masonry: is it not burnt brick and good?
>
> (Sandars 1972, 61)

Both sets of walls manifest security and control: protection against invading enemies and protection against the whims of the divine. In the temple, the gods are worshipped, placated, tamed, made more predictable—controlled. Ultimately, this very desire for agency leads Gilgamesh and his friend to tragedy. Having demonstrated their mastery over nature by slaying the guardian of the forest, they are punished for their pride, their hubris. The gods cannot allow humanity such an elevated stature, and Enkidu is struck down with plague as an expression of divine displeasure and of human vulnerability. This inspires Gilgamesh to embark on his famous search for everlasting life, a quest that ends in failure. However, despite the ultimate frustration of its heroes, the *Epic of Gilgamesh* expresses a deep desire of Mesopotamian humanity, the longing for control.

A similar theme may be found in the *Enuma Elish*, a Babylonian creation myth. Dating from the Babylonian or Kassite periods (perhaps 1800–1500 BCE), the story speaks of the mating of Tiamat and Apsu (saltwater and fresh water) and the generations of gods that result from it. Ultimately, Apsu grows tired of his offspring and plans to do away with them. He is restrained by Tiamat. However, Ea, god of waters, hears of Apsu's plot and kills him to prevent it. When Tiamat learns of the fate of her lover, her form and her demeanor are transformed; she assumes the shape of a terrible dragon and creates a band of monsters to aid her in avenging Apsu. Stricken with fear, her previous offspring—the gods—accept Marduk, son of Ea, as their protector, champion, and ruler. Marduk and Tiamat clash in deadly combat, and he emerges as the victor.

At this point, the energetic story a acquires a new dimension. Marduk divides the body of Tiamat and uses it to form earth and heaven; he takes the other gods—water, storm, sky—and sets them in place in the world he has built.

> Then the lord rested, gazing upon her dead body,
> While he divided the flesh of the . . ., and devised a cunning plan.
> He split her up like a flat fish into two halves;
> One half of her he stablished as a covering for heaven.
> He fixed a bolt, he stationed a watchman,
> And bade them not to let her waters come forth.
> He passed through the heavens, he surveyed the regions (thereof),
> And over against the Deep he set the dwelling of Nudimmud.
> And the lord measured the structure of the Deep,
> And he founded E-<u>sh</u>ara, a mansion like unto it.
> The mansion E-<u>sh</u>ara which he created as heaven,
> He caused Anu, Bêl, and Ea in their districts to inhabit.
>
> (King 1902, Book IV, lines 135–146)

In short, he imposes order and control on the cosmos, a cosmos constructed of the body of Tiamat. Further, that order is expressed in precise, quantitative terms:

> He (Marduk) made the stations for the great gods;
> The stars, their images, as the stars of the Zodiac, he fixed.
> He ordained the year and into sections he divided it;
> For the twelve months he fixed three stars.
> After he had [. . .] the days of the year [. . .] images,
> He founded the station of Nibir to determine their bounds;
> That none might err or go astray . . .
>
> (King 1902, Book V, lines 1–7)

Marduk produces an orderly universe, mediated by regular, mathematically precise patterns. While it is true that the *Enuma Elish* is a story of the gods, not humanity, it is equally true that the ordered cosmos established at the end of the epic is one that may be comprehended and described by the human mind. After all, humans wrote the story, and the mathematical patterns imposed on the heavenly bodies by Marduk represent some of the knowledge accumulated by Babylonian astronomers at the time of its composition. The *Enuma Elish*, like the *Epic of Gilgamesh*, demonstrates a Mesopotamian emphasis on mastery, on the ability to order or control one's surroundings, or in this case, on the capacity to understand and predict the motions of the heavenly bodies with precision.

The *Epic of Gilgamesh* and the *Enuma Elish* are both important works of literature that treat issues of divinity, creation, nature, and immortality. However, some more mundane sources reveal even more about Mesopotamian cultural priorities. The lion's share of cuneiform tablets that survive, from Sumer, Akkad, Babylon, or Assyria, describe the material transactions of complex societies and economies. Tax records, government expenditures, inventories, business transactions, land sales—all are recorded in great detail in quantitative terms. Many of these transactions require complex calculations. Such records, so essential for civilization, law, and government, show how numeration and mathematics served as potent agents of control. Number allowed for the precise tallying of wealth, for the application of power, for the distribution of resources. It enabled Mesopotamian humanity to plan for the future. The economic records of the Third Dynasty of Ur vividly illustrate such activities. The Third Dynasty ruled the city of Ur from 2112 to 2004 BCE, at a time when Sumerian and Akkadian peoples had blended to form a composite population. Thousands of clay tablets reveal the operations of a complicated government, intent on recording its myriad transactions with precision. In addition to records of taxes and human labor, records of commodity prices survive, based on silver as a means of exchange.

These documents, and many others like them, indicate a practical approach to mathematics; a notion of number as associated with concrete objects and applications. One might argue that the capacity to record and manipulate numerical quantities was an essential element of Mesopotamian society. Such a view complements an emphasis on control and planning as central to civilization and government. The evolution of written numbers themselves in ancient Mesopotamia further demonstrates the practical aspect of mathematics there. Early Sumerian numbers were sexagesimal, that is, based on the number 60 (see the later entry). Initially, the symbols were pictographic or curviform; these evolved into cuneiform, or wedge-shaped, characters by the end of the third millennium BCE (around 2000 BCE). The early system of pictographic symbols does not include a notion of place-value or a concept of zero; symbols for 1, 10, and 60 are repeated as many times as necessary to represent a given amount, as shown in Figure 1.3.

Figure 1.3 Early Sumerian representation of 275.

The written numbers resemble the actual process of counting physical objects in groups of 10 and 60. This is made even more apparent by the resemblance between the pictographic (curviform) symbols and Sumerian calculi, or clay counting tokens (see later entry). It is quite possible that early Sumerian number symbols are representations of physical objects used in early accounting (Ifrah 1998, 192–197).

Cuneiform mathematical symbols for the Sumerian sexagesimal system became dominant in Mesopotamia around 2600 BCE and were far less cumbersome than the older characters. Combinations of wedge-shaped marks could be produced on a clay tablet quite quickly with a stylus (Figure 1.4).

Figure 1.4 Cuneiform representation of 275.

However, with the decline of the Sumerians and the rise of different Semitic peoples in the region (Akkadians, Babylonians, Assyrians), new systems of numeration developed. Sumerian cuneiform characters were combined with Semitic decimal numbers based on powers of 10. In addition, by the 19th century BCE, Babylonian scholars had developed a sexagesimal system with a truly positional character, one that facilitated complex operations. Eventually, it included a symbol for zero as well.

Educated scribes, skilled in writing and mathematics, served as the primary agents of administration in Mesopotamian governments. They developed sophisticated methods of problem solving and adopted innovative techniques to speed their calculations, including the use of mathematical tables of predetermined values (very much like the use of trigonometric tables in modern mathematics). Much of their work had a practical character, linked to accounting or engineering; indeed, a preference for concrete rather than abstract mathematics complemented a Mesopotamian view of number as associated with mastery of the real world. Babylonian astronomy applied mathematical knowledge to the motions of the heavenly bodies for the purposes of timekeeping and religious observance—a practical concern, if one believes that stellar and planetary motion reveals the intentions of the gods. Babylonian mathematical proofs had a decidedly practical character as well and were often confirmed by trial and error. However, the complexity of some texts suggests abstract mathematical work as well. For example, Plimpton 322, a Babylonian tablet from the period around 1800 BCE, contains a series of Pythagorean triples; that is, numbers that represent the lengths of the sides of different right triangles (Figure 1.5). Such triples demonstrate Babylonian mastery of the Pythagorean theorem centuries before Pythagoras lived. The Yale tablet discussed earlier dates from the same period and contains a geometrical problem that also requires knowledge of the Pythagorean theorem. The Yale tablet appears to be a school exercise, part of a scribe's training, and it is likely that knowledge of triangular geometry served a practical purpose, especially with regard to the planning of significant irrigation and engineering projects. However, it is difficult to imagine a similar practical application for the knowledge exhibited in Plimpton 322. While scholars disagree on details, the tablet appears to contain a method for generating Pythagorean triples, sets of numbers that fulfill the pattern of the theorem. The same method may be applied to the solution of quadratic equations. The tablet seems to be an expression of pure mathematics, an abstract investigation of the links between quantity and physical space. Mesopotamian mathematics, grounded in the control of the real world, ultimately transcended that world as an expression of human imagination and creativity.

Figure 1.5 Plimpton 322 is a mathematical tablet in cuneiform from the period around 1800 BCE. It contains a series of Pythagorean triples and may represent an example of pure mathematical speculation. (agefotostock/Alamy Stock Photo)

SUMERIAN COUNTING TOKENS

Sumerian counting tokens, or calculi, illustrate the practical character of Mesopotamian mathematics at its early stages of development. Dating from the fourth and third millennia BCE, they represent an early approach to counting, mathematical recordkeeping, and rudimentary calculation. The term "calculation," in fact, has a common root with "calculi" or "calcification"—all refer to the counting stones once used by the ancient Greeks and Romans.

In early Sumerian civilization, counting tokens were small figures of dried clay (Figure 1.6). Differently shaped tokens represented different quantities: 1, 10, 60, 600, etc. One should note that the tokens resembled the early pictographic written characters for Sumerian numbers, suggesting a link.

Such tokens provided an effective, highly tangible means of counting, figuring, and storing quantitative information—a means that did not require actual literacy. When counting commodities—sheaves of grain, livestock, chariot wheels—a single token would correspond to every item counted. Groups of 10 would be represented by a token for "10"; groups of 60 by a token for "60." The tokens corresponded to the sexagesimal

Figure 1.6 Mesopotamian counting tokens represent an early form of keeping quantitative records. The clay vessel, or *bulla*, would contain the tokens representing a given record or account. (Adam Ján Figel/Alamy Stock Photo)

system of the Sumerians. A group of 596 measures of grain would be represented by:

9 tokens for 60

5 tokens for 10

6 tokens for 1

The practical value of the technique is clear: groups of tokens representing quantities of commodities could be used to record commercial transactions, inventories, taxes, or government expenditures. Rudimentary calculations could be carried out with tokens as well, particularly addition and subtraction. Ultimately, the use of counting tokens as a means of manipulating quantities gave way to the Sumerian or Babylonian version of the abacus. It is likely that such a device, composed of moveable wooden counters, found its roots in the tokens themselves (Ifrah 1998, 237–244).

Archaeologists have recovered many examples of counting tokens from the fourth to the second millennia BCE. Often, tokens are contained in hollow clay capsules inscribed with cuneiform characters: *bullae*. A token-filled bulla represents a single transaction or a single inventory. For example, an official might count the jars of olive oil in a royal storehouse by placing the appropriate number of tokens of proper value in a bulla, closing it, and sealing it with his personal cylinder seal (a common means of personal identification for the elite). The bulla in question would serve as a tangible record of available olive oil. Sumerian counting tokens demonstrate the quantitative ingenuity of the early Sumerians, and indeed, of humanity in general: many other cultures around the world have used similar techniques to represent quantitative data in lieu of writing. Ultimately,

Sumerian tokens would give way to more sophisticated cuneiform records. In addition, they illustrate the concrete foundations of Sumerian mathematics. The link between numbers and actual physical things is clear when *numbers are* actual physical things. For the Sumerians, math begins as a path to stability, a means to exert control over the real world of objects.

SUMERIAN SEXAGESIMAL SYSTEM

Note: *Sumerian culture is credited with the invention of writing, and the earliest written words consisted of pictographic symbols pressed into clay. Number symbols had a similar pictographic character. However, by the late third millennium BCE, stylized wedge-shaped characters—cuneiform—had replaced the earlier pictograms. The following discussion addresses Sumerian cuneiform numbers.*

As early as the fourth millennium BCE, the ancient Sumerians developed a numerical system based on the powers of 60, a sexagesimal system. Historians have suggested a number of reasons for the choice of 60 as a base. Different methods of hand counting could have led to an emphasis on 60, just as counting on one's fingers in the simplest manner could obviously lead to a numeration system based on 10. Sixty may also have served as the base of Sumerian numeration because of its many divisors—many numbers divide into it evenly. A system based on 60 greatly facilitates the calculation of fractions. Sixty could also have religious or ideological significance.

Regardless of its roots, a sexagesimal system posed Sumerian scribes with a significant challenge. In a modern decimal-based system, nine symbols are needed to represent integers smaller than 10; at that point, the first set of ten is represented by the figure "1" and a "0." Counting continues with the same nine symbols. Thus, the symbol "14" represents:

$$\begin{array}{r} 1 \times 10 \\ +4 \\ \hline \end{array}$$

Further, modern numbers are based on a place-value notation, where a symbol's placement in a number represents its multiplication by a particular power of 10. So "4,985" represents the following:

$$\begin{array}{r} 4 \times 1000 = 4000 \\ 9 \times 100 = 900 \\ 8 \times 10 = 80 \\ +\ 5 \times 1 = 5 \\ \hline 4985 \end{array}$$

Finally, this place-value system includes the symbol "0" to represent the absence of value at a given place in the number. Hence, "107" actually means:

$$\begin{array}{rcl} 1 \times 100 & = & 100 \\ 0 \times 10 & = & 0 \\ + \; 7 \times 1 & = & 7 \\ \hline & & 107 \end{array}$$

The ancient Sumerians did not develop a notation based on place-value, nor did they discover the concept of zero. In addition, their choice of 60 as the basis of their numeration system would seemingly require fifty-nine separate counting symbols before the appearance of a symbol representing the first set of 60. Such a system would be quite demanding on any scribe. They addressed this difficulty by using the number 10 as a secondary base. A single cuneiform mark—a vertical wedge—was used to represent 1; that wedge was repeated twice for 2, thrice for 3, and so on, up to 9. The wedges were placed in a consistent, recognizable pattern, rather like the pips on a pair of dice or a playing card (Figure 1.7).

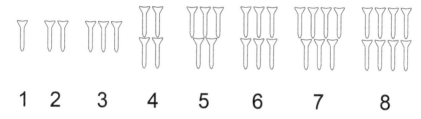

Figure 1.7 Grouping of cuneiform unit symbols.

"Ten" was represented as a horizontal chevron, and later numbers combined symbols for 10 and 1 (Figure 1.8).

Figure 1.8

Thus, the Sumerians could write the numbers from 1 to 59 with combinations of just two symbols (Figure 1.9).

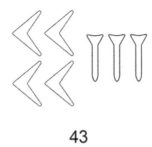

43

Figure 1.9

The number 60, the base for the system, took the form of a large single wedge or stroke. Numbers beyond 60 included that symbol and those for 1 and 10. Multiple large wedges indicated multiple sets of 60 (Figure 1.10).

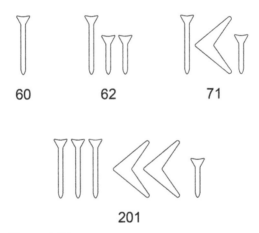

Figure 1.10

It is important to point out once more that this Sumerian system did not include real place-notation or a concept of zero. Symbols were included as many times as necessary to represent the number in question. In the same way that 10 served as an intermediate base between 1 and 60, so 600 served as an intermediate between 60 and 3600 (60^2) (Figure 1.11).

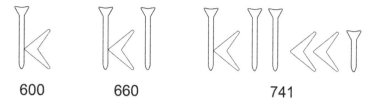

|600|660|741|

Figure 1.11

As one can see, the number 741 is represented by one symbol for "600," two for "60," two for "10," and one for "1."

Just as a decimal system is based on powers of 10 (10, 100, 1,000, etc.), the Sumerian sexagesimal system is based on powers of 60 (60, 3,600, 216,000, etc.). The intermediary bases, 10 and 600, aid in the written representation of numbers, but they do not affect the nature of the system. Higher powers of 60 have their own symbols (Figure 1.12).

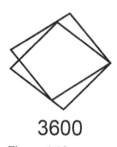

3600

Figure 1.12

So, the number 8,003 would be represented as shown in Figure 1.13.

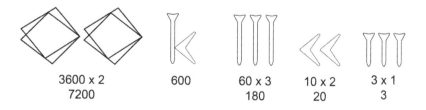

| 3600 × 2 | 600 | 60 × 3 | 10 × 2 | 3 × 1 |
| 7200 | | 180 | 20 | 3 |

8003

Figure 1.13

In conclusion, the practical character of the Sumerian system must be stressed. For the most part, Sumerian numerals represented numbers of objects (goats, sheaves of grain, jars of olive oil) or the dimensions of physical spaces. The written symbols and their manner of representing quantity make that clear. Sumerian numeration may be seen as a means of exerting control over the physical world, of imposing rule and order upon it.

LATER APPROACHES TO NUMBERS IN MESOPOTAMIA
Semitic Numeration

In the late third millennium BCE, groups of Semitic-speaking people challenged the Sumerians for control of much of Mesopotamia. The Akkadian Empire rose around 2350 BCE and lasted for two centuries. It was followed by a period of Sumerian resurgence, but by 2000 BCE the Semitic peoples regained power. Different groups—Babylonian, Kassite, Assyrian, Chaldean—would rule the region until the 6th century BCE. However, these dominant groups were heavily influenced by Sumerian culture, and they adopted many ideas and institutions: from Sumerian religion and mythology to cuneiform writing. In addition, the Semitic peoples combined the Sumerian sexagesimal numeration system with their own decimal numbers to produce a mathematical synthesis. Traditionally, Semitic peoples used a counting system based on the powers of 10. By creating cuneiform symbols for 100 and 1,000, Semitic scribes created a truly compound mathematical system that employed two different bases: 10 and 60.

As a result, during the second millennium BCE, in addition to traditional sexagesimal numbers, scribes could employ a synthesis of the two. In this way, the mathematics and recordkeeping methods of the region reflected the rise and fall of empires and the mixing of cultures and ethnicities.

Positional Numeration

In the 19th century BCE, elite scholars in Babylon developed a truly positional sexagesimal number system. While scribes employed the various systems described earlier for everyday calculations, the new positional approach was apparently applied to such exalted subject matter as complex mathematics or astronomy. The positional system used just two symbols: a vertical stroke or wedge for 1 and a horizontal chevron for 10. For small numbers, this worked much like the earlier sexagesimal system, combining the two symbols to represent values (Figure 1.14).

Figure 1.14

However, for numbers greater than 60, the value of symbols depended on their position within the mathematical expression. Thus, the number 134 was written as shown in Figure 1.15.

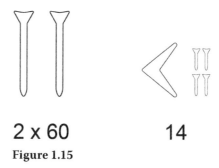

Figure 1.15

Here, the first two strokes represent 2 × 60, while the second group of symbols indicates 14. Further placements would represent further powers of 60. The system resembles a modern decimal positional system, where digits represent powers of 10 based on their position. Using this positional sexagesimal approach, the number 4,292 would look like Figure 1.16.

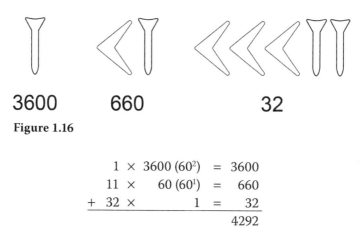

Figure 1.16

$$
\begin{array}{rrcr}
1 \times & 3600\,(60^2) & = & 3600 \\
11 \times & 60\,(60^1) & = & 660 \\
+ \ 32 \times & 1 & = & 32 \\
\hline
 & & & 4292
\end{array}
$$

This positional approach to representing quantity reflects the increasing sophistication of Mesopotamian mathematics during the period of the Babylonian Empire, for it allowed scribes to handle complex problems and computations in a streamlined manner. Indeed, the sources described earlier, the Yale tablet and Plimpton 322, both employ such numbers. Still, the system lacked a notion of "zero"—the absence of value at a given placement in a mathematical expression. Without such a null placeholder, calculations could be unclear. Mesopotamian scribes addressed this problem by leaving an empty space to represent a lack of value for a given power of 60. However, a more effective solution to the problem arose with the introduction of the concept of "zero" in the second half of the first millennium BCE. Characters for zero appear in astronomical texts from the Seleucid period, after the conquest of Mesopotamia by Alexander the Great in the late 4th century BCE. However, it is reasonable to assume that the concept developed earlier, and it may be argued that Mesopotamian mathematicians were the first to conceive of it (Ifrah 1998, 297–300).

MESOPOTAMIAN CALCULATION METHODS

Modern numbers, based on powers of 10 and expressed in a place-value system with a symbol for zero, may be used both to represent quantity and to perform mathematical operations. In other words, one may easily calculate using the written symbols themselves. However, numeration systems without place-value do not allow for such easy written calculations. Sumerian and early Babylonian texts suggest a number of different techniques for mathematical operations, and many of these highly effective techniques reflect the practical, applied character of Mesopotamian mathematics. However, with the rise of a place-value system of numeration in the early second millennium BCE, more sophisticated calculations were possible, and these suggest an abstract element in the mathematical thought of the region.

As stated earlier, the early Sumerians employed counting stones, or calculi, for basic operations of addition, subtraction, and multiplication, as well as inventory and recordkeeping. The use of physical objects as calculational tools may have led to the development of a sexagesimal abacus in the third millennium BCE. While no physical examples of such a device have been discovered by archaeologists, several cuneiform tablets seem to refer to it. Such an abacus would replace clay counting stones with wooden beads or sticks, arranged on a frame for easy manipulation. With sticks representing values of 1, 10, 60, 600, and 3,600 (60^2) arranged in parallel rows, a skilled scribe could perform sexagesimal operations quite rapidly. With the introduction of Semitic decimal numeration in the late third millennium BCE, an abacus based on powers of 10 may also have been employed. The fragile nature of these devices would prevent their preservation, but they may have played a central role in Mesopotamian culture throughout the ancient period (Ifrah 1998, 242–259).

Other techniques allowed for more complex operations. Mesopotamian scribes lacked a method for long division, but achieved the same results by multiplying the dividend by the reciprocal of the divisor.

Geometrical problems required values for squares, cubes, and square roots, and it appears that scribes, scholars, and engineers used prefigured tables of values to speed their calculations. Such tables of reciprocals, exponents, and roots could be produced through laborious work with an abacus or through different forms of estimation. Once produced, they could be copied *en masse* and employed in a wide variety of practical or scholarly settings. A scribe working on financial problems might use a table of reciprocals for division; an engineer laying out a structure would need a table of squares to calculate area. The Yale tablet (see earlier) may indicate the utilization of such tables. The tablet shows a square with a side length of 30. It also includes a value for the square root of 2 and a figure for the product of these two numbers. If, as believed, the tablet represents a school problem, then the student would have recognized the operation necessary to find the diagonal of a square and then consulted a table of square roots to get the values necessary to complete the problem. Such practical techniques indicate the substantial institutional resources dedicated to the solution of mathematical problems in ancient Mesopotamia, for the production and duplication of mathematical tables must have been costly. This suggests a role for government and implies that easy and accurate mathematical calculations were a priority.

A numeration system based on true place-value arose in Mesopotamia in the early second millennium, during the Old Babylonian Empire. Such a system allowed easier calculation. However, it appears that older methods persisted as well, and versions of the abacus probably continued to be employed.

BABYLONIAN ASTRONOMY

The ancient Babylonians established a tradition of astronomical observation and prediction that lasted for almost 2,000 years, from the time of the Old Babylonian Empire (19th to 16th century BCE) to that of the Seleucids (4th to 1st century BCE). Despite the rise and fall of the varied hegemonies that dominated Mesopotamia during this vast stretch of time, the study of the heavens persisted as a feature of Babylonian culture. That same culture played a crucial role in the development of mathematics in Mesopotamia. Indeed, Babylonian astronomy represents an important application of mathematics to the natural world, as scholars, priests, and scribes recorded the cycles and patterns of the cosmos in quantitative terms.

The role of priests in this endeavor must be emphasized, for Babylonian astronomy had a significant, even dominant, religious element. Several major Babylonian gods were associated with the heavenly bodies: Shamash with the sun, Sin with the moon, Ishtar with Venus, and Marduk with Jupiter. For the Babylonians of the second millennium BCE, the motion of these bodies could indicate the gods' intentions at a given time. In other words, the heavens served as a source of omens, a way to interpret divine influence on the present or the future. Furthermore, these omens had a precise character; measurement of planetary positions over time yielded a mathematical understanding of heavenly motion, and therefore, of godly power. Because most astronomical motions observable with the naked eye are cyclical in nature, Babylonian priest-astronomers could find intricate mathematical patterns in their observations, regular cycles that they used to foretell the fertility of the land, the fate of kings, the time for war. Such practices were a form of astrology; indeed, while the Babylonians may also have been interested in the heavens for the purposes of timekeeping or pure curiosity, one cannot separate astronomy from astrology in Babylonian culture. By the time of the Neo-Babylonian or Chaldean Empire (7th c BCE), astrology had evolved beyond the interpretation of omens. Priests now believed that the general state of the heavens at a particular time affected earthly events (North 1995, 34). The notion of individual horoscopes arose fairly late in the Babylonian tradition, and the earliest one to appear in written records dates from the early 5th century BCE. The religious or astrological dimension of Babylonian astronomy reflects a larger cultural emphasis on human agency or control; it is quite practical in its own way. Over many centuries, priests and scholars systematically recorded and analyzed astral observations in an attempt to grasp, and perhaps benefit from, the influence of the gods on their world.

Babylonian priest-astronomers amassed an impressive body of astronomical observations over 1,500 years. They paid particular attention to solar and lunar eclipses, to the motions of the sun and moon, and to the positions of the five planets visible to the naked eye: Mercury, Venus, Mars, Jupiter, and Saturn. Planetary positions were often recorded in terms of the dates of their appearance in the night sky, or their disappearance from it. The Venus Tablet of Ammisaduqa serves as an excellent example of such practice. The tablet contains a set of observations of the planet Venus made during the reign of Ammisaduqa in the 17th or 16th century BCE. The Babylonians linked the movements of Venus to Ishtar, goddess of fertility. These observations span twenty-one years and record the dates of Venus's appearances and disappearances. The tablet was copied and recopied many times over the centuries and survived in a version produced in the 7th century BCE and preserved in the Assyrian library at Nineveh. It formed part of the *Enuma Anu Enlil*, a collection of astronomical tablets dating back to the time of the Old Babylonian Empire. The importance of this archive to the Assyrians of the 7th century testifies to the extraordinary continuity of the Babylonian astronomical tradition.

During the Neo-Babylonian period (8th c to 6th c BCE), records took the form of "astronomical diaries," yearly records of important celestial events recorded in quantitative terms (North 1995, 34–36). The diaries record a great deal of astronomical data and reflect the Babylonians' knowledge of important cyclical phenomena: periodic appearances of the planets, regular occurrence of eclipses, and the movements of the sun and moon. Analysis of these phenomena led to important achievements in calendar development. For example, in the 6th century BCE, the Babylonians defined the Metonic Cycle, the nineteen-year period that reconciled the lunar month with the solar year. Specifically, they found that 235 lunar cycles corresponded closely to 19 solar cycles. It is important to point out that while this increasingly sophisticated astronomical knowledge involved significant mathematical agility, the Babylonians did not develop geometrical models of the universe based on their observations. Rather, they based their knowledge of celestial cycles on patterns in their measurements themselves. In addition, their astronomical work continued to have great religious significance; it remained a priestly endeavor.

Babylonian astronomical activity continued under the Persian Empire after the late 6th century BCE and under the rule of the Seleucids, successors to Alexander the Great. During the Seleucid period (4th c to 1st c BCE), the tradition thrived, drawing on records going back almost 2,000 years. Most notably perhaps, Babylonian astronomers continued their work on planetary motions, creating a detailed mathematical understanding of the planets' cyclic behavior. Their work, and that of their predecessors, was combined with Greek and Egyptian views to produce the intricate astronomical systems of the Hellenistic age. The records of the Babylonians played a major role in the construction of such systems, for it was they who first applied number to the heavens in a systematic way.

CONCLUSION

Mesopotamian scribes and scholars developed one of the most efficient and sophisticated mathematical traditions of the ancient world. Their choice of a sexagesimal number system remains influential in the contemporary world and lives on in modern timekeeping and geography, where minutes and seconds are still based on 60. Using math as a means of imposing order on the world, they employed effective practical techniques for accounting and calculation. In the early years of the second millennium BCE, Mesopotamian scholars, probably from Babylonia, introduced a true system of place-value numeration and used it to perform complex operations in arithmetic, geometry, and algebra. They were familiar with the Pythagorean theorem long before Pythagoras, and their work suggests theoretical as well as practical mathematical thought. Finally, their astronomical traditions represented an ongoing effort to describe natural—and divine—phenomena in precise quantitative terms. In short, their work served as an important foundation for the later achievements in Mediterranean, Middle Eastern, and European mathematics.

PRIMARY TEXT: ENUMA ELISH, TABLET V

Here, Marduk orders the cosmos and assigns constellations ("stars") to mark the regular divisions of the year. "Nebir," or Nebiru, may refer to the planet Jupiter or some other point of reference in the heavens. Marduk uses the moon to regulate the month and describes its movements and phases ("horns") in terms of mathematical cycles.

He. (i.e., Marduk) made the stations for the great gods;

The stars, their images, as the stars of the Zodiac, he fixed.

He ordained the year and into sections he divided it;

For the twelve months he fixed three stars.

After he had [. . .] the days of the year [. . .] images,

He founded the station of Nibir 1to determine their bounds;

That none might err or go astray,

He set the station of Bêl and Ea along with him.

He opened great gates on both sides,

He made strong the bolt on the left and on the right.

In the midst thereof he fixed the zenith;

The Moon-god he caused to shine forth, the night he entrusted to him.

He appointed him, a being of the night, to determine the days;

Every month without ceasing with the crown he covered(?) him, (saying):

"At the beginning of the month, when thou shinest upon the land,

"Thou commandest the horns to determine six days,

"And on the seventh day to [divide] the crown.

"On the fourteenth day thou shalt stand opposite, the half [. . .].

"When the Sun-god on the foundation of heaven [. . .] thee,

"The [. . .] thou shalt cause to . . ., and thou shalt make his [. . .].

"[. . .] . . . unto the path of the Sun-god shalt thou cause to draw nigh,

"[And on the . . . day] thou shalt stand opposite, and the Sun-god shall . . . [. . .]

"[. . .] to traverse her way.

"[. . .] thou shalt cause to draw nigh, and thou shalt judge the right."
King, William. *Seven Tablets of Creation, Book V.*
London: Luzac and Co., 1902: lines 1–24.

FURTHER READING

Dalley, Stephanie. 1989. *Myths from Mesopotamia*. Oxford: Oxford University Press.

Ifrah, Georges. 1998. *The Universal History of Numbers: The World's First Number Systems*. London: The Harvill Press.

King, William. 1902. *Seven Tablets of Creation*. https://www.sacred-texts.com/ane/stc/index.htm

North, John. 1995. *The Norton History of Astronomy and Cosmology*. New York: W.W. Norton and Co.

Oates, Joan. 1986. *Babylon*. London: Thames and Hudson.

Robson, Eleanor. 2008. *Mathematics in Ancient Iraq*. Princeton: Princeton University Press.

Sandars, N.K. (translator). 1972. *The Epic of Gilgamesh*. London: Penguin Books.

Shekoury, Raymond. 2010. *Mesopotamians: Pioneers of Mathematics*. Createspace.

2

Numbers in Ancient Egypt

Chronology of Ancient Egypt

Early Dynastic Period	3100–2686 BCE
Old Kingdom	2686–2181 BCE
First Intermediate Period	2181–2025 BCE
Middle Kingdom	2025–1700 BCE
Second Intermediate Period	1700–1550 BCE
New Kingdom	1550–1069 BCE
Third Intermediate Period	1069–664 BCE
Late Period	664–525 BCE

CULTURAL ICON: ANCIENT EGYPTIAN RULE FOR THE AREA OF A CIRCLE

Our most important source for ancient Egyptian mathematics, the Rhind Papyrus, dates from the 16th century BCE. It contains a varied set of problems that illustrate different Egyptian mathematical techniques and capabilities, as well as an impressive table of fractions of the number 2 intended as an aid in calculation (see later entry). Several problems treated in the papyrus deal with geometry, and two of the most striking present the reader with an approach to finding the area of a circle.

The basic Egyptian unit of length was the royal *cubit* (20.6 inches; 52.3 centimeters). A *khet*, a unit of land measure, was 100 cubits (about 57 yards). One square khet was known as a *setat*. Problem 50 in the Rhind Papyrus shows the reader how to find the area of a circle with a diameter of 9 *khet*. Such a circle would represent a substantial piece of land, more than 40 acres in modern terms (Gillings 1982, 139). The author of the problem instructs the reader as follows:

Take away $\frac{1}{9}$ of the diameter, namely, 1.

The remainder is 8.

Multiply 8×8.

It makes 64.

Therefore, it contains 64 *setat* of land.

(Gillings 1982, 140)

In short, the Egyptian method is to take $\frac{8}{9}$ of the circle's diameter and to square that figure to find the area. It is important to point out that this approach to the problem does not employ the figure of π, the ratio of the circumference of a circle to its diameter. A modern solution to the same problem would multiply the square of the circle's radius by π to produce an answer of 63.6174. The Egyptian answer to the problem in the Rhind Papyrus (64 *setat*) is thus accurate to within less than 1 percent.

Problem 50 illustrates some important features of ancient Egyptian mathematicians: their ingenuity; their focus on practical applications and measurement; and their concern with quadrature, the interpretation of curved figures in terms of those bounded by straight lines (Clagett 1999, 20). Most of the problems in the Rhind Papyrus deal with issues of measurement: the division of loaves of bread, the strength of beer, or in the case of Problem 50, the division of land. The relatively small degree of error matters little when considering such questions; this demonstrates the Egyptian desire for workable solutions to real-world problems. In addition, the problem represents an interesting approach to the quadrature of the circle; it equates the area of a circle to that of a square with a side $\frac{8}{9}$ of the circle's diameter (Figure 2.1).

One may argue that the author of this problem is the first person in history to suggest a solution to the problem of relating the geometry of the circle to that of the square.

It is not clear how the Egyptians discovered this rule for the area of a circle, but Problem 48 in the Rhind Papyrus may shed some light on its origins. The problem asks the reader to compare the area of an octagon with the area of a square in which it is inscribed (Figure 2.2).

The length of the side of the square is 9 Egyptian *khet*; the octagon is inscribed within it. Some historians suggest that the octagon in question is intended to serve as an approximation of the area of a circle; both would be nine units wide at the point where they touch the sides of the square. By finding the area of the inscribed octagon and relating it to the square, one may approximate the area of a circle inscribed in a square. If the square is divided into nine equal boxes, then each small box must have a side length of 3 *khet* and an area of 9 *setat* (square *khet*). The octagon is made up of 7 such boxes: 5 whole boxes and 4 halves. That makes the area of the inscribed octagon 63 *setat* (7 boxes × 9 *setat* each). This figure may be accepted as the area of the inscribed circle. The error involved—the difference in area between the circle and the octagon—is negligible. As a further approximation, one may note that 63 *setat* is very close to the figure of 64 *setat* and that 64 *setat* is the area of a square with a side length of 8 *khet*. Thus, the area of a circle with a diameter of 9 units is approximately the area of a square with a side length of 8 units. To speak in more general terms, the area of any circle is equivalent to that

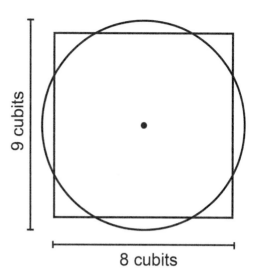

Figure 2.1 In Problem 50 of the Rhind Papyrus, the area of a circle 9 units in diameter is equated to that of a square with a side of 8 units.

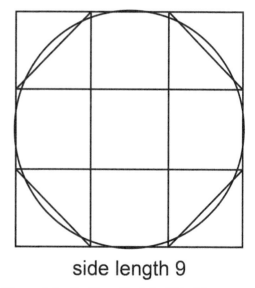

Figure 2.2 Problem 48 of the Rhind Papyrus: area of an octagon inscribed in a square.

of a square with a side length $\frac{8}{9}$ of the diameter of the circle, a workable solution (Gillings 1982, 141–146).

Other historians reject the relationship between Problem 50 and Problem 48. They suggest that the Egyptian rule for the area of a circle probably was developed from the practice of measuring the volume of a cylindrical granary, and therefore the amount of grain it could contain (Clagett 1999, 79). In order to do so, some approximation of the area of a circle would be required. In any case, the circular geometry exhibited in Problem 50 of the Rhind Papyrus serves to illuminate the practical, even pragmatic, character of Egyptian mathematics. This pragmatism extended to the question of proof as well. While later geometrical proofs would be based on logical necessity, Egyptian proofs depended upon testing. In other words, if a given approach produced workable results when applied, it was accepted. The Egyptian rule for the area of a circle satisfied these conditions, and it served its purpose for more than 2,000 years.

CULTURAL COMMENTARY: MEASUREMENT AND SOCIETY

In the valley of the Nile lay the ancient kingdom of Egypt, renowned for its wealth and power. For more than 3,000 years, Egypt occupied a central place in the politics and culture of the Eastern Mediterranean, and it challenged the empires of Mesopotamia and Asia Minor for supremacy in the region. Culturally, it had a profound influence, affecting areas from Nubia to Greece. The artifacts and monuments of Pharaonic Egypt still capture the modern imagination, and three of these reveal a great deal about the place of mathematics in ancient Egyptian culture: *The Book of the Dead*, the Great Pyramid of Khufu, and the Palermo Stone.

The Book of the Dead played an important role in the funerary rites of prominent Egyptians during the period of the New Kingdom, from the 16th to the 11th centuries BCE. Compiled from earlier images, rites, and customs, the book contained an intricate set of prayers and spells intended to guide a deceased person along the difficult paths to the next world. Wealthy Egyptians were frequently entombed with a copy of the *Book of the Dead*, written on papyrus and often richly illustrated. Some surviving examples of the *Book of the Dead* include an image of the deceased appearing for judgement before Osiris, god of the underworld (Figure 2.3).

Anubis, the jackal-headed god of death and the afterlife, weighs the heart of the dead person in a balance against a white feather, representing *ma'at*—truth or order. If the heart and *ma'at* are in balance, then the person is allowed into the afterlife. If not, the heart is thrown to Ammit, the

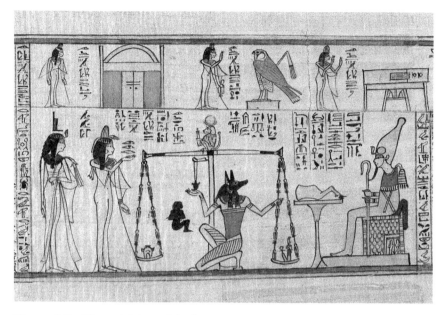

Figure 2.3 The weighing of the heart is an important scene from the ancient Egyptian *Book of the Dead*. Here, the god Anubis weighs the heart of a woman as the god Osiris looks on. (The Metropolitan Museum of Art/Rogers Fund, 1930)

eater of the dead—a terrible beast resembling a blend of crocodile, lion, and hippopotamus. Thoth, god of wisdom, is often depicted with a stylus to record the outcome of the weighing. The image presents a fascinating example of the Egyptian view of life as an ethical test, concluding in an act of divine judgement. Further, one is struck by the fact that this particular episode of divine judgement is essentially an act of measurement, complete with instruments and the recording of results. In a society deeply concerned with the afterlife, one's ultimate fate is determined by a balance, a scale. One may argue that the *Book of the Dead* reflects the importance of quantitative measurement in Egyptian culture.

An older and more iconic monument to Egyptian beliefs about life and death, the Great Pyramid dates from the time of the Old Kingdom (27th to 22nd centuries BCE). The largest of the pyramids, it was built during the reign of the Pharaoh Khufu in the 26th century BCE (Figure 2.4).

Historians suggest that a massive workforce, including thousands of skilled laborers, completed the project in ten to fourteen years. The builders exhibited an astounding degree of precision. Flinders Petrie, the British archeologist who surveyed the Great Pyramid in 1880, wrote that the original baselines measured 440 Egyptian royal cubits (an Egyptian royal cubit was slightly less than 21 inches). He found that the largest variance or error of the

Figure 2.4 The Great Pyramid of Khufu was built during the 26th century BCE. Its precise design and construction illustrate the engineering skill of the ancient Egyptians. (Vladyslav Siaber/Dreamstime.com)

baselines was only 7 inches, in an average length of 755 feet, 9 inches—less than .07 percent (Romer 2016, 384). The right angles of the walls were accurate to with 58 seconds of arc, and the heights of the four sloping faces of the pyramid varied by only 2 centimeters—less than an inch in a total height of 455 feet, or 280 cubits (Schneider 2013, 172). Such accuracy in the execution of the design demonstrates the great skill of ancient Egyptian engineers and their capacity for accurate measurement. John Romer, in his *History of Ancient Egypt*, suggests that the Egyptian builders designed the emerging structure itself to be a gauge of the accuracy of the construction work, using the sight lines and diagonals of the growing pyramid to indicate any irregularities (Romer 2016, 384). The Great Pyramid of Khufu thus embodied the rigor of ancient Egyptian practical mathematics.

The Palermo Stone also dates from the time of the Old Kingdom. It is a fragment of a large stone monument, or stele, carved with the written annals of the first five dynasties of ancient Egypt. The intact stele listed the rulers of these dynasties, along with notable events during each year of their respective reigns (Figure 2.5).

Prominent among the events preserved on the Palermo fragment are accounts of the yearly flooding of the Nile River. For example, the Palermo Stone tells that in the fourteenth year of the reign of King Sneferu, the Nile rose to a level of 8 cubits, 1 palm, and 1 finger, that is, 8 feet $10\frac{4}{5}$ inches (Romer 2016, 94). The accumulation of data on the Nile, from the

Figure 2.5 The Palermo Stone contains records from the early dynasties of ancient Egypt. It features precise measurements of the flooding of the Nile River. (World of Triss/Alamy Stock Photo)

Palermo Stone and elsewhere, reveals a great deal about the priorities of ancient Egyptian civilization and the role of measurement, number, and quantification in pursuing those priorities.

The civilization of ancient Egypt brings to mind images of great cities, extraordinary monuments, and striking works of art and architecture. Yet as the Nile measurements from the Palermo Stone indicate, all of these things ultimately rested upon the extraordinary fertility of the Nile Valley and the agriculture that it supported. Egypt contained some of the most productive agricultural land in the ancient Mediterranean and Near East; centuries later, Rome would still depend on surplus Egyptian grain to help feed its empire. That fertility, that extraordinary production, proceeded from the yearly flood of the Nile during the Egyptian season of *Akhet*, (the inundation): from June to September. The Nile flooded predictably every year and left behind a layer of fresh silt that made Egyptian agriculture possible. The Greek historian Herodotus, writing of his travels in Egypt in the 5th century BCE, writes:

> ... for as soon as the river has risen and watered the land and then retreated again, every man sows his field, and having put in his swine to tread in the seed, he waits for the harvest ...

(Herodotus 1958, 97)

The precision of the measurements from the Palermo Stone indicates how the Egyptians took great care to record the behavior of the river, perhaps the single most important factor that affected agriculture and food supplies. Similarly, the Egyptian calendar represents a practical application of careful quantitative measurement. Herodotus credits the Egyptians with developing a sophisticated calendar based on the measurement of the motion of the stars, as opposed to the phases of the moon. Significantly, that calendar was based upon the rising of the star Sirius, an event that usually coincided with the flooding of the Nile. It is quite likely that the Egyptian calendar arose from measurements and observations linked to the fluctuations of the river. In short, for the Egyptians, the Nile was the foundation of their society, and they sought to describe it quantitatively.

Thus, the ancient Egyptians apparently viewed mathematics in largely practical terms, particularly as a means to manage agriculture, food supplies, and engineering through the act of measuring and recording. Egyptian numeration and mathematics exhibited great consistency over 2,000 years; traditional systems of numeration and methods of computation changed relatively slowly as far as available sources indicate. Hieroglyphic and hieratic numbers appeared in the late fourth millennium BCE and remained in use for centuries. Much computation was done by techniques of doubling (see later)—a direct and effective method with its roots in everyday counting and inventory. Egyptian geometry also had a practical character; it appeared to be focused on the surveying of farmland and the guidance of engineering projects, as the Great Pyramid demonstrates. The Egyptian method for finding the area of a circle suggests an emphasis on concrete applications as well (see earlier). Water supplies, agriculture, engineering, surveying—all reveal the importance of measurement and quantification as means of managing a prosperous society. Indeed, the Rhind Papyrus, the most useful surviving mathematical text from ancient Egypt, contains a series of problems, probably meant for students, that deal largely with the even division of loaves of bread among recipients, the determination of areas and volumes, measures of grain, and the design of granaries and pyramids. Certainly, one might conclude that for the ancient Egyptians, mathematics and measurement served as a path to stability and order in this life, and perhaps the next.

EGYPTIAN NUMERATION

The ancient Egyptians used a system of numeration based on powers of 10, and they employed two different sets of symbols for mathematical purposes. While the symbols and conventions for representing numbers evolved somewhat over 3,000 years, the general system remained surprising stable over such a vast period. Egyptian numeration corresponded to Egyptian writing; as a general rule, the ancient residents of the Nile Valley used *hieroglyphic* symbols for formal and monumental purposes and *hieratic* characters for ordinary recordkeeping.

Hieroglyphic writing was the older of the two systems and appears in Egyptian culture before 3000 BCE. Hieroglyphics consisted of a series of pictorial symbols, read from right to left or from left to right. The characters faced towards the start of the line, so right-facing characters were read from right to left and vice versa. Hieroglyphic characters certainly had a pictographic character; that is, the character indicated what it depicted. An image of an eye represented an eye; an image of a bird represented a bird. However, the same symbols could be used as ideograms to represent concepts, and mathematical symbols fell into this category. Finally, hieroglyphic characters could also stand for specific syllables in spoken language, so a series of them could be used to denote words or names. While hieroglyphic writing was quite complex, the representation of numerical quantities had a simple character. Seven symbols were employed to denote specific powers of 10: a vertical stroke for 1, an inverted "U" for 10, a coil for 100, a lotus flower for 1,000, a bent rod or finger for 10,000, a tadpole for 100,000, and a kneeling man for 1 million (Figure 2.6).

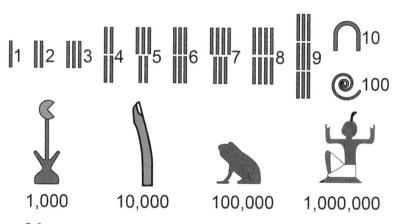

Figure 2.6

Egyptian numbers were not expressed in terms of place-value and did not include a symbol for zero. Quantities were expressed by writing the appropriate powers of 10 as many times as necessary. For example, 30,415 was written as shown in Figure 2.7 (read from the left).

34   Numbers

Figure 2.7

While 1,100,000 would look like Figure 2.8.

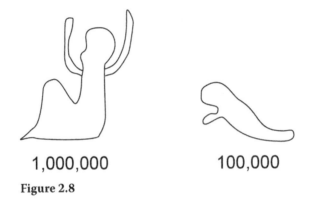

Figure 2.8

Such numbers could be read left to right or right to left, depending on the orientation of the characters (Figure 2.9).

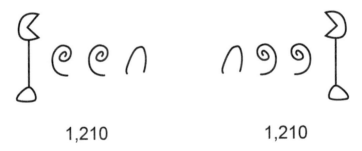

Figure 2.9

Over time, writing conventions developed that made numbers easier to read. Specifically, signs of a given magnitude were clustered together into easily recognizable quantities (Ifrah 1998, 324). For example, the number 659 contained three distinct groupings (Figure 2.10).

Numbers in Ancient Egypt   35

659

Figure 2.10

While hieroglyphics were employed in formal situations, hieratic script represented a more compact and flexible means of writing. First developed between 3200 and 3000 BCE, hieratic script was used for administrative, scholarly, and religious purposes. Written from right to left, usually on papyrus, it was also the script applied to most mathematical work in ancient Egypt. Hieratic numerals probably evolved from hieroglyphics; hieroglyphic characters were blended together to create more easily written figures. However, it was more demanding to master; separate hieratic characters were developed for the numbers 1 to 9 and all their multiples of 10 (up to 9,000) (Figure 2.11).

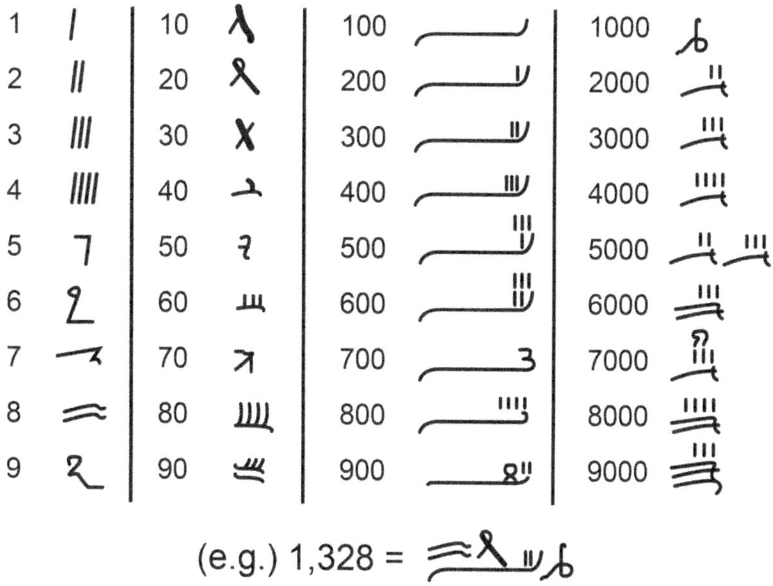

Figure 2.11 Egyptian hieratic numerals.

Clearly, such a system allowed for a more streamlined manner of mathematical expression; numbers could often be written with far fewer characters than with hieroglyphics. The figure 7,641 may be written with just four characters (Figure 2.12).

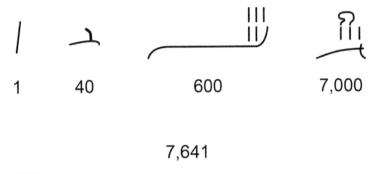

7,641

Figure 2.12

On the other hand, an individual scribe needed to work with thirty-six separate signs for the numbers up to 10,000 (as opposed to the ten signs used in modern mathematics for most operations: 0, 1, 2, 3, 4, 5, 6, 7, 8, 9). A high level of training would be required for such facility, and Egyptian scribes were clearly elite in this sense.

The ancient Egyptians had strict conventions for dealing with rational numbers, or fractions. They expressed these quantities in terms of unit fractions, that is, fractions with the number "1" as the numerator: $\frac{1}{5}$, $\frac{1}{12}$, $\frac{1}{20}$. Any other fraction was reduced to a series of unit fractions, written in a sequence of decreasing value. No unit fraction would be repeated, and no sequence ran longer than four terms. For example, the fraction $\frac{3}{5}$ would be written as follows:

$$\frac{1}{2} + \frac{1}{5} + \frac{1}{20}$$

The only nonunit fractions allowed in a sequence were the special terms $\frac{2}{3}$ and $\frac{3}{4}$. The Egyptians proved extraordinarily skilled in finding the simplest possible sequences for any given fraction.

In hieroglyphics, fractions were written with the mouth-shaped symbol for "part" over a number. Thus, $\frac{1}{12}$ would be written as shown in Figure 2.13, reading right to left.

For larger expressions, the "part" symbol would be placed over a portion of the number. So, $\frac{1}{322}$ would look like Figure 2.14.

Figure 2.13 1/12 in hieroglyphics.

Figure 2.14 1/322 in hieroglyphics.

Hieratic fractions replaced the "part" symbol with a line or a dot. The mathematical tables of fractions that play an important part in Egyptian calculation were written in hieratic characters.

EGYPTIAN CALCULATION

Egyptian methods of calculation followed a series of mathematical conventions, probably derived from practical necessity and preserved for centuries in a highly conservative intellectual culture. While these conventions and techniques appear quite labor-intensive to a modern reader, training, repetition, and the use of mathematical tables as aids to problem-solving probably facilitated operations to a great degree. Hieroglyphic and hieratic numbers were not based on place-value, and were thus difficult to manipulate when doing complex mathematical operations. However, Egyptian scribes, engineers, and bureaucrats developed effective strategies to satisfy the mathematical requirements of their society.

Even without a place-value system, the ancient Egyptians could perform addition and subtraction by combining and carrying. In addition, they probably employed a form of abacus or counting device similar to those in use in Mesopotamia; the Greek historian Herodotus describes such a device in his account of his Egyptian travels. For problems involving multiplication and division, the Egyptians adopted a practical, direct method: doubling and adding. In other words, they would double and redouble the number to be multiplied and then add the appropriate results to yield the desired product. As an illustration of the method, consider the problem 17 × 35. By doubling the factor 35 successively, one gets the following result:

$$
\begin{array}{rcl}
35 \times 1 & \quad 35 & 1 \\
35 \times 2 & \quad 70 & 2 \\
70 \times 2 & \quad 140 & 4 \\
140 \times 2 & \quad 280 & 8 \\
280 \times 2 & \quad 560 & 16 \\
\end{array}
$$

Once the doubling is completed, it is a simple matter to complete the problem. The doubling factors (1, 2, 4, 8, 16) are combined to reach the factor of 17. The corresponding results of the doubling are added to yield the answer, or product.

$$
\begin{array}{r}
560 \ (16 \times 35) \\
+ \ 35 \ (1 \times 35) \\
\hline
595 \ (17 \times 35) \\
\end{array}
$$

The method is just as effective if one doubles 17 successively:

$$
\begin{array}{rcl}
17 \times 1 & \quad 17 & 1 \\
17 \times 2 & \quad 34 & 2 \\
34 \times 2 & \quad 68 & 4 \\
68 \times 2 & \quad 136 & 8 \\
136 \times 2 & \quad 272 & 16 \\
272 \times 2 & \quad 544 & 32 \\
\end{array}
$$

Again, doubling is followed by adding:

$$
\begin{array}{r}
544 \ (17 \times 32) \\
34 \ (17 \times 2) \\
+ \ 17 \ (17 \times 1) \\
\hline
595 \ (17 \times 35) \\
\end{array}
$$

Obviously, the Egyptians would use hieratic numbers, but the essence of the process would be unchanged. The method worked for numbers of any size and would be facilitated considerably by the creation of tables of pre-calculated doubles.

A similar doubling process can be applied to division problems. The divisor is doubled sequentially, and then the appropriate doublings are added to find the quotient. For example, to solve $\frac{943}{23}$, one starts by doubling the divisor, 23:

$$
\begin{array}{rcrrr}
23 & \times & 1 & 23 & 1 \\
23 & \times & 2 & 46 & 2 \\
46 & \times & 2 & 92 & 4 \\
92 & \times & 2 & 184 & 8 \\
184 & \times & 2 & 368 & 16 \\
368 & \times & 2 & 736 & 32
\end{array}
$$

Next, the doubling results are combined to reach the dividend, 943:

$$
\begin{array}{rr}
736 & 32 \\
184 & 8 \\
\underline{23} & \underline{1} \\
943 & 41
\end{array}
$$

By adding the doubling factors that correspond to the appropriate doubling results, one reaches the quotient, 41.

The doubling method had a pragmatic character; it functioned as an effective means of solving the problem at hand. Such calculating methods reinforce the view of Egyptian mathematics as focused on practical problems. That is not to say that such mathematics were rudimentary, for Egyptian scribes could deal with a wide variety of calculations. While doubling uses the number 2 as a multiplier, the Egyptians would also use 10, $\frac{1}{2}$, or $\frac{2}{3}$. This significantly increased the speed of their work.

As in Mesopotamia, the Egyptians employed mathematical tables to aid in calculation, just as modern mathematicians did in the days before computers and electronic calculators. This allowed scribes or engineers to draw on prefigured values for often-used quantities. One can imagine the value of a table of doubles for the process outlined earlier (although such a table would be tremendous). The Egyptian Mathematical Leather Roll, a text from approximately 1650 BCE, contains tables of fractions. Tables in other texts address the multiplication of fractions or the divisions of the *heqat*, the principal Egyptian unit of volume (Clagett 1999, 19). However, the best-known tables are found in the Rhind Papyrus. In fact, that hieratic document serves as the most important source for understanding Egyptian mathematical techniques and goals.

RHIND MATHEMATICAL PAPYRUS

The Correct method of reckoning, for grasping the meaning of things, and knowing everything—obscurities and all the secrets.

(MacGregor 2010, 107)

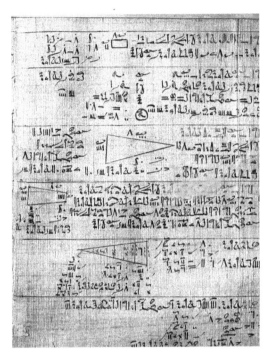

Figure 2.15 The Rhind Papyrus reveals a great deal about ancient Egyptian mathematics. It consists of a series of problems, perhaps intended for educational purposes. (Album/Alamy Stock Photo)

So begins the Rhind Papyrus, the best surviving source for ancient Egyptian mathematics (Figure 2.15). The document dates from the 16th century BCE, during the reign of the Hyksos king Apophis. However, the papyrus indicates that it was actually copied from a far older source, from the time of King Amenemhet III (1844–1997 BCE). Named for Henry Rhind, the Englishman who purchased it in the mid-19th century, the Rhind Papyrus is part of the collection of the British Museum, although a small fragment of it resides in the Brooklyn Museum in New York. It contains two tables of fractions, presumably for use in calculation, and eighty-four problems that illustrate Egyptian mathematical methods and priorities quite effectively. The opening passage of the Rhind Papyrus suggests the elite status of mathematical knowledge in ancient Egypt; such knowledge represents "the meaning of things" and contains the key to "obscurities and all the secrets." These secrets were mastered by only an exalted few. However, many of the problems in the text deal with the measurement and distribution of food and other commodities; others treat useful techniques in calculations that might be applied to administration or engineering. "The meaning of things" mentioned at the start of the papyrus could refer to the special mathematical skills needed to manage the complex society of the Nile Valley.

The first part of the Rhind Papyrus is perhaps the most significant: a table of fractions representing the divisions of the number 2 by all the odd numbers from 3 to 101. In typical Egyptian fashion, the figures are expressed as sums of unit fractions, that is, fractions with the number 1 as the numerator. For example, $\frac{2}{15}$ is represented as $\frac{1}{10} + \frac{1}{30}$. The table is probably intended as an aid to other mathematical operations, as a tool for scribes and engineers. Such scribes could draw on its values when solving problems, much like a trigonometric table today. It must have been produced through painstaking calculation, and modern historians debate the methods that may have been employed in its construction. The table exhibits genuine economy and elegance; while the values that it contains could be indicated by various combinations of fractions, the sequences listed are usually the shortest ones possible without repeating any particular unit fraction. The entire table testifies to the mathematical skill of the authors and to the strict conventions observed by ancient Egyptian mathematicians, rules that are consistently followed over centuries.

The papyrus continues with a second table that shows the divisions of the numbers 1 to 9 by 10 and then moves on to a series of problems. Many of these have a practical character. Some explore the sharing of different numbers of loaves of bread among ten men. Solutions consist of a series of unit fractions representing the proper division:

Divide 7 loaves among ten men.

Do the multiplication of $\frac{2}{3}$,

$\frac{1}{30} \times 10$;

the result is 7.

Total: 7 loaves, which is the correct total for 10 men after each man receives $\frac{2}{3}$; one-tenth of it.

(Clagett 1999, 135)

One wonders if the Egyptians actually divided loaves into such small pieces or if these problems are meant to demonstrate a general process for the division of resources. Other problems deal with *pefsu* (or *pesu*), that is, the relative amount of grain in bread or beer. In other words, bread or beer could be concentrated or weak, with obvious effects on its nutritive value. They ask, for example:

$3\frac{1}{2}$ heqat of meal is made into 80 loaves of bread. Make known to me the amount of meal in each loaf and their pefsu.

Or

From 1 des-jug of beer $\frac{1}{4}$ has been poured off and then the jug has been refilled with water. What is the pefsu of the diluted beer?

(Clagett 1999, 174–176)

These problems employ the Egyptian units particular to the measurement of food volume and potency and resemble modern exercises in percentages.

Aha problems address the finding of unknown quantities, given some information about the quantity:

A quantity with $\frac{1}{5}$ of it added to it becomes 21.

What is the quantity?

The quantity is $17\frac{1}{2}$ and $\frac{1}{5}$ of it is $3\frac{1}{2}$ and the total is 21.

(Clagett 1999, 142)

The Egyptians solved these problems through an interesting process of estimation and testing. Some of their techniques anticipate those of Arab algebra, developed twenty-eight centuries later.

The Rhind Papyrus also treats geometrical problems. Some seek the volumes of differently shaped granaries; others treat the areas of different polygons. The Egyptians knew the formulae for the areas of rectangles, triangles, and trapezoids and had an effective approximation for the area of a circle (see earlier). By combining these, they could find the areas of different irregular shapes. These ideas demonstrate a sophisticated geometry that could also be applied to issues of land, agriculture, and engineering (see next entry).

For a modern reader, the Rhind Papyrus serves as a compendium of Egyptian mathematical capabilities in the early second millennium BCE. It appears to be an educational text, perhaps a guide to problem-solving for young scribes, complete with useful tables for future calculations. It reveals some of the priorities of Egyptian scribal training and illustrates the practical quantitative character of ancient Egyptian administration.

ANCIENT EGYPTIAN GEOMETRY

> ... this king (Sesostris) made an equal division of all the land in Egypt and allotted a square piece to every man; and he fixed a certain rent to be paid for it every year ... yet if any man were robbed by the river of a part of his allotment he was at liberty to go before the king and declare the fact, and the king would send some to verify and measure the loss ... I think it was in this way that geometry came to be discovered ...
>
> (Herodotus 1958, 94)

In the second book of his *History*, written in the mid-5th century BCE, the Greek author Herodotus credits the Egyptians with the discovery of geometry. More specifically, he claims that this knowledge arose from the measuring of farmland along the Nile River. Herodotus traveled through Egypt long after the height of its power, and much of what he says in his book is fanciful. However, an examination of the Rhind Papyrus and the Moscow Mathematical Papyrus supports the view that Egyptian geometry was focused largely on the measurement of land and the planning of structures.

The Moscow Mathematical Papyrus dates from the mid-19th century BCE, around the same period as the parent document of the Rhind Papyrus. Like the Rhind Papyrus, it contains a series of problems, as well as tables to aid in calculation. The sixth problem of the Moscow Papyrus indicates that the Egyptians knew the process for finding the area of a rectangle, multiplying its length by its width. The Rhind Papyrus also has a number of geometrical problems dealing with area. Problem 51 shows the reader a method for finding the area of a triangle, multiplying the height by $\frac{1}{2}$ of the base. Problem 52 seeks the area of a trapezoid, and Problem 50, as shown earlier, treats the area of a circle. It important to point out that all of these area problems from the Rhind Papyrus deal explicitly with the measurement of land. While some problems ask about geometry on its own (Problem 48, as discussed earlier, looks at the area of a circle and its circumscribing square, using an octagon as an approximation), it appears that the management of land played a crucial part in Egyptian mathematics.

The Rhind Papyrus also treats three-dimensional geometry. Problem 41 asks for the volume of a cylindrical granary with a diameter of 9 units and a height of 10 units. The text instructs the reader to find the area of the circular cross-section of the granary and to multiply that by the height of the structure. The cross-sectional area is found through the method discussed in Problem 50. In fact, some scholars have suggested that the Egyptian approach to finding the area of a circle grew from solving such problems. The Rhind Papyrus has a number of problems, some quite challenging, that explore the storage of grain as measured in their volumetric units: *heqats* and *khar*.

Pyramids are perhaps the most iconic features of ancient Egyptian culture, and both the Rhind and the Moscow papyri address the geometry of pyramids, presumably in a way that related to the building process. These texts both date from a time when pyramids were still built to serve as tombs for Egypt's rulers. Neither papyrus relates the formula for the volume of a whole pyramid, but they do examine the volume of truncated pyramids, that is, partial pyramids with the apex removed. In addition, the Rhind Papyrus has a number of *seqed* problems. *Seqed* is related to the modern notion of slope (as the cotangent of the slope angle) and may be defined as "the inclination of any one of the four faces of a pyramid to the horizontal plane of its base" (Gillings 1982, 212). It was measured in terms of horizontal run and vertical rise and must have been a crucial factor in the design of any pyramidal structure.

In short, the available sources indicate that the geometry of the ancient Egyptians was largely directed towards practical purposes: land measurement, engineering, and the design of buildings. While the Rhind and Moscow texts do treat some questions of pure geometry, these appear to be building skills necessary for the solution of more complex practical problems. Egyptian geometry thus appears to be the most sophisticated part of a highly conservative tradition of applied mathematics.

EGYPTIAN CALENDAR

> ... they all agreed that the Egyptians were the inventors of the year and the division of the course of the seasons in twelve, and they said they found out how to do it from the stars. ... the Egyptians make each of the twelve months thirty days and then add five days to make up the year, so that the cycle of the seasons is completed in every year.
>
> (Herodotus 1958, 94)

The Greek historian Herodotus toured Egypt in the 5th century BCE, and in his account of his travels, he credits the Egyptians with "inventing the year" and basing it on stellar observation; he goes on to describe the nature of the Egyptian calendar. The ancient Egyptian civic calendar represents the earliest example of a 365-day yearly cycle based on stellar observation. As such, it serves as a potent example of the Egyptian emphasis on measurement, recordkeeping, and practical quantification. Most ancient calendars were lunar in nature and based their months on the 29-day cycle of the phases of the moon. The Babylonian, Greek, and early Egyptian calendars all shared this feature; indeed, the Egyptians continued to use the lunar calendar for religious purposes for centuries after their development of the 365-day cycle. However (to speak in modern terms), the lunar cycle does not correspond to the orbit of the earth around the sun, for twelve lunar (or synodic) months do not make up a 365-day year (they add up to 348 to 354 days). As a result, lunar calendars became progressively detached from astronomical and seasonal events over time. In order to address this issue, the Babylonians (and other cultures) periodically added a thirteenth month to the year, in a process known as intercalation. Ultimately, the Babylonians developed the Metonic calendar, composed of twelve 12-month years and seven 13-month years. This nineteen-year cycle, based on lunar months, proved to be quite accurate.

The Egyptians developed a more sophisticated calendar early in the third millennium BCE. The calendar was originally based on the heliacal rising of the star Sirius, the date on which the star first appears above the horizon at dawn. The rising of Sirius probably struck the Egyptians as significant for two reasons: Sirius is the brightest star in the sky, and it appears at the time when the crucial Nile floods begin. They noted that heliacal risings of Sirius were separated by a period of 365 days (again, in modern terms, the periodic rising of Sirius is a consequence of the earth's orbit). The Egyptians thus created an impressive stellar, or sidereal, calendar and linked it to the important agricultural events that punctuated their existence. They divided the 365-day cycle into twelve 30-day months, with five feast days added. The months were grouped into three seasons: *Akhet* (flood), *Peret* (growing season), and *Shemu* (harvest).

The Egyptians had a detailed understanding of the movement of the night sky, and they marked the hours of the night by the sequential rising of small groups of stars called *decans*. Like Sirius, each decan also had a yearly heliacal rising, that is, the date on which it first appears in the sky at sunrise. These heliacal events were used to divide the 30-day months into 10-day weeks: a new decan rose heliacally every 10 days. One must note that the Egyptians tied their calendar closely to their measured observations of the stars. The calendar also had an important religious component; Egyptian deities were associated with the stars, and the five feast days at the end of the year were devoted to the great gods Osiris, Horus, Set, Isis, and Nephthys. The movement of the stars, the seasons of the year, the tasks of agriculture, and the worship of the divine all came together in the structure of the calendar.

The earth actually orbits the sun every $365\frac{1}{4}$ days. As a result, like the lunar calendars discussed earlier, the Egyptian 365-day calendar became detached from astral and seasonal events, although at a much slower rate. Originally, the heliacal rising of Sirius heralded the beginning of the year, on the first day of the season of *Akhet*. Over time, that celestial event worked slowly backwards through the calendar year. Because of the extraordinary longevity of Egyptian civilization, Egyptian astronomers recognized that Sirius would rise again on the first day of *Akhet* after 1,461 years—one *Sothic Cycle*.

While Pharaonic Egypt had been conquered before in its long history, after the 7th century BCE it would come under the control of a series of foreign powers: first the Persians, then the Macedonians and Greeks, and finally the Romans. In the 1st century BCE, the Romans extended the Julian calendar to Egypt, adding an extra day every fourth year and correcting the $\frac{1}{4}$ day error. However, the ancient Egyptian calendar stands out as the first truly sidereal yearly cycle in history, and the later calendars of the Mediterranean all felt its influence.

CONCLUSION

Over many centuries, the ancient Egyptians developed a tradition of measurement and applied mathematics. They recorded the behavior of their local environment and of the heavens in quantitative terms, and they employed their mathematical knowledge in the administration and distribution of resources and the design of structures. The surviving mathematical texts from ancient Egypt seem to have served as repositories of such knowledge, intended for the training of scribes and engineers. The culture of ancient Egypt exhibited great stability over the millennia, as is demonstrated by the persistence of hieroglyphic and hieratic writing. The Rhind and Moscow papyri, and the other Egyptian mathematical sources that remain, appear to be a manifestation of that stability. Some scholars have suggested that more abstract mathematics evolved in ancient Egypt, as it did in Mesopotamia. That is possible, but no evidence of such work is known. What is clear is that the Egyptians used mathematics as a tool to manage their society and their world, from the strength of the bread they baked to the flooding of the Nile River.

FURTHER READING

Clagett, Marshall. 1999. *Ancient Egyptian Science, A Source Book, vol 3*. Philadelphia: American Philosophical Society.

Gillings, Richard. 1982. *Mathematics in the Time of the Pharaohs*. New York: Dover Publications.

Herodotus. 1958. *History, Book II*. New York, Heritage Press.

Ifrah, Georges. 1998. *The Universal History of Numbers: The World's First Number Systems*. London: The Harvill Press.

Imhausen, Annette. 2016. *Mathematics in Ancient Egypt*. Princeton: Princeton University Press.

MacGregor, Neil. 2010. *History of the World in 100 Objects*. London: Viking Press.

Romer, John. 2016. *A History of Ancient Egypt*. New York: St. Martin's Press.

Schneider, Thomas. 2013. *Ancient Egypt in 101 Questions and Answers*. Ithaca: Cornell University Press.

Shaw, Ian. 2004. *Oxford History of Ancient Egypt*. Oxford: Oxford University Press.

3

Numbers in Ancient Greece

Note: *In this chapter, and in succeeding ones, the term "abstract" refers to things that are separated from worldly reality. "Abstract truth," then, is pure idea—idea that does not proceed from the experiences of the physical world. Many thinkers consider pure mathematics to be a form of abstract truth.*

Chronology of Major Figures

Thales of Miletus	(approximately 624–546 BCE)
Pythagoras of Samos	(approximately 570–495 BCE)
Plato	(427–347 BCE)
Eudoxus of Cnidus	(390–337 BCE)
Aristotle	(384–322 BCE)

CULTURAL ICON: THE SPHERICAL COSMOLOGY OF EUDOXUS

In *The Discarded Image*, C.S. Lewis writes of the earth-centered view of the universe dismissed by Galileo, Kepler, and their contemporaries in the 17th century. This model of the cosmos featured a central earth surrounded by a series of concentric, solid, transparent spheres, rather like the layers of an onion. The sun, moon, stars, and planets were embedded in these spheres, and as the spheres rotated, they carried the heavenly bodies in circular paths around the earth (Figure 3.1).

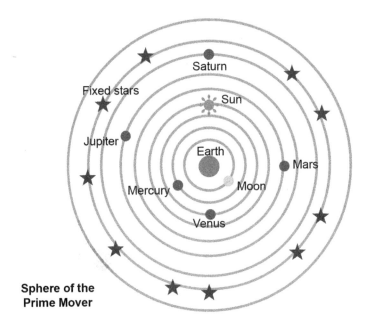

Figure 3.1 Medieval Europe adopted the Greek idea of an earth-centered system of spheres carrying the heavenly bodies in circular paths.

The scheme proceeded largely from the ideas of the ancient Greek philosopher Aristotle (384–322 BCE), and it dominated European cosmology in the Middle Ages and the Renaissance. However, Aristotle actually borrowed the concept from Eudoxus, a philosopher and mathematician of the early 4th century BCE. Eudoxus's cosmology helps to illustrate the place of mathematics, particularly geometry, in the intellectual culture of Classical Athens.

Aristotle argued that a series of actual spheres made up the universe, and he sought to explain their actual nature. His system addresses the material composition of the spheres, as well as their physical interactions with each other. On the other hand, his predecessor Eudoxus considered the heavenly spheres in a more theoretical sense, as a complex geometrical solution to the problem of celestial motion. Eudoxus wished to account for the observed movements of the heavenly bodies: the cycle of the stars and the individual paths of the sun, moon, and five visible planets against the star background. However, he worked from some basic assumptions. First, for him the earth lay at the center of the universe—a reasonable view based on the fact that people do not perceive the earth as moving. In addition, he would only use circular or spherical geometry in his explanation, because he accepted the idea that the heavenly motions were eternal and therefore had to be based on the most perfect and changeless of geometrical forms, the circle. In adopting this concept, Eudoxus was following the teaching of his contemporary, the Athenian philosopher Plato, and his system reflected Platonic philosophical values.

Bound by the assumptions that celestial movement must be circular, unchanging, and earth centered, Eudoxus suggested that the motions of the heavenly bodies could be explained by the combined rotations of a series of concentric spheres. He proposed four spheres to produce the movement of each planet. Each sphere rotated in a constant way, and the planet's visible path across the sky was a combination of the four different spherical (or circular) components (Figure 3.2).

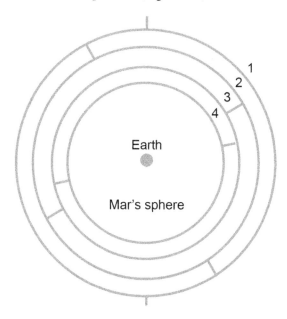

Figure 3.2 For Eudoxus, the movement of each planet resulted from the combined motions of four spheres associated with it.

The sun and moon needed three spheres each to explain their movements, while the fixed stars required just one. The entire system employed 27 spheres.

Eudoxus's system did not predict actual planetary positions. Rather, he sought to show how the complex observed motions of the planets, sun, moon, and stars could be produced by the combination of regular circular or spherical movements. His spheres served as mathematical models to explain observed motion; he probably did not conceive of the spheres as actual objects. Aristotle, with his emphasis on material nature, definitively took that step, giving the system real solidity and making it even more complicated.

The thought of Eudoxus demonstrates many of the features of Greek mathematics in the Hellenic or Classical Period, that is, from about 500 to 323 BCE. Like the Egyptians and the Mesopotamians before them, the Greeks certainly used mathematics in a variety of practical ways: in commerce, for example. However, many Greek thinkers preferred to treat mathematics as an abstract discipline, an exercise in pure thought. Geometry especially fascinated them, and Eudoxus's system is essentially an exercise in geometry. Yet the Greeks also wanted to explain the workings of the actual physical universe. It is therefore not surprising that Eudoxus, an important figure in the history of Greek mathematics, would seek to explain the movements of the observed cosmos with spherical geometry. His system, so influential for so long, reflects the fascinating confluence of Greek mathematics and philosophy.

CULTURAL COMMENTARY: PHILOSOPHY AND NATURE

Ancient Greek culture had its roots in the Bronze Age; the great rulers of Mycenaean Greece were the contemporaries of Hammurabi of Babylon (18th c BCE) and Ramses of Egypt (13th c BCE). However, most of the well-known achievements of the ancient Greeks—the architecture of the Parthenon, the statues of Olympus, the drama of Sophocles—date from much later. The Archaic Period of Greek history stretched from 800 to 500 BCE, and the Classical Period from 500 to 323 BCE, ending with the death of Alexander the Great. During these centuries, the Greeks produced art, literature, and political systems that would have a profound effect on the world. In particular, Greek philosophy represented new approaches to the definition of knowledge and the explanation of nature, approaches that were often intimately associated with mathematical thought.

Plato and Aristotle stand at the foundations of Western philosophy, just as they stand together at the center of Raphael's Renaissance masterpiece, *The School of Athens* (Figure 3.3). Raphael sought to depict eight centuries of ancient Greek thinkers in a single image. Yet despite the dozens of luminaries that populate his fresco, it is still the central group that catches the eye of the viewer. Plato stands lightly on his feet, pointing upwards, holding a book—his *Timaeus*—with his other hand. To his left is Aristotle, feet solidly on the ground, gesturing downwards and grasping his *Ethics*.

The postures give the viewer a brief sense of the intellectual priorities of the two figures. Plato stressed the abstract nature of pure truth—truth that existed beyond the physical world. For example, the idea of a circle does not depend on the existence of real circular objects; one can conceive of a circle in entirely intellectual terms. For him, truth was perfect, eternal, and unchanging, and thus could not originate in a physical universe that changed constantly and where nothing lasted forever. He argued that truth existed in a realm of pure idea, an eternal plane untainted by material existence. Because human minds can know some truth (however incompletely or imperfectly), they must have some capacity to reach that realm. The goal of Plato's philosophy was to help the mind overcome the limitations and distractions of the material world and to reach the truth through acts of disciplined thought and contemplation.

Figure 3.3 Painted during the Italian Renaissance, Raphael's *School of Athens* illustrates the priorities of Plato and Aristotle. Plato points upward to indicate ideal thought, while Aristotle gestures downwards to indicate the importance of concrete thinking. (Pytyczech/Dreamstime.com)

Aristotle, perhaps Plato's greatest student, disagreed with his former teacher. He denied the existence of Plato's ideal realm of truth and instead stressed the systematic study of the physical world itself. For Aristotle, Nature acted consistently, and truth might be found in the consistencies of Nature. Thus, if one wanted to learn about trees, one could observe them and take note of their consistent character. One could learn that a particular tree with a particular kind of leaf always has a particular kind of wood—in that case, one would know the *truth* about the tree through observation and reason. However, the discovery of such truth required discipline and ordered inquiry, and Aristotle wrote extensively on the intellectual processes of observing the universe, constructing valid categories, and making legitimate generalizations. In his two books of *Analytics*, he explored different ways to combine these elements to produce logical arguments. The ideas of Plato and Aristotle had a huge effect on the development of later philosophy in Europe, the Middle East, and North Africa, but it is important to point out that they both were influenced by earlier ideas. For example, both philosophers had important intellectual relationships with the mathematical tradition in Archaic and Classical Greece (that is, Greece from about 800 to 300 BCE).

In Plato's *Meno*, the author wishes to make an argument about the universal character of human virtue. The work is structured as a dialogue between Socrates and Meno, and Socrates presumably expresses the opinions of Plato, the author. Socrates argues that there is a concept of goodness—pure virtue—which people can strive to reach. Meno asks how one may possibly seek for something when one does not yet know its nature. Socrates responds by claiming that the human mind or soul is immortal and can remember truths from its previous existence. In other words, through the voice of Socrates, Plato argues that the human mind can recall ideas from the realm of pure truth (in this case, the notion of virtue).

Socrates demonstrates this by posing a problem in geometry to one of Meno's servants, a young boy who lacks any formal education. He draws a square on the ground and asks the boy the following: given a square with a side length of two units and an area of four units, how can one find the length of the side of another square with twice the area of the first? The boy initially suggests doubling the length of the side, an incorrect solution:

> **Socrates:** Now then, try to tell me how long each of its sides will be. The present figure has a side of two feet. What will be the side of the double-sized one?
>
> **Boy:** It will be double, Socrates, obviously.
>
> **Socrates:** You see, Meno, that I am not teaching him anything, only asking. Now he thinks he knows the length of the eight-foot square.
>
> (Plato 1961, *Meno* 82e)

Socrates proceeds to asks him a series of guiding questions, and ultimately the boy arrives at the correct solution. To double the area of a square, one must draw a diagonal, dividing the square into two triangles. Four of these triangles may be arranged to show that a second square, with a side equal to the diagonal of the first square, would have twice the area of that square (Figure 3.4).

In this part of the dialogue, Plato uses a geometrical

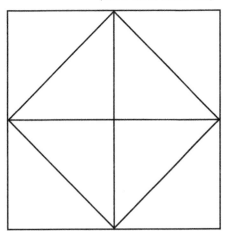

Figure 3.4 Socrates's solution to doubling the area of a square in Plato's *Meno*.

demonstration to illustrate his view of pure truth and human knowledge. For Plato, truth exists in a realm of idea. If human minds have some sense of the truth, then they must have once existed in this realm, with access to truth, before they were embedded in material bodies and thus corrupted. The boy's ability to learn geometry supports Plato's general approach to human understanding:

> **Socrates:** And if the truth about reality is always in our soul, the soul must be immortal, and one must take courage and try to discover—that is, to recollect—what one doesn't happen to know, or more correctly remember, at that moment.
>
> (Plato 1961, *Meno* 86b)

Philosophy may be seen as the effort to master the limits of the physical body and to revive the mind's relationship with pure truth—to recollect the truths with which the mind once had familiarity. This is Plato's explanation for the kind of intuitive knowledge often employed in mathematics—the fact that mathematical ideas just make sense to people. In the verbal exchange concerning the area of a square, Socrates's questions awakened the young boy's recollection of forgotten truth. Socrates (and Plato) goes on to argue that just as the boy may be guided to recall the truths of mathematics, so philosophers may aspire towards the most exalted of all eternal truths, that of pure virtue.

Plato thus uses mathematical knowledge to represent the kind of pure, abstract ideas central to his entire philosophy. Indeed, for him, mathematics is perhaps the most accessible form of truth, the form that requires the least discipline to attain. Plato refers to mathematics many times in his other works; for example, in the *Republic*, he maintains that training in arithmetic, geometry, and harmonics is essential to the education of young guardians, the future philosopher-kings whose capacity to rule depends on their capacity to understand truth. Again, he stresses the abstract character of mathematics as its most significant feature. In addition, in the *Timaeus* (see later discussion), he tells the story of the creation of the physical universe. For him, the universe was created as an expression of eternal ideas, with special emphasis on geometrical truths. In short, the cosmos is based largely on pure geometry.

Plato's emphasis on ideal mathematics suggests a link with the earlier thought of the Pythagorean school (see later discussion). Pythagoras of Samos worked in the late 6th century BCE; his followers maintained his teachings for centuries after that. The Pythagoreans had a particular interest in examples of mathematical harmony in the universe. For them, pure number served as a pattern or a template for the physical world. Some historians suggest that the Pythagoreans went so far as to claim that worldly objects themselves were numbers in material form. Their ideas clearly went beyond mathematics to numerology and number mysticism. However, the notion of pure number as the foundation of physical existence clearly influenced Plato's concept of abstract idea.

Plato's school, the Academy, played an important part in the evolution of Greek mathematics in the 4th century BCE. Plato's respect for mathematics—particularly geometry—as a form of pure thought attracted scholars with similar interests. Indeed, many of the most prominent mathematicians of Classical Greece studied or worked at the Academy, including Theodorus of Cyrene, Theaetetus, and Eudoxus of Cnidus. Eudoxus certainly adopted Platonic values, and his cosmology reflected those views. He explained complex celestial motions in terms of "ideal" circular and spherical geometry, drawing upon Platonic—and perhaps Pythagorean—concepts of mathematical truth as a basis for explaining visible realities. However, the ideas of Eudoxus also serve as a bridge to the consideration of Aristotelian thought, for his system of the universe ultimately gave rise to Aristotle's scheme of the cosmos, an intellectual approach that moved away from abstract truth.

Aristotle sought to explain nature systematically and logically, and his logical methodology exhibits a fascinating relationship with the mathematical culture of Classical Greece. Aristotle is often hailed as the originator of formal logic; his six basic works on logical argumentation are referred to as the *Organon*, or instrument. Aristotle sought to establish rules for constructing arguments that were rationally true, that *must be so*. He stressed the derivation of necessary conclusions from carefully constructed and arranged statements, or premises. In other words, he defined his terms and ordered his statements in such a way that only one result was logically possible. Some of his arguments were based on self-evident principles; for example, that the whole is equal to the sum of its parts. Others included empirical elements, based on observation of the world. Aristotle modeled many different forms of logical arguments, or syllogisms, but the following serves as a basic example:

All men are mortal.

Socrates is a man.

Therefore, Socrates is mortal.

Here, Aristotle defines a category (all men) and links, or *predicates*, a characteristic to that category—that all men will ultimately die. The statement is based on observation of consistent nature—all men and women die eventually. The second statement identifies an individual, Socrates, and places him in the category previously defined. The conclusion proceeds from the two statements: what is true of the category must be true of every member of the category, so Socrates will die one day.

While the example seems simple, the technique involved is quite powerful, for it allows the logician to infer new truths from established ones. It is important to point out that while the content of the previous argument is based on observations of the world, its structure represents a form of self-evident reasoning: if all members of group X have the characteristic Y, and an individual Z is in group X, then Z must have the characteristic Y. The link of such thinking to mathematical or geometrical reasoning is clear.

The works of the *Organon* discuss the varied components of logical reasoning: the construction of valid categories, the nature of definitions, the structure of different kinds of proofs or demonstrations, and the character of self-evident statements. Aristotle intended his logic to be a tool for the analysis of the world around him. He applied his logical methods to a variety of subjects in his many works: the behavior of matter, the motion of objects, the operation of the cosmos, the nature of ethical behavior, the basis of successful government. While Aristotle's interests were focused on nature and society, the deductive character of his logic had much in common with the kind of abstract mathematical and geometrical discourse that was developing in ancient Greek culture before and during his lifetime. His definitions and self-evident statements resembled the definitions and axioms of Greek geometry in particular. It is probable that Aristotle's ideas and those of the Classical mathematical tradition influenced each other.

In short, while Plato's thought related to the abstract character of Classical Greek mathematics and geometry, Aristotle's logic reflected its emerging methodology, a process of rigorous reasoning that would be expressed in a mature form in Euclid's *Elements*, written about twenty years after Aristotle's death.

Greek mathematics and philosophy grew together during the Archaic and Classical periods, and they were closely linked as a result. However, there was also a practical side to Greek views of number. The ancient Greeks had a strong seafaring tradition and a wide-ranging commerce, and mathematics played a major role here. Like the Romans, the Greeks employed a simple numeration system without a notion of place value; alphabetical symbols were used to represent specific quantities, and they were repeated as many times as necessary to make up a given number. It is notable that the Greeks employed special numerical symbols for commercial transactions; trade clearly had a big effect on methods of calculation.

Greeks traded throughout the Mediterranean; they certainly had contact with the civilizations of Egypt and Mesopotamia. It is important to consider the influence of the older traditions of Egyptian and Babylonian mathematics on Classical Greek thought. According to ancient sources, many Greek mathematicians were said to have traveled in

Egypt, including Thales, Pythagoras, and Eudoxus. Egyptian knowledge of measurement, practical geometry, and engineering was transmitted to Greece through such trade and travel. In addition, the Babylonians had developed a highly sophisticated mathematics and geometry, and such knowledge probably influenced the pursuit of mathematics in Hellenic Greece. However, one must point out that during this period, the Greeks developed deductive methods of abstract geometrical proof that do not appear elsewhere, and they used these methods to great effect. The emergence of Greek philosophy and its interactions with mathematics and geometry would have a transforming effect on ancient thought in the Mediterranean.

THALES OF MILETUS

Thales of Miletus worked in the early 6th century BCE. He may be considered the earliest known figure in the ancient Greek philosophical tradition, and he is associated with a number of specific mathematical ideas. His work demonstrated, in a rudimentary form, the Hellenic Greek interest in abstract views of geometry. In addition, ancient accounts of his life suggest a link between his ideas and the geometry of ancient Egypt.

The city of Miletus played a significant role in the history of Greek philosophy and mathematics. Located on the eastern edge of the Aegean Sea, in what is now modern Turkey, ancient Miletus was part of the Greek-speaking world. Thales represents the earliest of a series of Milesian thinkers who stood at the foundations of Greek philosophy, including Anaximenes and Anaximander.

Thales's thought apparently covered a wide spectrum of subjects, from mathematics to the nature of matter. While none of his writings survive, his ideas were briefly described by Aristotle, Herodotus, and other Greek authors. According to Aristotle, Thales claimed that all matter was derived from water as a first principle and that the earth floated upon water. It is notable that the Milesian thinker argued as a natural philosopher, explaining nature in rational or observational terms without reference to mythological entities. Other authors write of Thales's interest in astronomy, his predictions of eclipses, and his knowledge of the sun's apparent movements. His geometrical achievements supposedly included a discussion of a triangle inscribed within a circle. According to tradition, Thales proved that if one of the sides of an inscribed triangle was the diameter of the circle, then the angle opposite that side must be a right angle. This is true regardless of the measures of the other angles of the triangle (Figure 3.5).

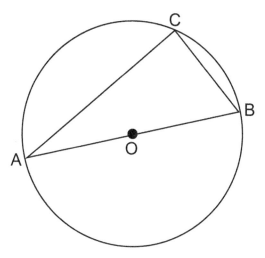

Figure 3.5 Thales's Theorem.

The Greek philosopher Eudemus credited Thales with the theorem that triangles with one side and two adjacent angles in common are identical. This account survives in the work of Proclus, a philosopher of the 5th century CE:

> Eudemus in his treatise on Geometry attributed to Thales this theorem (that triangles which are equal with respect to one side and its two adjacent angles are equal in all respects); arguing that Thales must have employed the theorem in computing, as he is said to have done, the distance of ships at sea.
>
> (Wheelwright 1966, 49)

Thales is also said to have traveled to Egypt and learned geometry there. While in Egypt, he measured the height of the pyramids with relation to their shadows. He noted the time of day when his own shadow was equal to his height because of the angle of the sun, and he argued that at that same time of day, the length of the shadow of a pyramid was equal to its height. In the process, he revealed his knowledge about the similar nature of all right isosceles triangles.

Because Thales's work is only known through the writings of other figures, some composed centuries after his death, his life and accomplishments have a legendary quality to them. Indeed, many historians question the accounts in ancient sources, including those that discuss travels to Egypt. The mathematical ideas attributed to him may have been discovered by others: Babylonians, Egyptians, or other Hellenic Greeks. However, the very respect paid to him by later writers indicates that he probably played some foundational role in Greek mathematics and Greek natural philosophy. Further, the reported link between his ideas and those of ancient Egypt is fascinating. Whether Thales went to Egypt or not, whether he measured the pyramids or not, the ancient authors who recounted such stories accepted some connection between the geometrical traditions of ancient Greece and Egypt. Perhaps such intellectual influences were real, or perhaps they were a fiction of the later Hellenistic period, a time when Macedonian Greeks ruled Egypt and cultural interchange was common. In any case, to the later Greeks, Thales represented a vehicle for the transmission of knowledge.

PYTHAGORAS OF SAMOS AND THE PYTHAGOREANS

Pythagoras of Samos lived and taught in the 6th century BCE. Perhaps the most well-known of the pre-Socratic philosophers, he had a significant impact on later Greek thinkers. In addition to his influence on the philosophy of Plato, his work was discussed by a variety of other Hellenic authors, including Aristotle and Herodotus. Such writers gave Pythagoras a semi-mythical status and credited him with ideas and accomplishments that were probably not his own. Indeed, modern historians disagree about the scope of his work and cannot really determine whether key concepts originated with Pythagoras or with his followers. While he is best known for the theorem that bears his name, Pythagoras's actual interests, and those of his intellectual disciples, probably ranged from mysticism and numerology to mathematics and astronomy.

At the age of forty, Pythagoras left his birthplace and traveled to Croton in southern Italy, founding an intellectual community of sorts. The Pythagoreans placed great emphasis on the notion of the eternity of the soul, or spirit, and its transfer from one body to another at the time of death. Their interest in metempsychosis, or reincarnation, was accompanied by a focus on harmony, the ordered interaction or confluence of parts. The soul was viewed as the source of harmony for the body, its life force. Such beliefs might have served as the basis of the vegetarian practices attributed to the Pythagoreans by other ancient authors; the eating of flesh may have been associated with the destruction of living harmony and therefore forbidden.

In a related vein, the Pythagoreans were credited by later Greek authors with discoveries in the field of harmonics. One of these concerned the mathematical link between the length of a string and the note it produced when plucked. The Pythagoreans observed that the pitch of a note was proportional to the string's length: a string half as long as another would produce a note twice as high (a full octave). Another insight concerned the notes in a consonant musical chord, that is, a chord in which the notes produced beauty and harmony. Such notes were separated by specific mathematical ratios. In other words, there was an apparent link between mathematical patterns and the beauty of sound. Many modern historians dispute the actual discovery of such phenomena by the Pythagoreans. However, it does appear that the Pythagoreans were interested in all physical expressions of mathematical harmony: real things that showed numerical patterns; or numerous components coming together to produce something unified. Such experiences led them to conclude that the physical world was ruled by number—that number lay behind physical existence. The notes produced by a set of strings had a mathematical relationship to each other because the strings themselves were governed by number in some way, or perhaps—a striking assertion—because the strings *were* actually numbers in another form. Aristotle refers to such a view in his discussion of the Pythagoreans in the *Metaphysics*:

> Furthermore they observed empirically that the properties and ratios of harmonious musical tones depend upon numbers. Since they found, in short, that everything else, too, in its intrinsic nature, seemed to be essentially numerical, and thus that numbers seem to be the ultimate meaning of everything that exists, they concluded that the elements of numbers must be the elements of everything, and that the visible heavens in their entirety consist of harmony and number.
>
> (Aristotle, Metaphysics 985 b 22, in Wheelwright 1966, 213)

Philolaus, a later member of the Pythagorean school, claimed that even the souls of living things could be considered an expression of numerical harmony, the ordered expression of multiple parts. Soul (or life) proceeds from the harmonious actions of the parts of the body:

> The soul is established in the body through number, which is to say, through immortal and incorporeal harmony.
>
> (Wheelwright 1966, 234)

Clearly, the Pythagoreans did not understand mathematical natural laws in modern terms. Rather, they pursued a form of number mysticism, seeking the hidden numerical patterns and harmonies that lay behind physical or visual phenomena. This led to an interest in numerology, in numbers with mystical or sacred character. The Pythagoreans shared the general Greek reverence for the number 1, the monad, the source of things. They added other numbers of note: 2—the dyad—or 7—the number of heavenly bodies in the sky that move against the background of the stars. Pythagoras is said to have revered the number 10 above all: as the sum of the first four integers and as the basis of the Tetractys, a figure

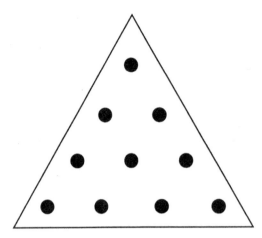

Figure 3.6 The Tetractys reflected the numerical and geometrical significance of the number 10.

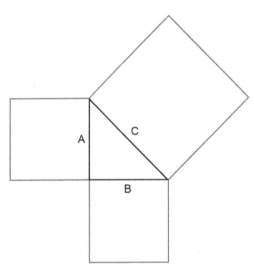

Figure 3.7

revered by his followers (Figure 3.6).

The Pythagorean Theorem may be seen as an example of such harmony in the field of mathematics. The theorem is usually represented as $a^2 + b^2 = c^2$, where "c" is the length of the hypotenuse of a right triangle and "a" and "b" the lengths of the other two sides. Pythagoras did not in fact discover this relationship: the concept was well-known by the ancient Mesopotamians. However, ancient Greek authors credit him with an actual proof of the theorem based on the areas of squares constructed from the sides of the triangle in question (Figure 3.7).

If one constructs squares based on the hypotenuse and the sides of a right triangle, one may prove that the area of square "a" (literally, side "a" *squared*) added to the area of square "b" equals the area of square "c." This is the proof of the Pythagorean Theorem that later appeared in Euclid's *Elements*, written around 300 BCE.

Again, modern historians doubt Pythagoras's discovery of this formal proof. They point out that the kind of deductive geometrical arguments used in the proof were not developed in Greece until after his death. What does seem possible is that the Pythagoreans knew of the relationship between the sides of a right triangle and understood it as yet another example of unity and concord arising from multiple parts. The sides of the triangle related to each other mathematically and formed a harmonious whole, an additional sign that the universe was based on the underlying relationships of number.

The Pythagoreans thus believed that number was embedded in physical phenomena and in the figures of geometry. They stressed the harmonious, exact relationships between whole numbers (integers) and viewed such relationships in a mystical way. Because of these mystical beliefs, the Pythagoreans struggled with the concept of irrational numbers and their occurrence in geometry. Irrational numbers are numbers that cannot be expressed as the ratio of two integers (rational numbers may expressed as such ratios, as in $\frac{4}{1}$ or $\frac{2}{3}$). They cannot be understood exactly. They include the square roots of most counting numbers other than the perfect squares, as well as several important geometrical relationships. For example, the value of pi, the ratio of the circumference of a circle to its diameter, is irrational; it cannot be expressed in exact numerical form. Another irrational geometrical value is the length of the diagonal of a unit square (a square with a side length of one unit) (Figure 3.8).

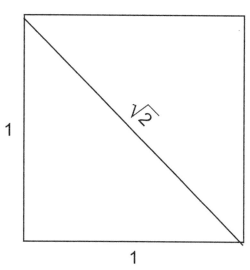

Figure 3.8 The diagonal of a square shows that irrational numbers are inherent in basic geometrical figures.

Based on the Pythagorean theorem, the length of the diagonal of such a figure would be the square root of two, which is an irrational number. Because the length cannot be expressed precisely as a ratio of whole numbers, its relationship to the length of the side cannot be known exactly.

For the Pythagoreans, such irrational values in geometry presented a problem. They felt that irrational numbers were violations of the mathematical harmony that they sought in the world, and they found it difficult to accept that such basic geometrical figures as circles and squares could have irrational relationships between their component parts (Wheelwright 1966, 207). How could the world be harmonious if the relationship between the circumference of a circle and its diameter could not be known exactly?

In modern terminology, two quantities related by an irrational number—that is, two quantities that cannot be described precisely in terms of a common unit—are said to be *incommensurable*. Thus, the circumference and diameter of a circle are incommensurable, as are the side and diagonal of a square. The problem of incommensurability would continue to confront Greek mathematicians throughout antiquity.

Pythagoras and his followers developed an ideology based on the notion of number as the basis of physical existence. His legend grew, and the later Greeks attributed a variety of specific discoveries to him and his school. Some of these may be genuine, some not. More significantly, Pythagorean ideas on number and reality had a profound influence on later Greek philosophical and mathematical developments, particularly the Platonic notion of abstract truth as the basis of physical nature.

PLATO'S *TIMAEUS* AND THE PERFECT SOLIDS

Plato (427–347 BCE) gives an account of the creation of the physical universe and its relationship with abstract truth in his work *Timaeus*. True to the rest of his philosophy, the author begins with abstract, eternal ideas. These ideas serve as the basis for the ordered universe; a creator, or *demiourgos*, shapes the physical world in their image. Thus, spherical objects—the moon or a pearl—are based on the idea of a perfect sphere. Plato explains the stages of creation, stressing the spherical shape of the universe and the circular character of its motions—ideas that probably influenced Eudoxus's cosmology (see earlier).

Over the course of his discussion, Plato embraces a number of older Greek philosophical traditions. For example, he describes the divisions of the world soul, the spirit of the universe, in terms of harmonic ratios, very much in the spirit of the Pythagoreans. While Plato is usually viewed as a rational thinker, his account of the creation of the world soul is quite obscure and reveals the influence of Pythagorean number-mysticism:

> Next he proceeded to fill up both the double and triple intervals, cutting off further parts from the original mixture and placing them between the terms. . . . These links produced intervals of 3/2 and 4/3 and 9/8 within the original intervals. And he went on to fill in all the intervals of 4/3 (fourths) with the interval 9/8 (the tone), leaving over in each a fraction.
>
> (Plato 1965, 27)

Plato also draws on the ideas of the 5th-century philosopher Empedocles, who maintained that matter was composed of four elements: earth, water, air, and fire. However, Plato adds a unique component to his account of the elements in the *Timaeus*. According to him, the elements are the physical embodiments of four of the five perfect solids. The perfect solids may be defined as regular polyhedra in which all sides are congruent regular polygons. The Greek mathematician Theaetetus described the five figures in the early 4th century BCE, although some were undoubtedly known of earlier. They include the tetrahedron (four sides, all equilateral triangles), the cube (six sides, all squares), the octahedron (eight sides, all equilateral triangles), the dodecahedron (twelve sides, all regular pentagons), and the icosahedron (twenty sides, all equilateral triangles) (Figure 3.9).

Figure 3.9 The perfect, or "Platonic," solids.

In the *Timaeus*, Plato claims that the four material elements proceed from the forms of the perfect solids. Earth is based on the cube, water on the icosahedron, air on the octahedron, and fire on the tetrahedron. Some of these pairings appear to link physical properties with the shape of the perfect solid in question: fire seems pointed, like a tetrahedron, while earth is stable and unmoving, like a cube. Yet the most striking feature of Plato's conception is his definition of the material elements themselves as the physical expressions of ideal geometrical forms. His approach to the elements serves as a perfect example of the Hellenic Greek fascination with idealized mathematics, as well as his own emphasis on the realm of eternal ideas as the basis for an imperfect, constantly changing, material world.

Plato's concept of the *actual* relationship between the ideal perfect solids and the four elements is not clear. Some scholars suggest that Plato believed the elements to be composed of particles in the shape of the corresponding solid: earth would be made up of cubical particles, fire of tiny tetrahedra. This would link Plato to Democritus and other Greek theorists who argued for a particulate understanding of matter. Unfortunately, the text does not treat the issue explicitly.

The *Timaeus* was not Plato's most important work; at one point in the text he refers to his account of creation as only a "likely story." However, it illustrates the close relationship between developing Greek mathematical thought and an important strain of Greek philosophy. When Plato conceived of the universe as an expression of ideal truths, he drew some of the most fundamental of those truths from mathematics. Because of their role in the Timaeus, the five perfect solids are often called the *Platonic* solids.

EUDOXUS OF CNIDUS

While he is best known for his attempt to explain planetary motion in geometrical terms, Eudoxus of Cnidus was also the most accomplished Greek mathematician of the Classical period. Active during the first half of the 4th century BCE, he studied at Plato's Academy and, according to some sources, traveled to Egypt as well. Eudoxus's thought stressed geometrical reasoning over mathematical calculation; his background in Platonic philosophy may have influenced him in this respect. His most significant work addressed the problem of incommensurability that particularly troubled Greek intellectuals in antiquity.

Two numbers, figures, or quantities are said to be incommensurable if they cannot be related to each other in precise terms. Irrational numbers, like the square roots of two, or three, or five, cannot be expressed exactly in terms of integers, and so they cannot be related precisely to integers—such relationships or ratios are incommensurable. Many basic geometrical figures give rise to incommensurable ratios: the length of the diagonal of a square is incommensurable with the length of its side, for example. Similarly, the diameter of a circle is incommensurable with its circumference. Because such figures were fundamental to the discipline of geometry, the issue of incommensurability was a source of philosophical distress for many Greek thinkers. For example, the Pythagoreans struggled with irrational numbers and incommensurability; such concepts challenged the central Pythagorean belief in the universe as an expression of numerical or arithmetical harmony. Incommensurability, by its very nature, seemed to be an expression of disharmony.

Eudoxus confronted the problem by formulating his theory of proportionality. He adopted the view that while one could not relate incommensurable quantities to each other exactly, one could treat the ratio *between* those quantities as a functional whole and use it in statements of proportion. For example, the circumference C of a given circle is incommensurable with its diameter D, but the ratio C/D may be used in mathematical statements without being resolved arithmetically—it is represented by the symbol π. While pi is often represented as 3.14 in computations, this is an imprecise estimate. However, π may be used in geometrical statements without being expressed in exact numerical terms. Thus, the area of a circle is πr^2.

Eudoxus recognized that many geometrical problems required a demonstration of proportionality, that is, a demonstration that two different ratios were equal: $a/b = c/d$. His theory of proportionality served to establish that equality. His approach involved using common multipliers to show that the ratios were equal, a technique related to the modern algebraic method of cross-multiplication. However, his procedure resulted in logical proof, not arithmetical solution: he could prove that the relationship a/b was the same as the relationship c/d without resolving these quantities numerically. This made his theory of proportionality crucial to pure geometry, for it allowed Greek mathematicians to explore the relationships between geometrical figures without using numerical values. For example, Eudoxus established that the areas of different circles were proportional to the squares of their radii and that the volumes of different spheres were proportional to the cubes of their radii. The actual dimensions of individual circles and spheres did not matter; what mattered was the proof of the relationships involved. Further, Eudoxus's method addressed the problem of incommensurability, for it could prove a proportional relationship between two incommensurable ratios without concern for the exact value of those ratios.

Eudoxus added to the capabilities of Greek geometry in other important ways as well. He refined the so-called "method of exhaustion" (a modern term) that allowed for the estimation of the area of curved spaces by using polygons of increasingly similar dimensions. He noted that a hexagon inscribed in a circle covered much of the area of the circle. By replacing the six-sided hexagon with a twelve-sided dodecagon, he increased that similarity, cutting the difference between the area of the circle and that of the inscribed polygon in half (Figure 3.10).

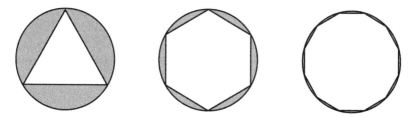

Figure 3.10 The method of exhaustion served as a means of approximating the area of a circle.

By progressively doubling the number of sides of the inscribed polygon, he could produce increasingly accurate estimates. His approach here resembles the modern use of limits and infinitesimals.

Eudoxus's work helped to set the trajectory of Classical Greek mathematics. The Greeks would direct their energies largely towards geometrical pursuits, despite the difficulties of incommensurability. As they explored different techniques for proving theorems and propositions, geometrical proof came to replace Pythagorean arithmetical harmony as an ideal. The geometrical synthesis of Euclid proceeded from this trend, and several of the most important proofs in Euclid's synthesis of geometry, *The Elements*, are ascribed by historians to Eudoxus.

THEAETETUS

An important example of Plato's interest in mathematics is his dialogue *Theaetetus*, named for an Athenian mathematician of the early 4th century BCE. A contemporary and an acquaintance of Plato, the actual Theaetetus was credited with the proof that there are exactly five perfect solids, the figures that play such a significant part in Plato's *Timaeus*. In addition, he was fascinated with irrational numbers and their role in geometry. Irrational numbers may be defined as numbers that cannot be described as an exact ratio between integers. Plato's dialogue uses this subject to explore the boundaries of philosophy, to consider what can and cannot be known. The work goes on to ask if truth is based on human sense experience or on pure thought.

Plato undoubtedly chose Theaetetus as a character for the dialogue because of his mathematical work. The *Theaetetus* takes place when the title character is only fifteen, and it relates a conversation between Theaetetus and Socrates, along with several other characters. Socrates asks Theaetetus to define knowledge, and the young man struggles with the question. As he explores what can and cannot be known, he describes the problem of a rectangle with an area of 17 units. Such a rectangle can be easily understood or constructed; imagine any rectangle and divide it into 17 equal parts with vertical lines—each part is one unit. The problem lies in relating the area of the rectangle to the length of its sides. The area of a rectangle is found by multiplying its length by its width, yet 17 is a prime number and therefore cannot be the precise product of any numbers other than itself and 1. As a result, one cannot know the precise length of the sides of the rectangle in question—those lengths are irrational numbers. Plato uses young Theaetetus's discussion to point out the boundaries of human knowledge, to illustrate Socrates's argument that part of wisdom is knowing that some things cannot be known. Here, geometry is employed to point out human limitations, rather than human certainty (Cropsey 1995, 28–29, 34–36).

The *Theaetetus* goes on at length to examine the qualities of both pure thought and sense perception as approaches to what may be known. Socrates is interested in challenging the notion of "man as the measure of all things," as stated by the philosopher Protagoras, for the initial discussion of irrational numbers serves to point out that some things are beyond human understanding. The dialogue also reflects the ongoing Greek fascination, and frustration, with irrational quantities in geometry.

The problem of incommensurability, that is, of quantities that cannot be related to each other in exact terms, was one of the questions that defined Hellenic mathematics; it flew in the face of the assumption that mathematics and geometry were systems of pure intellectual harmony. Plato's *Theaetetus* was probably written before Eudoxus of Cnidus addressed the problem in part with his theory of proportion (see earlier discussion). The issue remained a significant one in Greek mathematics and philosophy, and as Plato points out, it served as an important source of philosophical humility in a culture characterized by intellectual optimism.

CONCLUSION

The interaction of philosophy and mathematics in Greece from the 7th to the 4th centuries BCE produced important trends in the thought of the region, trends that would have long-standing influence on the history of Western and world ideas. The Pythagoreans sought numerical harmony in the universe and saw nature as an expression of number. Plato stressed the abstract character of truth and viewed mathematics, and especially geometry, as examples of such abstraction. His ideas showed some Pythagorean influence, for he maintained that matter itself was an expression of geometrical idea. Eudoxus's work helped to move Greek mathematics towards an emphasis on geometrical proof as opposed to arithmetical calculation, a move that exhibits Platonic influence. His spherical system of planetary motion serves to illustrate the convergence of many of these intellectual and cultural elements: he explained visible nature in terms of pure geometry, the complex motions of perfect spheres. Finally, Aristotle developed rigorous rules for logical reasoning, rules that shaped, and were shaped by, the growing tradition of geometrical analysis in 4th-century Greece. Greek mathematical thought in the classical period certainly drew on older Babylonian and Egyptian influences, but it developed a speculative character of its own. At the close of the 4th century, the geometrical synthesis of Euclid would systematically codify the work of the mathematicians of Classical Greece. Euclid lived and worked in Alexandria for part of his life, and the rise of Alexandria as the intellectual hub of the eastern Mediterranean signaled a new era in the cultural history of Greece, Egypt, and the Middle East. Indeed, the intellectual—and mathematical—traditions of these cultures would combine in fascinating new ways during the Hellenistic Age.

PRIMARY TEXT: PLATO'S *MENO*

> Soc. . . . The soul, then, as being immortal, and having been born again many times, and having seen all things that exist, whether in this world or in the world below, has knowledge of them all; and it is no wonder that she should be able to call to remembrance all that she ever knew about virtue, and about everything; for as all nature is akin, and the soul has learned all things; there is no difficulty in her eliciting or as

men say learning, out of a single recollection—all the rest, if a man is strenuous and does not faint; for all enquiry and all learning is but recollection. And therefore we ought not to listen to this sophistical argument about the impossibility of enquiry: for it will make us idle; and is sweet only to the sluggard; but the other saying will make us active and inquisitive. In that confiding, I will gladly inquire with you into the nature of virtue.

Men. Yes, Socrates; but what do you mean by saying that we do not learn, and that what we call learning is only a process of recollection? Can you teach me that?

Soc. I told you, Meno, that you were a rogue, and now you ask whether I can teach you, when I am saying that there is no teaching, but only recollection; and thus you imagine that you will involve me in a contradiction.

Men. Indeed, Socrates, I protest that I had no such intention. I only asked the question from habit; but if you can prove to me that what you say is true, I wish that you would.

Soc. That is no easy matter, but I will try to please you to the utmost of my power. Suppose that you call one of your numerous attendants, that I may demonstrate on him.

Men. Certainly. Come hither, boy.

Soc. He is Greek, and speaks Greek, does he not?

Men. Yes; he was born in the house.

Soc. Attend now to the questions which I ask him, and observe whether he learns of me or only remembers.

Men. I will.

Soc. Tell me, boy, do you know that a figure like this is a square?

Boy. I do.

Soc. And you know that a square figure has these four lines equal?

Boy. Certainly.

Soc. And these lines which I have drawn through the middle of the square are also equal?

Boy. Yes.

Soc. A square may be of any size?

Boy. Certainly.

Soc. And if one side of the figure be of two feet, and the other side be of two feet, how much will the whole be? Let me explain: if in one direction the space was of two feet, and in other direction of one foot, the whole would be of two feet taken once?

Boy. Yes.

Soc. But since this side is also of two feet, there are twice two feet?

Boy. There are.

Soc. Then the square is of twice two feet?

Boy. Yes.

Soc. And how many are twice two feet? count and tell me.

Boy. Four, Socrates.

Soc. And might there not be another square twice as large as this, and having like this the lines equal?

Boy. Yes.

Soc. And of how many feet will that be?

Boy. Of eight feet.

Soc. And now try and tell me the length of the line which forms the side of that double square: this is two feet-what will that be?

Boy. Clearly, Socrates, that will be double.

Soc. Do you observe, Meno, that I am not teaching the boy anything, but only asking him questions; and now he fancies that he knows how long a line is necessary in order to produce a figure of eight square feet; does he not?

Men. Yes.

Soc. And does he really know?

Men. Certainly not.

Soc. He only guesses that because the square is double, the line is double.

Men. True.

Soc. Observe him while he recalls the steps in regular order. *(To the Boy.)* Tell me, boy, do you assert that a double space comes from a double line? Remember that I am not speaking of an oblong, but of a square, and of a square twice the size of this one—that is to say of eight feet; and I want to know whether you still say that a double square comes from double line?

Boy. Yes.

Soc. But does not this line become doubled if we add another such line here?

Boy. Certainly.

Soc. And four such lines will make a space containing eight feet?

Boy. Yes.

Soc. Let us describe such a figure: is not that what you would say is the figure of eight feet?

Boy. Yes.

Soc. And are there not these four divisions in the figure, each of which is equal to the figure of four feet?

Boy. True.

Soc. And is not that four times four?

Boy. Certainly.

Soc. And four times is not double?

Boy. No, indeed.

Soc. But how much?

Boy. Four times as much.

Soc. Therefore the double line, boy, has given a space, not twice, but four times as much.

Boy. True.

Soc. Four times four are sixteen – are they not?

Boy. Yes.

Soc. What line would give you a space of eight feet, as this gives one of sixteen feet; do you see?

Boy. Yes.

Soc. And the space of four feet is made from this half line?

Boy. Yes.

Soc. Good; and is not a space of eight feet twice the size of this, and half the size of the other?

Boy. Certainly.

Soc. Such a space, then, will be made out of a line greater than this one, and less than that one.

Boy. Yes; that is what I think.

Soc. Very good; I like to hear you say what you think. And now tell me, is not this a line of two feet and that of four?

Boy. Yes.

Soc. Then the line which forms the side of eight feet ought to be more than this line of two feet, and less than the other of four feet?

Boy. It ought.

Soc. Try and see if you can tell me how much it will be.

Boy. Three feet.

Soc. Then if we add a half to this line of two, that will be the line of three. Here are two and there is one; and on the other side, here are two also and there is one: and that makes the figure of which you speak?

Boy. Yes.

Soc. But if there are three feet this way and three feet that way, the whole space will be three times three feet?

Boy. That is evident.

Soc. And how much are three times three feet?

Boy. Nine.

Soc. And how much is the double of four?

Boy. Eight.

Soc. Then the figure of eight is not made out of a line of three?

Boy. No.

Soc. But from what line? Tell me exactly; and if you would rather not reckon, try and show me the line.

Boy. Indeed, Socrates, I do not know.

Soc. Do you see, Meno, what advances he has made in his power of recollection? He did not know at first, and he does not know now, what is the side of a figure of eight feet: but then he thought that he knew, and answered confidently as if he knew, and had no difficulty; now he has a difficulty, and neither knows nor fancies that he knows.

Men. True.

Soc. Is he not better off in knowing his ignorance?

Men. I think that he is.

Soc. If we have made him doubt, and given him the "torpedo's shock," have we done him any harm?

Men. I think not.

Soc. We have certainly done something that may assist him in finding out the truth of the matter; and now he will wish to remedy his ignorance, but then he would have been ready to tell all the world that the double space should have a double side.

Men. True.

Soc. But do you suppose that he would ever have inquired or learned what he fancied that he knew and did not know, until he had fallen into perplexity under the idea that he did not know, and had desired to know?

Men. I think not, Socrates.

Soc. Then he was the better for the torpedo's touch?

Men. I think that he was.

Soc. Mark now the farther development. I shall only ask him, and not teach him, and he shall share the inquiry with me: and do you watch and see if you find me telling or explaining anything to him, instead of eliciting his opinion. Tell me, boy, is not this a square of four feet which I have drawn?

Boy. Yes.

Soc. And now I add another square equal to the former one.

Boy. Yes.

Soc. And a third, which is equal to either of them?

Boy. Yes.

Soc. Suppose that we fill up the vacant corner?

Boy. Very good.

Soc. Here, then, there are four equal spaces?

Boy. Yes.

Soc. And how many times larger is this space than this other?

Boy. Four times.

Soc. But it ought to have been twice only, as you will remember.

Boy. True.

Soc. And does not this line, reaching from corner to corner, bisect each of these spaces?

Boy. Yes.

Soc. And are there not here four equal lines which contain this space?

Boy. There are.

Soc. Look and see how much this space is.

Boy. I do not understand.

Soc. Has not each interior line cut off half of the four spaces?

Boy. Yes.

Soc. And how many spaces are there in this division?

Boy. Four.

Soc. And how many in this?

Boy. Two.

Soc. And four is how many times two?

Boy. Twice.

Soc. And this space is of how many feet?

Boy. Of eight feet.

Soc. And from what line do you get this figure?

Boy. From this.

Soc. That is, from the line which extends from corner to corner?

Boy. Yes.

Soc. And that is the line which the learned call the diagonal. And if this is the proper name, then you, Meno's slave, are prepared to affirm that the double space is the square of the diagonal?

Boy. Certainly, Socrates.

Soc. What do you say of him, Meno? Were not all these answers given out of his own head?

Men. Yes, they were all his own.

Soc. And yet, as we were just now saying, he did not know?

Men. True.

Soc. And yet he had those notions in him?

Men. Yes.

Soc. Then he who does not know still has true notions of that which he does not know?

Men. He has.

Soc. And at present these notions are just wakening up in him, as in a dream; but if he were frequently asked the same questions, in different forms, he would know as well as any one at last?

Men. I dare say.

Soc. Without any one teaching him he will recover his knowledge for himself, if he is only asked questions?

Men. Yes.

Soc. And this spontaneous recovery of knowledge in him is recollection?

Men. True.

Soc. And this knowledge which he now has must he not either have acquired or always possessed?

Men. Yes.

Soc. But if he always possessed this knowledge he would always have known; or if he has acquired the knowledge he could not have acquired it in this life, unless he has been taught geometry; for he may be made to do the same with all geometry and every other branch of knowledge. Now, has any one ever taught him? You must know that, if, as you say, he was born and bred in your house.

Men. And I am certain that no one ever did teach him.

Soc. And yet he has the knowledge?

Men. The fact, Socrates, is most certain.

Soc. But if he did not acquire the knowledge in this life, then he must have had and learned it at some other time?

Men. That is evident.

Soc. And that must have been the time when he was not a man?

Men. Yes.

Soc. And if there have been always true thoughts in him, both at the time when he was and was not a man, which only need to be awakened into knowledge by putting questions to him, his soul must have always possessed this knowledge, for he always either was or was not a man?

Men. That is clear.

Soc. And if the truth of all things always existed in the soul, then the soul is immortal. Wherefore be of good cheer, and try to recollect what you do not know, or rather do not remember.

> Jowett, B., trans. *The Dialogues of Plato, Volume 1.*
> New York: Charles Scribner's Sons, 1908: 255–261.

FURTHER READING

Cropsey, Joseph. 1995. *Plato's World*, Chicago: University of Chicago Press.

Lloyd, G.E.R. 1974. *Early Greek Science: Thales to Aristotle.* New York: W.W. Norton and Company.

Plato. 1961. *The Collected Dialogues.* Edited by Edith Hamilton and Huntington Cairns. Princeton: Princeton University Press.

Plato. 1961. *Meno.* In *The Collected Dialogues.* Edited by Edith Hamilton and Huntington Cairns. Princeton: Princeton University Press.

Plato. 1961. *Theaetetus.* In *The Collected Dialogues.* Edited by Edith Hamilton and Huntington Cairns. Princeton: Princeton University Press.

Plato. 1965. *Timaeus.* New York: Everyman's Library.

Wheelwright, Philip. 1966. *The PreSocratics.* New York: Macmillan Publishing Company.

4

Numbers in the Hellenistic Mediterranean

Chronology of Hellenistic Authors

Euclid	325–270 BCE (approximate)
Archimedes	287–211 BCE (approximate)
Eratosthenes	276–195 BCE (approximate)
Apollonius	240–190 BCE (approximate)
Hipparchus	190–120 BCE
Ptolemy	100–170 CE (approximate)
Diophantus	200–284 CE (approximate)
Hypatia	c. 360–415 CE

CULTURAL ICON: ERATOSTHENES OF CYRENE AND THE CIRCUMFERENCE OF THE EARTH

Eratosthenes was born in the Greek city of Cyrene in North Africa in the year 276 BCE, at a time when the region was controlled by the Macedonian Ptolemaic rulers of Egypt. He studied in Athens and demonstrated his talents in a number of fields: mathematics, history, and poetry. His reputation grew, and the ruler of Egypt, Ptolemy III, sought him out as a teacher for his son and for a position in the *Mouseion* (Museum) of Alexandria, a royally supported group of scholars and authors. The Great Library at Alexandria had been founded in the early 3rd century BCE for the use of the scholars of the Mouseion, and Eratosthenes rose to become chief librarian. His academic career mirrored the culture of Alexandria itself during its golden age; he was inquisitive, energetic, and eclectic. He played a central role in the vibrant scholarly community of the Mouseion, and he had access to a wide range of different intellectual resources. Throughout his long life, he continued to pursue many interests. He did significant work in geography and astronomy and developed a method for identifying prime numbers, the so-called "Sieve of Eratosthenes." However, he is best known for his measurement of the circumference of the earth. His method for measuring the earth's circumference combined Greek geometry with Egyptian knowledge of practical land measurement.

Eratosthenes knew that in the city of Syene at the time of the summer solstice, the rays of the sun were vertical at noon; that is, the sun stood at the very apex of the sky. One may define the vertical rays of the sun as rays that are perpendicular to a plane that is tangential to the earth at a particular point. Syene is located quite near to the Tropic of Cancer and that explains why the rays of the sun would be vertical on the day of the summer solstice (June 21). In practical terms, at noon on the summer solstice, a vertical stick driven into the ground would cast no shadow at Syene. On the other hand, at the same moment in time, a similar stick would indeed cast a shadow at Alexandria, approximately 500 miles to the north. Eratosthenes used this phenomenon as the basis of his calculation. He assumed the earth to be spherical and theoretically "extended" each of the two sticks to the center of the earth. He noted that they each then formed a radius of the sphere (Figure 4.1).

He also assumed that the rays of the sun were parallel. In this case, at noon on the day of the summer solstice, the ray striking the stick at Syene would be vertical; the stick would cast no shadow. However, a parallel ray striking the stick at Alexandria would do so at an angle, and a shadow would result. That angle could easily be measured (Figure 4.2).

Eratosthenes recognized that if the two rays of light were parallel, then the radius formed by the stick at Alexandria formed a *transversal*, that is, a straight line that crosses two parallel lines. As a result, the angle formed by the light ray and the stick at Alexandria and the angle formed by the two radii at the center of the earth were *alternate interior angles*, and they had to be congruent (both are labeled "A" in Figure 4.2). Thus, by measuring the angle at Alexandria, one would know the size of the angle made by the two radii at the center of the earth.

Numbers in the Hellenistic Mediterranean 77

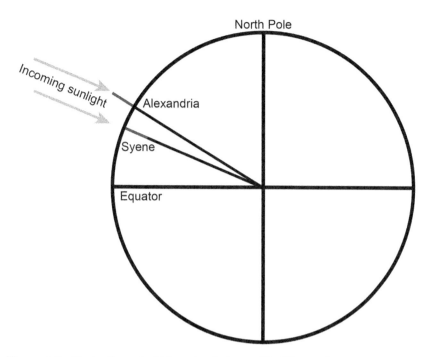

Figure 4.1 Eratosthenes's sticks extended as radii of the earth.

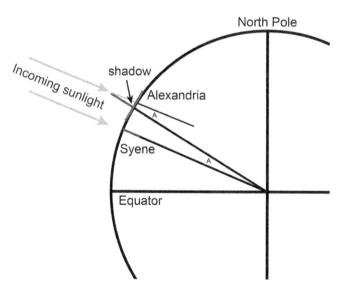

Figure 4.2 Geometry of Eratosthenes's measurement of the earth.

Further, angle "A" was a central angle of a circle (or sphere, in this case), and a central angle intercepts an arc, a portion of the circle's circumference, that is equal to itself (for example, a 90-degree angle intercepts a quarter of the circle's circumference, a 90-degree arc). Eratosthenes thus found the size of the arc represented by the distance between Syene and Alexandria, or to be more precise, the fraction of the earth's circumference that was represented by that arc. If the arc was 7 degrees, then it was about 1/50th of the circumference of the earth ($360 \div 7 = 51.43$). By using the linear distance between the two cities, he could then calculate the circumference of the earth. Eratosthenes did not actually use a 360-degree measure for a circle; rather, he divided it into sixty equal parts for the purposes of his calculations. His approach clearly proceeded from his knowledge of Mesopotamian sexagesimal mathematics.

Eratosthenes used *stadia* as his units; one stadium measures approximately 0.175 kilometers or 0.109 miles. His result in miles has an error of about 10 percent when compared to a modern figure for the earth's circumference, 24,500 miles. Eratosthenes's error arose from a number of factors. Syene does not lie precisely on the Tropic of Cancer and therefore does not receive perfectly vertical rays of the sun on the summer solstice. The earth is not in fact a perfect sphere, but rather is slightly oblate: flattened at the poles and bulging at the equator. His tools and methods of measurement were crude by modern standards. However, despite all these issues, he achieved a remarkable degree of accuracy. His determination of the earth's circumference illustrates the sophistication of Hellenistic mathematical thought and the ability to apply that thought to the solution of practical problems. Like many of his fellow thinkers in the scholarly community at Alexandria, Eratosthenes harnessed the intellectual power of building upon the varied mathematical traditions of earlier antiquity and combining them to produce fascinating new work.

CULTURAL COMMENTARY: COLLECTION AND DISCOURSE

At the British Museum on Great Russell Street in London, the object that receives the most attention from visitors is the Rosetta Stone. For a long time, it stood just inside the main doorway; it was literally the first thing that people saw when they entered the museum, and they could walk up and touch it. It was placed behind protective glass in the 1990s. Discovered by the French in 1799, the stone dates from the Hellenistic Period, when Macedonian Greek kings governed Egypt. The text on the stone appears in three different written languages: Egyptian hieroglyphics, Egyptian Demotic, and Hellenistic Greek or *Koine*. In 1822 Jean-Francois Champollion used the Greek part of the inscription to decipher the hieroglyphics and establish the principles for reading ancient Egyptian monumental writing. This ultimately allowed historians and archaeologists to unlock the 3,000-year history of the Egypt of the pharaohs.

However, the Rosetta Stone also reveals a great deal about the time when it was created, a time when Egypt was part of a larger Eastern Mediterranean cultural sphere dominated by Greek-speaking Macedonian dynasties. The stone was part of a larger monument, or stele, engraved with a decree in honor of Ptolemy V Epiphanes, the fifth ruler of the Ptolemaic dynasty. Dated to 196 BCE, the trilingual character of the writing represents a unique cultural moment, a period when Greek language and culture were imported into Egypt and combined with Egyptian culture. One might view this as cultural domination; clearly the Macedonians were establishing a Greek identity in the lands that they had conquered. Yet it is also apparent that the Ptolemaic rulers of Egypt came to adopt Egyptian customs when dealing with their subject peoples, such as using Egyptian hieroglyphics on public monuments. The Rosetta Stone survives as a physical manifestation of the mixing of cultures and peoples during an age when the regions of the eastern Mediterranean were bound together by the Greek language.

Empires and dynasties are transient by nature; the realities of power lead to the eventual decline of rulers and the domains they govern. Yet some empires leave an enduring cultural legacy. The Macedonian king Philip II defeated the free city-states of Greece in the mid-4th century BCE and unified the Greek peninsula. His son, Alexander the Great, went on to conquer vast territories, stretching from Egypt in the west to the borders of India in the east. He died in Babylon in the year 323 BCE at the age of thirty-three. After his death, his officers fought bitterly over his conquests, and three dynasties—the Antigonids in Greece, the Seleucids in Mesopotamia and Asia Minor, and the Ptolemies in Egypt—came to dominate the eastern Mediterranean and the Middle East. Like Alexander, all of these ruler-generals were Macedonian. Immersed in Greek culture, Alexander's varied successors in power sought to spread that culture throughout their new domains. The export of Greek culture transformed the intellectual atmosphere of Egypt and the other areas under Macedonian control in the three centuries after Alexander, a time known as the Hellenistic Period.

The term "Hellenistic" comes from the Greek word *Hellas*, referring to Greece itself. Two major trends characterize the intellectual and cultural life of the period. On the one hand, Hellenistic intellectuals responded to the thought of Classical, or *Hellenic*, Greece. Thus, Hellenistic philosophers reacted to the central figures of Plato and Aristotle, while mathematicians built upon the work of Eudoxus and his contemporaries. The work of Euclid serves as an excellent example; around the year 300 BCE he wrote a systematic treatment of Greek geometry, organizing and clarifying the ideas of the Greek mathematicians of the preceding two centuries.

On the other hand, Hellenistic culture was affected by the determined efforts of the Macedonian rulers of Egypt and Mesopotamia to foster Greek ideas and artistic styles in their kingdoms. This resulted in significant royal support for communities of thinkers, writers, and artists and also led to the mingling of Greek ideas with the far older traditions of the ancient Middle East. The Ptolemaic rulers of Egypt had the most profound cultural ambitions. Named for Ptolemy I Soter, a close companion of Alexander and the founder of the dynasty, they worked to establish Greek language, literature, architecture, art, and philosophy in their realm, particularly at their new capital at Alexandria. Ptolemy I established the Mouseion, a collection of scholars and authors paid for by a generous royal endowment (Casson 2001, 33), and his successor, Ptolemy II Philadelphus, founded the Great Library as a resource for those scholars. The resulting cultural energy lasted for centuries. Alexandria lay at the epicenter of that energy and produced extraordinary achievements in engineering, the arts, astronomy, and . . . mathematics (Figure 4.3).

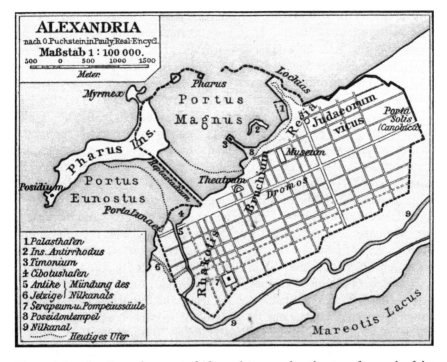

Figure 4.3 The city and seaport of Alexandria served as the stage for much of the mathematical thought of the Hellenistic Mediterranean. The rulers of Ptolemaic Egypt actively patronized intellectual pursuits. (*F. W. Putzgers Historischer Schul-Atlas*, 1901)

The Mouseion was located in the precincts of the royal palace of Alexandria, and its resident academicians were housed, fed, and paid for their efforts. The institution represented a significant royal commitment to Greek culture, and its rewards attracted thinkers from across the Greek world. The assembly of scholars formed a company of discourse, a fertile atmosphere of exchange and criticism. Some monarchs clearly had personal intellectual interests. Ptolemy I famously asked Euclid if there was an easier path to mathematical knowledge than the rigors of his proofs and was told that "there was no royal road to geometry." Ptolemy III's devotion to literature led him to pursue Eratosthenes as a member of the Mouseion and as tutor to his son. Greek language and learning lay at the center of the enterprise, and this was especially evident in the organization of the Great Library.

The library of Alexandria was associated with the Mouseion and situated in an adjacent area of the palace grounds. Its royal patrons sought to build as great a collection of Greek books as possible. They purchased books in Greece, or they borrowed them and had copies made. Books found on ships visiting Alexandria's bustling harbor were appropriated and copied; the library often kept the originals and returned the copies (Casson 2001, 35). Books in Alexandria took the form of papyrus scrolls, formed of sheets sewn together; the sheets were rolled and labeled with the name of the author. The early directors of the library—Zenodotus and Callimachus—confronted the great task of organizing and cataloging the growing collection. Some sources claim that at its largest extent, the library held almost half a million rolls; a smaller library located at the Temple of Serapis (the Serapeum) had another 40,000.

In addition to Greek works, the library acquired significant texts from other cultures and translated them into Greek. No one knows the list of titles involved, but it probably included religious, astronomical, and mathematical texts from Egypt, Mesopotamia, and Asia Minor. The Hebrew Bible was among them; Ptolemy II asked Jewish scholars to translate it from Hebrew to *Koine* Greek in the 3rd century BCE. Koine was a form of Greek that crystallized during the early Hellenistic period, and it rapidly became the common language of the eastern Mediterranean. The translation of these Egyptian and Mesopotamian books into Koine, in Alexandria and elsewhere in the Hellenistic world, had important cultural effects. It made them available to the scholars of the Mouseion, and it allowed ideas that were formerly available only in Mesopotamian cuneiform or Egyptian hieratic to circulate more easily throughout the Mediterranean world.

The main focus of study in the Mouseion and the Great Library was Greek literature, especially the poetry of Homer. However, the intellectual atmosphere of Hellenistic Alexandria also stimulated the development of mathematics and related disciplines, particularly astronomy. The community of scholars in Alexandria provided a forum for the discussion of mathematical and astronomical ideas, and the holdings of the Great Library gave scholars easy access to the work of their predecessors, as well as the concepts and data produced by other intellectual traditions. Historians speak of a "golden age" of Hellenistic mathematics, roughly the years from 300 to 120 BCE. This period encompassed the work of Euclid, Eratosthenes, Archimedes, Apollonius of Perga, and Hipparchus of Nicaea, as well as dozens of lesser figures. All contributed to a Mediterranean-wide atmosphere of inquiry, bound together by a common literary language and by the collections of the Library at Alexandria—and the other libraries that built on its example. In particular, the Library of Pergamum in Asia Minor grew to be an important rival.

The works of these figures had a profound impact on the development of mathematical thought. Euclid systematized Greek geometry during his time in Alexandria. Eratosthenes's studies on geography and the circumference of the earth built on that geometry and included Egyptian and Babylonian influences. Archimedes of Syracuse, the great scholar of solid geometry, corresponded with Eratosthenes and spent time in Alexandria as well—presumably to take advantage of the resources there. Apollonius of Perga studied in Alexandria and wrote a number of books on geometry and astronomy, but only his *Conics* survives. In that text, he explored the characteristics of single and double cones and demonstrated how basic curves—circles, ellipses, parabolas, and hyperbolas—could be derived by sectioning those cones with differently angled planes (Figure 4.4).

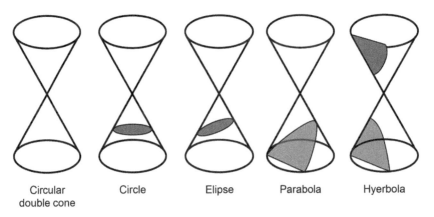

Figure 4.4 Characterization of different curves in the *Conics* of Apollonius.

Hipparchus of Nicaea spent most of his working life on the island of Rhodes. Perhaps the most accomplished astronomer of the age, he accurately described the precession of the equinoxes and the motions of the sun and moon, with attention to the irregularities of the latter. He drew significantly on Babylonian mathematics and celestial observations in his work and adopted the Mesopotamian division of the heavenly sphere into 360 degrees.

Hipparchus also pioneered the discipline of trigonometry and developed the first trigonometric tables for use in his astronomy—in his case, tables of chords intercepted by central angles in a circle. These chord values were crucial in explaining celestial motions. Thus, early trigonometry may be considered a form of applied mathematics, that is, mathematics used to analyze actual phenomena. Scholars who pursued such studies, such as Eratosthenes and Hipparchus, frequently exhibited significant Egyptian and Mesopotamian influences, probably because of the practical character of those particular mathematical traditions. Taken as a whole, the impressive body of work done from 300 to 120 BCE testifies to the existence of a vibrant community of mathematical ideas in the Hellenistic Period, a community of intellectual exchange centered on Alexandria but extending throughout the Mediterranean.

The vigor of Alexandrian scholarship declined after 145 BCE; in that year, royal rivalries led to the purging of the Mouseion and much of the city by Ptolemy VIII. Many of the resident scholars were killed or exiled, and the institution never regained its former luster.

The steady rise of Roman power in the Mediterranean ultimately had a significant effect on the city of Alexandria as well. The Romans subdued the Macedonian kingdoms of Greece and Asia Minor over the space of a 150 years, beginning around 214 BCE. Ptolemaic Egypt was the last remaining Hellenistic realm. During the Roman civil wars of the 1st century, the statesman Pompey the Great was defeated by Julius Caesar and fled to Egypt in the year 48 BCE, where he was assassinated by agents of the young king Ptolemy XIII. At the time, Ptolemy was engaged in a struggle for power with his sister, Cleopatra VII. When Caesar arrived in Alexandria, he objected to the murder of Pompey—his enemy, but a fellow Roman—by Ptolemy, and he supported Cleopatra's bid for the throne. According to the ancient historian Plutarch, Caesar's forces were surrounded in the ensuing conflict, and he set fire to Ptolemy's fleet. That fire raged out of control, and part of the city, including the Great Library, was destroyed.

However, other sources disagree. It is possible that the story in Plutarch referred to warehouses full of books near the Alexandrian waterfront that were consumed by fire and that the library survived. The geographer Strabo visited Alexandria between 30 and 25 BCE and briefly described the Mouseion in his account of his journey. Clearly, it was still operating twenty years after the time of Julius Caesar. Strabo does not mention the library explicitly: perhaps because he simply viewed it as a working part of the Mouseion, perhaps because it was gone. The latter seems unlikely, because the work of many Alexandrian scholars after that time suggests that they had continued access to a great variety of sources, indicating that the resources of the library were still available in some form. It is possible that only some of the Great Library's books were lost or that the smaller library of the Serapeum had absorbed its surviving volumes and grown to take its place.

In any case, after the year 30 BCE, Egypt came under the control of the Roman state, first as a protectorate and later as part of the Roman Empire. The kingdom of the Ptolemies passed into history, but the intellectual importance of Alexandria persisted. The city still possessed a long-established identity as a center of study, and it still boasted the commercial wealth to support that identity. In addition, the emperors and provincial governors of Rome admired Hellenistic scholarship. For example, the Emperor Claudius (ruled 41–54 CE) built an addition to the Mouseion/Library, another indicator that it continued to operate as an institution (Casson 2001, 111). The study of mathematics continued, but the focus of work shifted from geometry to applied mathematics. Hero of Alexandria lived in the 1st century CE and did substantial work in engineering, mechanics, and optics; his inventions demonstrate the existence of a technical culture that was lost in later antiquity. He is also noted for his method for finding the area of triangles based on the length of their sides. However, the most important mathematical figure of Roman Alexandria was Claudius Ptolemy. His *Almagest*, a great synthesis of ancient astronomy written in the 2nd century CE, incorporated the ideas of four centuries of Hellenistic astronomers, as well as Babylonian observations and mathematical techniques (perhaps through the intermediacy of Hipparchus of Nicaea). Ptolemy's access to the ideas of his predecessors, coupled with the spirited debates and commentaries inspired by his book, demonstrates the ongoing existence of an elite community of intellectuals in the city.

After 200 CE, Alexandrian intellectual life waned significantly. The Mouseion (and perhaps the surviving library) was probably destroyed during the suppression of a rebellion by the Romans in the year 272, but the Serapeum had already replaced it as the scholarly center of the city. Mathematical work continued. Historians write of a "silver age" of achievement—notable, but much less significant than the Hellenistic past. Diophantus of Alexandria produced the *Arithmetica* in the 3rd century, a collection of problems and solutions that had a significant effect on the later development of algebra in Arab and European mathematics. Pappus of Alexandria did some important geometrical work in the 4th century and wrote commentary on many of his predecessors. However, the rise of Christianity progressively altered the intellectual atmosphere of the eastern Mediterranean. Because of its prominence, Alexandria became an important center of the new religion. Christianity was legalized in the Roman Empire in the early 4th century and quickly grew dominant. The Serapeum, a pagan temple, was closed in 325 and destroyed in 391. Hypatia of Alexandria, a famous teacher and the author of several mathematical works, was brutally murdered by a mob in 415, probably because of her involvement in religious politics. The age of Alexandrian literary, philosophical, and mathematical scholarship finally came to an end in the 5th century, as inquiry was replaced by orthodoxy.

EUCLID'S *ELEMENTS*

Euclid's famous study of geometry, the *Elements*, embodied the synthetic character of Hellenistic Greek thought; it drew on a number of earlier Greek mathematical and philosophical traditions and combined them into a highly organized survey. It is not clear whether Euclid is the original source of any of the mathematical arguments or proofs in his book. However, most scholars agree that his great contribution lay in the arrangement of these arguments as a progressive set of deductions, where later proofs proceeded from earlier ones. In this sense, the *Elements* served as a textbook. Indeed, it played that role in the Mediterranean, the Middle East, and Europe for more than 2,000 years, making it one of the most influential works in the history of mathematics.

Little is known about Euclid himself. The late-antique philosopher Proclus (5th c CE) wrote that he was younger than the students of Plato and that he exchanged views with the first Ptolemaic ruler of Egypt (Lloyd 2010, 167). This would make him active around 300 BCE in Alexandria. Other sources also associate him with that city, but the site of his birth and education is not clear. His work exemplifies the active export of Greek language and culture throughout the eastern Mediterranean in the years following the conquests of Alexander the Great. It also illustrates the extent to which Hellenistic thought built upon, and reacted to, earlier Classical Greek foundations.

Euclid's *Elements* is a compendium of Hellenic Greek mathematical thought with respect to both specific proofs and general approaches to truth itself. The work deals mainly with geometry, number theory, and incommensurability—all topics of intense interest to the earlier Greeks. It does not include significant treatments of arithmetic, nor does it address practical problems. The abstract character of the work resembles the emphasis on pure thought typical of the Platonic philosophical tradition. However, it is not clear if Euclid actually embraced Platonic values or if he merely drew on the Greek mathematical ideas that also affected Plato's thought.

Euclid's mode of argument is deductive; he begins with a set of definitions, postulates, and axioms and derives the arguments in his various proofs from these in logical terms. His definitions establish the common language for his proofs: he describes the lines, angles, and figures that will be the focus of his work. His axioms form a set of statements, apparently self-evident, that he will draw upon in the course of his mathematical arguments. For example, he states that "things which are equal to the same thing are equal to one another" (Euclid 1933, 6). His postulates represent assumptions about the nature of geometry itself: "That a straight line may be drawn from any one point to any other point" and "That all right angles are equal." Thus, at the start of the *Elements* he provides the reader with the logical foundations and tools necessary for geometric analysis. His work is cumulative, as his later proofs build upon earlier ones. Euclid's logical approach resembles that of Aristotle, and it is possible that he was influenced by Aristotle's *Analytics*. Once again, however, he may have drawn primarily upon the tradition of mathematical and logical analysis that grew in Greek culture from the time of Pythagoras to the end of the 4th century BCE, a tradition that certainly shaped Aristotle's logical work as well.

Throughout the *Elements*, the author's debt to earlier Greek mathematics becomes clear. The first six books of the Elements deal with plane geometry; Book I ends with the proof of the Pythagorean theorem. Euclid shows that the areas of squares based on the two shorter sides of a right triangle add up to the area of a square based on the hypotenuse. Book V addresses the Greek fascination with the notion of mathematical proportion; Euclid draws heavily on the work of Eudoxus of Cnidus here (Lloyd 2010, 168). Books VII–IX deal with number theory rather than geometry, and Book X treats the problems of incommensurability that so troubled the Hellenic Greeks. Here, Euclid borrows from Theaetetus (see earlier discussion). The final three books of the *Elements* treat questions of solid geometry, and again the previous works of Eudoxus and Theaetetus play an important part.

Euclid wrote several other works, and four have survived to this day: the *Data*, the *Division of Figures*, the *Phaenomena*, and the *Optics*. The last is of particular interest, for it represents the application of mathematics to the real physical phenomena of light and vision. It would inspire a long tradition of work on optics in the Mediterranean, the Middle East, and western Europe. However, the *Elements* surpasses all of his other work in influence and importance. Culturally, it stands out as a synthesis of Greek mathematical thought at the dawn of the Hellenistic age, a time when Greek ideas would combine with those of Mesopotamia and Egypt in new and powerful ways. Historically, it would serve as a model of rational thinking from the time of its authorship until the modern era.

ARCHIMEDES OF SYRACUSE

The story is well-known: in the early 3rd century BCE, King Hiero II of Syracuse had an elaborate gold crown made in the shape of a laurel wreath. However, he grew suspicious of its composition; he thought that the maker of the crown had substituted less valuable metal for some of the gold. He asked the mathematician Archimedes, a resident of Syracuse (and possibly a kinsman), to determine the purity of the gold in the crown. While sitting in a bath, Archimedes realized that a body submerged in water displaced an amount of water equal to its own volume. Filled with excitement, he ran naked through the streets of the city shouting "Eureka" (I have it). He then proceeded to show that the crown displaced a greater volume of water than a lump of gold of the same weight would displace; its volume was larger than it would be if it were pure gold. In other words, the crown was not made of pure gold; it contained some lighter metal (silver, in this case).

Stated in more formal terms, Archimedes's principle states that a body submerged in water displaces a volume of water equal to its own volume and that it is acted upon by an upward force equal to the weight of that displaced water. This explains the buoyancy of any body in a fluid. Whether the story of the crown is true or not, Archimedes successfully described a physical phenomenon in precise mathematical terms and established a rule that could be used to predict the behavior of nature. However, his ideas on statics (buoyancy is an example of a static phenomenon) made up only a small part of his work; many historians view him as the most productive mathematical mind of the Hellenistic period.

Archimedes was probably born in 287 BCE, and he produced a remarkable number of works during his life, many of which have survived. A resident of Syracuse in Sicily, he is said to have traveled to Alexandria, and he corresponded with Eratosthenes, the chief librarian there. His interests ranged from geometry and astronomy to mechanics and statics. In his work *On the Measurement of the Circle*, he provided an estimate of the value of pi superior to those suggested by the Egyptians and the Babylonians, and he used that value to describe the area and circumference of a circle (Merzbach and Boyer 2011, 113). Archimedes also explained the geometry of spirals and parabolas in complex terms and did significant work in the field of solid geometry. His methods included formal proofs in the style of Euclid, as well as a more mechanical, intuitive approach to geometry described in his book *The Method*.

Throughout many of his works, he demonstrated an uncanny intuition about the nature of three-dimensional space. Pappus of Alexandria, a mathematician writing in the 4th century CE, credited him with discovering and describing the thirteen semiregular solids, that is, polyhedra whose external sides are regular polygons of different species—mixtures of equilateral triangles and squares, for example (Figure 4.5) (Merzbach and Boyer 2011, 121).

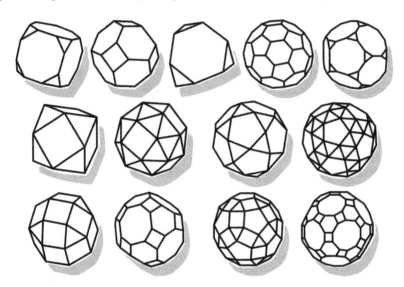

Figure 4.5 The semiregular solids of Archimedes.

Further, in *On the Sphere and Cylinder*, Archimedes proved a series of theorems on the surface area and volume of the sphere, and he examined the mathematical relationship between a sphere and a cylinder with the same diameter and height, that is, a sphere circumscribed by a cylinder (Figure 4.6).

One might argue that Archimedes's fascination with solid geometry related to his work on statics, mechanics, and other physical phenomena. In his work *The Sand Reckoner*, he considered the number of grains of sand necessary to fill the cosmos. The work was notable for its creative approach to the expression of very large numbers, despite the limitations of the Greek system of numeration. However, it also contained a physical estimate of the size of the

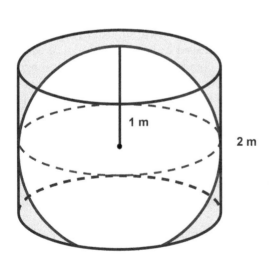

Figure 4.6 Archimedes's study of a sphere circumscribed by a cylinder.

three-dimensional universe, based on the sun-centered cosmology of Aristarchus. Archimedes did not actually accept Aristarchus's heliocentric model, but he used that system in *The Sand Reckoner* because it provided an interesting starting point for estimating the universe's volume. Archimedes's famous work on statics, *On Floating Bodies*, had a similar spatial quality: he explained the buoyancy of objects in terms of the volume of water displaced and the weight of that water. All of these works combined rigorous mathematics with a creative approach to physical space; they treated the physical world as an expression of solid geometry.

Archimedes is famous for other accomplishments: his work on levers and balances, his engineering, and his astronomy. He is credited with designing the largest ship of antiquity and with building a mechanism to effectively raise water (the screw of Archimedes). The Roman philosopher Cicero writes that Archimedes constructed a working spherical model that replicated the motions of the heavens. However, his first interest always lay with mathematics. According to the Roman historian Plutarch, Archimedes did not wish to be remembered for his practical designs and inventions. Rather, he wanted his tombstone adorned with the figure of a sphere circumscribed by a cylinder, a monument to solid geometry.

ANTIKYTHERA MECHANISM

Discovered in an ancient shipwreck, the Antikythera Mechanism illustrates the sophistication of mathematics, astronomy, and engineering during the Hellenistic period, as well as the fusion of Greek, Egyptian, and Mesopotamian intellectual traditions. It was found in a sunken ship in the year 1901, but detailed analysis of its functioning was not done until the second half of the 20th century. The shipwreck has been dated to around 70 BCE, but the mechanism may be significantly older and probably dates from between 150 and 80 BCE. While historians disagree on its exact functioning, it appears to be a mechanical device for modeling the motions of the sun, moon, and possibly the planets. It also traced the intersections of the lunar and solar calendars and indicated the dates of eclipses (Figure 4.7).

The mechanism consisted of over thirty-five interlocking gears linked to a series of pointers that moved over the star charts and calendars engraved on the faces of the instrument. Constructed of bronze, it was powered by a handcrank and encased in a wooden box. The gears moved the outward rings and pointers of the mechanism in an integrated manner, demonstrating astronomical motions over calendrical time. The clockwork technology involved required a high degree of metallurgical knowledge and skilled craftsmanship. The individual gears vary in size; the largest is about 14 centimeters wide.

Figure 4.7 The Antikythera Mechanism was a mechanical device that modeled the varied astronomical movements. Historians have worked to reconstruct the actual functioning of the device. (Have Camera Will Travel, Europe/ Alamy Stock Photo)

Examination of the teeth of the gears indicates that they were fashioned by hand in what must have been a labor-intensive and painstaking process. Such sophisticated clockwork would not be produced again until the European clocks of the Late Middle Ages; the Byzantines and Arabs made clockwork models of the heavens, but they were not nearly as intricate. The complicated character of the Antikythera Mechanism indicates that it could not have been a unique device; it must have had predecessors over years, even generations of development. The device illustrates the intellectual creativity and the technical prowess of Hellenistic culture, capabilities that proceeded from the vibrant intellectual atmosphere of the time and the combined knowledge of different regions of the Middle East and Mediterranean.

The inscriptions on the Antikythera Mechanism are in *Koine* Greek, and the device reflects the astronomical knowledge of Alexandria and the Hellenistic world at the time of its construction. For example, it incorporates the work done on the irregular motions of the moon done by Hipparchus of Nicaea in the 2nd century BCE. The engraved calendars on the device are expressions of the Egyptian Sothic calendar, but the names of the Egyptian months have been translated into Greek. One engraved face shows a star chart of the ecliptic, the path followed by the sun through the sky over the course of a year. The stars that lie along the ecliptic make up the twelve constellations of the Zodiac. The ecliptic chart is based on a circle of 360 degrees and thus clearly reflects the influence of Babylonian astronomy and sexagesimal mathematics. Also, according to the Babylonian practice, it is divided into twelve zones of 30 degrees, each associated with a zodiacal sign.

In short, the Antikythera Mechanism serves as a wonderful example of the culture that produced it. The observations and calculations that served as its foundation, the skill and expertise required for its construction, the creativity that lay behind its conception—all testify to the unique intellectual energy of the Hellenistic Mediterranean.

PTOLEMY'S *ALMAGEST*

Many scholars in Hellenistic or Roman Alexandria pursued both mathematics and astronomy; Eratosthenes, Apollonius of Perga, and Hipparchus of Pergamon all fall into this group. Claudius Ptolemy, writing in Alexandria during the 2nd century CE, created a grand synthesis of ancient Mediterranean astronomy, drawing on his Hellenistic predecessors and combining their work with Egyptian and Mesopotamian techniques of observation and measurement. Ptolemy also wrote a number of other works that applied mathematics to the observable world. His *Geography* included a coordinate system, his *Harmonics* explored the mathematical basis of music, and his *Optics* added to a growing body of work on the geometry of light and vision. However, he is best known for his great summary of astronomy, the Mathematical Treatise, or as medieval Arab scholars called it, the *Almagest.*

Ptolemy's work was a compendium of ideas; his work embraced many older assumptions of Classical Greek philosophy and Hellenistic astronomy. As discussed earlier, Hellenistic astronomy was *positional*: it sought to account for the movements of celestial bodies in mathematical or geometric terms. Ptolemy observed and recorded the positions of the stars and planets, and he probably used Babylonian astronomical records available to him in Alexandria as well. He worked in the geocentric, or earth-centered, astronomical tradition accepted by most of his Hellenic and Hellenistic predecessors. While the scholar Aristarchus had suggested a sun-centered view of the cosmos as early as the 4th century BCE, that view was rejected by most Hellenistic astronomers and by Ptolemy himself. Many factors contributed to the acceptance of a geocentric universe: contemporary religious or philosophical ideas, the lack of an observable parallax (see earlier), and the physics of Aristotle may all have played a role in his thought. Ptolemy also accepted the dominant Greek view, originating with the thought of Plato, that heavenly motions had to be circular and uniform. Because of the elevated status of the celestial regions, the motions of the stars and planets could only be characterized by the most perfect of geometrical figures, the circle. As a result, Ptolemy, like earlier astronomers in the Greek tradition, faced a daunting task: to account for the complex motions of the heavenly bodies using only circular geometry, assuming a central, stationary earth. (One must remember that, in fact, the earth and the planets move around the *sun* in *elliptical* paths—so Ptolemy faced a huge intellectual undertaking.)

Some discussion of observed celestial motions will serve to illustrate the complexity of the challenge facing Ptolemy and his Hellenistic predecessors. The stars and the sun exhibit a yearly motion (this is actually due to the earth's orbit around the sun). The moon and the planets move in their orbits. Finally, the planets periodically appear to move backwards against the star background. Such "retrograde" motion is not real; it actually results from the earth passing the planets as they all orbit the sun together. However, in an earth-centered universe, such motion can only be explained if the planets regularly move *backwards* in their paths around the earth. In Classical Greece, Eudoxus and Aristotle tried to explain these motions using a complex system of concentric spheres.

Ptolemy addressed these challenges by drawing on a number of geometrical techniques originated by earlier Alexandrian mathematicians and astronomers. Such techniques allowed him to construct a complex cosmological system that accounted for his observations of the planets and their movements. For example, he adopted the notion of eccentric orbits as a crucial feature of his astronomy. According to tradition, he kept the earth at the center of the universe, and he maintained that the paths of the sun, moon, and planets were circular. However, he did not place the earth at the exact *geometric* center of each planetary course (Figure 4.8).

Such eccentric geometry gave him the freedom to propose planetary motions that corresponded to his observations.

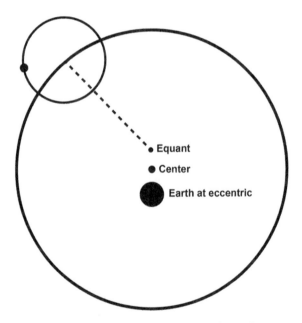

Figure 4.8 In an eccentric orbit, the earth does not lie at the geometrical center of the planets' path.

In addition, Ptolemy used the epicyclic orbits proposed by Apollonius of Perga in the 3rd century BCE. Apollonius had suggested that planets followed compound circular paths; instead of moving around the earth in a simple circle, a planet moved around a point that in turn moved around the earth. The small circle traced by the planet was called an epicycle; the larger circle was called a deferent. The combination of two circular motions produced a complex course of loops that could explain the occasional retrograde, or backward, motion of planets (Figure 4.9).

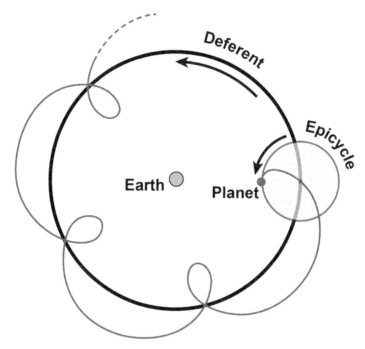

Figure 4.9 The geometry of epicycles described planetary movements in terms of two different circular motions.

Further, by varying the size and speed of the epicycle and the size and speed of the deferent, an astronomer had a great deal of geometrical flexibility in the attempt to account for observed planetary position. Finally, Ptolemy originated the notion of the equant, a point other than the earth that served as a basis of the planet's motion.

By combining epicycles, eccentrics, equant points, and other techniques of circular geometry, Ptolemy constructed a complex astronomical system that agreed with observed data within 3 degrees of error, a remarkable achievement. Modern readers tend to focus on the earth-centered nature of his work as a fundamental flaw. They neglect the rigorous and creative character of the mathematics involved in accounting for planetary position within the bounds of ancient assumptions about the heavens and circularity. Ptolemy's *Almagest* served as a synthesis of ancient thought on the heavens, a blend of mathematics, astronomy, and philosophy. His work also reveals something about the dynamic character of the intellectual community in ancient Alexandria. Clearly, there was an ongoing debate there about the nature of planetary motions; Ptolemy could draw on four centuries of previous Alexandrian work, as well as the accumulated ideas of Greece, Egypt, Asia Minor, and Mesopotamia. While his system was influential, it was by no means accepted by all. The older ideas of Eudoxus and Aristotle had many adherents, as did other views. Aristarchus, a contemporary of Euclid, had proposed a sun-centered astronomy, but the details are lost to us. In short, there was a spirited atmosphere of dispute and disagreement about astronomical issues in Alexandria. This was certainly true of mathematics, geography, philosophy, and literature as well.

In a larger sense, Ptolemy's great book treated the study of nature as an exercise in ideal circular mathematics, and it influenced views of the cosmos for 1,500 years in the Arab world and in medieval Europe. While Copernicus rejected Ptolemy's earth-centered model in the 16th century, he retained the Alexandrian scholar's data and some of his geometrical techniques. Indeed, Copernicus's picture of the universe, like Ptolemy's, included circular planetary paths, complete with eccentrics and epicycles. Not until the work of Galileo and Kepler was Ptolemy's vision truly dismissed.

HYPATIA OF ALEXANDRIA

While it certainly had a turbulent history, Alexandria retained its status as the center of Mediterranean mathematics and astronomy from the time of Euclid to the 5th century CE, spanning the Hellenistic, Roman, and Late Antique periods. Hypatia of Alexandria lived a life of scholarship and study in the tradition of her intellectual forbears. Educated in mathematics and philosophy, she wrote commentaries on the works of a number of Hellenistic mathematicians and astronomers. She also embraced the Neoplatonic philosophy of Plotinus, and she taught extensively. Caught up in the religious politics of the city in the 5th century, she was murdered by a Christian mob in the year 415 CE. One might argue that the death of Hypatia of Alexandria represents the final decline of the vibrant intellectual atmosphere that had thrived in the city for seven centuries.

Our sources on the life and work of Hypatia are sparse. None of her books survive in their original form, but segments of existing works have been attributed to her by scholars. Hypatia was probably born around the year 370. Her father, Theon of Alexandria, was a mathematician and astronomer in his own right, best known for editing Euclid's *Elements* and part of Ptolemy's *Almagest*, removing centuries of copying mistakes from the texts. The sources also refer to him as a member of the Mouseion, possibly a late antique version of the vanished Hellenistic institution. Hypatia was probably educated by her father and shared his interests in philosophy, mathematics, and astronomy, although she clearly surpassed him as a teacher and scholar.

Hypatia achieved great fame as a teacher of mathematics and philosophy, and she clearly played a central role in the intellectual life of the Alexandria of her day. She authored a number of mathematical books, but none have survived. She wrote a commentary on the *Arithmetica* of Diophantus, a collection of algebraic problems written in the 3rd century BCE. Some historians argue that parts of her work were incorporated into Diophantus's text itself and can be identified. She also produced a commentary on the *Conics* of Apollonius of Perga, and she edited the third book of Ptolemy's *Almagest*, and perhaps other parts as well. Philosophically, she embraced the ideas of Plotinus, a Neoplatonic thinker of the 3rd century CE. Plotinus adopted Plato's notion of pure, abstract truths, but he stressed the existence of the One, the Good, in which all truths reside. The material universe emanates necessarily from the One's existence, but the One does not create the universe through an act of will. Such ideas may have appealed to Hypatia as a mathematician and astronomer; they address the issue, so important to the Greeks, of why the universe exhibits mathematical patterns. At a time when Christianity was growing dominant in Alexandria, she adhered instead to a Neoplatonic view of pure Divine Mind.

As a prominent figure in Alexandrian society, Hypatia played a significant role in public life. She apparently acted as an adviser to Orestes, the Roman prefect of Alexandria (at this point in time, Alexandria was under the control of the Eastern Roman Empire). In the year 415, Orestes challenged the policies of Cyril, the new Christian bishop of Alexandria. Because of her association with the prefect, Hypatia was attacked in the streets by a group loyal to Cyril and brutally murdered. Her death has often been interpreted by historians as symbolic of the decline of the intellectual spirit of Alexandria in the 5th century in the face of the growing power of Christianity. For example, Edward Gibbon, the 18th-century author of the *Decline and Fall of the Roman Empire*, argued that her murder cast a shadow on the Church in Alexandria. Modern feminist scholars point out that her status as a female intellectual made her unacceptable to the male-dominated hierarchy of the Christian church. To say that Alexandrian philosophy and mathematics died with Hypatia in 415 is an oversimplification; schools of philosophy continued to operate there throughout the 5th century. However, her murder does serve as a potent sign that the intellectual focus of the city, and indeed, of the entire Mediterranean, was turning towards the Christian religion at the close of classical antiquity. The "Age of Faith" was dawning, and the era of classical philosophy and mathematics was drawing to a close.

CONCLUSION

The Antikythera Mechanism serves as an effective symbol of the intellectual culture and mathematical thought of the Hellenistic Mediterranean during the period of Alexandrian intellectual leadership. It was highly sophisticated and clearly drew on the accumulated work of an intellectual community. Its construction required great scholarship and skilled labor, and it was probably funded by royal wealth—rather like the Mouseion in Alexandria. Its Greek inscriptions illustrate the rise of Greek as an international language of commerce and scholarship during this time of Macedonian power. The mechanism's design reflected a combination of Greek geometry, Babylonian astronomical observations and mathematical techniques, and Egyptian calendars, all applied to the construction of a working model of the universe. It drew on some of the most recent innovations of the time, indicating the rapid spread of new ideas. Finally, like many of the books in Alexandria's library, the knowledge necessary to produce it was lost as Greco-Roman antiquity passed away. The Hellenistic period had an enduring influence on the development of mathematics, and the surviving works of Euclid, Archimedes, Apollonius, and Ptolemy would inspire new ideas around the world for centuries. Yet, so much is gone forever.

PRIMARY TEXT: EUCLID'S ELEMENTS, BOOK I
Definitions

Definition 1

A point is that which has no part.

Definition 2

A line is breadthless length.

Definition 3

The ends of a line are points.

Definition 4

A straight line is a line which lies evenly with the points on itself.

Definition 5

A surface is that which has length and breadth only.

Definition 6

The edges of a surface are lines.

Definition 7

A plane surface is a surface which lies evenly with the straight lines on itself.

Definition 8

A plane angle is the inclination to one another of two lines in a plane which meet one another and do not lie in a straight line.

Definition 9

And when the lines containing the angle are straight, the angle is called rectilinear.

Definition 10

When a straight line standing on a straight line makes the adjacent angles equal to one another, each of the equal angles is right, and the straight line standing on the other is called a perpendicular to that on which it stands.

Definition 11

An obtuse angle is an angle greater than a right angle.

Definition 12

An acute angle is an angle less than a right angle.

Definition 13

A boundary is that which is an extremity of anything.

Definition 14

A figure is that which is contained by any boundary or boundaries.

Definition 15

A circle is a plane figure contained by one line such that all the straight lines falling upon it from one point among those lying within the figure equal one another.

Definition 16

And the point is called the center of the circle.

Definition 17

A diameter of the circle is any straight line drawn through the center and terminated in both directions by the circumference of the circle, and such a straight line also bisects the circle.

Definition 18

A semicircle is the figure contained by the diameter and the circumference cut off by it. And the center of the semicircle is the same as that of the circle.

Definition 19

Rectilinear figures are those which are contained by straight lines, trilateral figures being those contained by three, quadrilateral those contained by four, and multilateral those contained by more than four straight lines.

Definition 20

Of trilateral figures, an equilateral triangle is that which has its three sides equal, an isosceles triangle that which has two of its sides alone equal, and a scalene triangle that which has its three sides unequal.

Definition 21

Further, of trilateral figures, a right-angled triangle is that which has a right angle, an obtuse-angled triangle that which has an obtuse angle, and an acute-angled triangle that which has its three angles acute.

Definition 22

Of quadrilateral figures, a square is that which is both equilateral and right-angled; an oblong that which is right-angled but not equilateral; a rhombus that which is equilateral but not right-angled; and a rhomboid that which has its opposite sides and angles equal to one another but is neither equilateral nor right-angled. And let quadrilaterals other than these be called trapezia.

Definition 23

Parallel straight lines are straight lines which, being in the same plane and being produced indefinitely in both directions, do not meet one another in either direction.

Postulates

Let the following be postulated:

Postulate 1
To draw a straight line from any point to any point.

Postulate 2
To produce a finite straight line continuously in a straight line.

Postulate 3
To describe a circle with any center and radius.

Postulate 4
That all right angles equal one another.

Postulate 5
That, if a straight line falling on two straight lines makes the interior angles on the same side less than two right angles, the two straight lines, if produced indefinitely, meet on that side on which are the angles less than the two right angles.

Common Notions (axioms)

Common Notion 1
Things which equal the same thing also equal one another.

Common Notion 2
If equals are added to equals, then the wholes are equal.

Common Notion 3
If equals are subtracted from equals, then the remainders are equal.

Common Notion 4
Things which coincide with one another equal one another.

Common Notion 5
The whole is greater than the part.

> Heath, T. L., trans. *The Thirteen Books of Euclid's Elements. Volume 1.* Cambridge: Cambridge University Press, 1908.

FURTHER READING

Berlinski, David. 2014. *The King of Infinite Space: Euclid and His Elements*. New York: Basic Books.

Casson, Lionel. 2001. *Libraries in the Ancient World*. New Haven: Yale University Press.

Euclid. 1933. *Elements*. Edited by Isaac Todhunter. London: J.M. Dent and Sons, Everyman's Library.

Hirshfeld, Alan. 2009. *Eureka Man: The Life and Legacy of Archimedes*. New York: Walker and Company.

Jones, Alexander. 2017. *A Portable Cosmos*. New York: Oxford University Press.

Lloyd, G.E.R. 1970, 1973, 2010. *Greek Science*. New York: W.W. Norton & Company. London: Folio Society.

Merzbach, Uta C. and Boyer, Carl B. 2011. *A History of Mathematics*. Hoboken: John Wiley and Sons.

Pollard, Justin and Reid, Howard. 2006. *The Rise and Fall of Alexandria*. New York: Viking Penguin.

Watts, Edward. 2017. *Hypatia, The Life and Legend of an Ancient Philosopher*. New York: Oxford University Press.

5

Numbers in Traditional China

Chronology of Chinese Dynasties

The Chinese Imperial government lasted for more than 3,000 years. While there were many significant disruptions during this vast period, one may refer to a largely continuous tradition of Imperial rule characterized by dynasty or ruling family. The major Chinese dynasties are as follows:

Xia Dynasty	(2070–1600 BCE)
Shang Dynasty	(1600–1050 BCE)
Zhou Dynasty	(1046–256 BCE)
Qin Dynasty	(221–206 BCE)
Han Dynasty	(206 BCE–220 CE)
Six Dynasties Period	(220–589)
Sui Dynasty	(581–618)
Tang Dynasty	(618–906)
Five Dynasties Period	(907–960)
Song Dynasty	(960–1279)
Southern Song Dynasty	(1127–1279)
Jin Dynasty	(1115–1234)
Yuan (Mongol) Dynasty	(1279–1368)
Ming Dynasty	(1368–1644)
Qing Dynasty	(1644–1911)

CULTURAL ICON: THE *SUAN SHU SHU* (BOOK ON NUMBERS AND COMPUTATION)

In 1983 a remarkable book was discovered in the tomb of a government official from the opening years of the Western Han Dynasty. The book was written around 200 BCE, and it consists of 200 long bamboo strips covered with written characters. Originally, the strips were connected with string as in a window blind; the entire book could be rolled up for storage and unrolled for use. Over the course of 2,200 years, the string had decomposed, but 180 of the bamboo strips remained intact. The book is the oldest surviving work of mathematics from ancient China, the *Suàn shù shū* (*Book on Numbers and Computation*) (Figure 5.1).

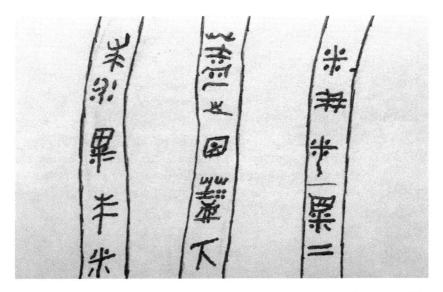

Figure 5.1 The text of the *Suàn shù shū* was written on 200 bamboo strips. The illustration shows a small portion of three of those. (Courtesy of the author)

The *Suàn shù shū* illustrates some of the priorities of ancient Chinese mathematics, as well as the level of mathematical sophistication present in China at the start of the 2nd century BCE. The work consists of sixty-nine problems, most of which have a practical character. The processes in the book include basic calculation, operations with fractions, ratios and proportions, methods of finding areas and volumes—all readily applicable to the work of a government functionary. Clearly, the *Suàn shù shū* had an instructional purpose; it may have served as a handbook, a set of model problems that demonstrated different kinds of problem-solving techniques useful in the act of government administration.

The book began with a list of prefigured results for a number of multiplication problems, some involving units of measurement, others involving fractions or powers of 10. This section bears a striking similarity to the beginning of the Rhind Papyrus in Egypt and probably served the same purpose: it provided scribes or officials with a quick reference of useful values that would speed calculations. Next, the *Suàn shù shū* discussed the rules for working with fractions; this section had a more theoretical character and gave the reader general concepts that might be applied to problem-solving (Dauben 2007, 212). The book also made references to the Chinese counting board, and one can imagine that the two were often used in close conjunction in the solution of problems.

The model problems discussed in the *Suàn shù shū* all included a solution and a method for reaching it. Many had a financial character and dealt with issues of taxation, customs duties, and the charging of interest. For example:

> A fox, raccoon, and hound go through customs, and together pay tax of 111 *qian*. The hound says to the raccoon, and the raccoon says to the fox: since your fur is worth twice as much as mine, then the tax you pay should be twice as much. How much should each one pay?

> (Dauben 2007, 208)

These problems reveal the role of mathematics in the ordinary transactions of government and commerce in the early Han period. Other questions dealt with the division and pricing of cloth or the value of grain. Geometrical problems often had an agricultural character and addressed the dimensions of fields or the volume of grain storage:

> If the length of a field is 30 *bu*, how wide is it if it is a field 1 *mu* in area? The answer says: the width is 8 *bu*. The method says: use 30 *bu* as the divisor, and use 240 square *bu* as the dividend. Finding the length may also be done in this way.

> (Dauben 2007, 211)

Here, 1 *mu* is equivalent to 240 square *bu*, so 240/30 = 8 bu.

While the *Suàn shù shū* clearly had a practical purpose, the text also reflects the operations and problem-solving techniques typical of Chinese mathematics in the early 2nd century BCE, as well as the place of mathematics in Han administration. It was probably a prized possession of the official who owned it, something that lent definition to his life and work. Perhaps that explains why it accompanied him to his grave.

THE FIVE CHINESE CLASSICS

The Five Classics occupied a special place in traditional Chinese thought. They were texts with great authority that influenced many later authors. These works were said to date back to the early centuries of the Zhou Dynasty, perhaps the 9th or 8th century BCE. However, while the Classics probably contained material from that era, most also incorporated later ideas and were clearly compiled over time. One, the *Yijing*, did not reach its final form until the 4th century BCE. They include the *Shijing* (Book of Poetry), the *Shujing* (Book of Documents), the *Liji* (Book of Ritual), the *Yijing* (Book of Changes), and the *Chunqiu* (Annals of Spring and Autumn).

CULTURAL COMMENTARY: CALCULATION AND COSMOLOGY

China has one of the oldest continuous cultures on earth; its written records date back more than 3,200 years, and its artistic traditions are even older. The earliest examples of written Chinese script are inscriptions on oracle bones and turtle shells from the 13th century BCE. These inscriptions included characters for numbers and were used in rituals of divination—attempts to predict future events. Over the centuries of the first millennium BCE, the ancient Chinese developed a sophisticated understanding of mathematics for practical purposes: measurement, commerce, taxation, timekeeping, and calendars.

Ancient Chinese numbers were decimal in character and incorporated a system of place-value, even though the Chinese did not have a character for zero until the 13th century CE. Elements of the system are still in use today. Calculations were performed with moveable rods on a counting board, and the Chinese developed techniques for the quick solution of complex problems. They also employed a working knowledge of geometry in their engineering projects, although they did not pursue the notion of geometrical proof. Chinese books of mathematics in antiquity and the Middle Ages emphasized problem-solving and were closely tied to worldly applications. The *Suàn shù shū* was a collection of sixty-nine problems written on bamboo strips, dating from around 200 BCE. The *Jiu zhang suan shu* (*Nine Chapters on the Mathematical Art*), written in the 1st century CE, included 246 problems and informed Chinese mathematical studies for more than 1,000 years.

A more mysterious view of number appeared in some of the systems of Chinese cosmology and philosophy that emerged during the period from the 6th to the 3rd century BCE. This was a time of great intellectual ferment in China; the age of the "Hundred Schools of Thought." The Chinese believed in a fundamental order that united heaven, nature, and human society, but the varied schools of Chinese philosophy understood this concept of order in different ways. For some, it was rooted in traditional values and authorities, reinforced by the realm of Heaven. Others understood order as the harmonious operation of the cosmos and linked this harmony to numerology—the study of numbers with hidden significance. In most, cosmic order included the spiritual realm, but it did not proceed from the conscious will or design of a god. A story from the *Shujing* (Book of Documents), one of the Five Chinese Classics, serves as an example of an ancient view of cosmic order:

> Early in the period of the Zhou dynasty in China (12th century BCE), the ruler of the realm fell seriously ill. His brother, the Duke of Zhou, went with some servants to a ritual place and performed the necessary rites in honor of his ancestors. He then prayed for his brother's recovery, and offered his own life in exchange. His requests were written on a piece of paper and locked in a metal-bound box, and the servants were sworn to secrecy. Unfortunately, the ruler died of his illness, and his son rose to the throne. The Duke of Zhou did all he could to aid his nephew and served as an advisor. Over time, the friends of the young ruler grew jealous of the Duke's influence, and convinced the king to banish him from court. In the period that followed, the realm suffered; a great storm blew across the land and flattened all the growing grain beyond recovery. Finally, the servants of the Duke of Zhou, who had been silent so long, came forward with the metal-bound box. The king opened it and read the paper within. Convinced of the Duke's loyalty and impressed with his devotion to the family, the king restored him to his rightful place. As a result, another storm came and blew the grain in the fields upright again.
>
> (Ebrey 1993, 6–7)

The story is fascinating because it demonstrates the existence of an order that pervades nature and that can be affected by human actions. No god is mentioned; the Duke makes a request after observing appropriate ritual. His request demonstrates his proper devotion to his family and his ruler. When his nephew later treats him with disrespect or impropriety, the order of nature is disturbed, and disaster results. When proper order is restored in the kingdom, the order of nature is also restored, and disaster is averted. This is not a story of divine judgement; rather, it reflects a belief that human beings exist as part of a larger cosmos and that the order or harmony of that cosmos links humanity with physical nature and with the mystical realm.

Ancient Chinese schools of thought disagreed on the essential character of this order. A survey of these schools provides an important backdrop for the understanding of number in Chinese cosmology. Kong fuzi (Confucius) lived from 551 to 479 BCE, at a time when the ruling Zhou Dynasty suffered from considerable internal strife. He stressed issues of ethics and morals and taught that society could only recover from its difficulties through adherence to rules and conventions of proper behavior rooted in tradition. Tradition served as the foundation of order; for Kong fuzi, it represented the bridge between contemporary human society and the wisdom of the past, a time when humanity lived in harmony. He especially stressed the correct relationships between parents and children, or rulers and subjects. Kong fuzi stressed the study of the Chinese Classics as a source of wisdom on propriety and conventions. Ritual and custom transcended individual desires and served as the guide to a stable society. Confucianism was one among many schools of thought in ancient China until the 2nd century BCE, when it was elevated to special status by the Han emperors.

The philosophy of Mohism challenged many of the values embodied by Confucians. Based on the teachings of Mozi (470–391 BCE), Mohism maintained that order proceeded from the rational actions of people rather than from reverence for tradition. Such rational action should be the basis of society and its institutions and values. Yet this emphasis on reason did not rule out acceptance of the supernatural; Mozi accepted the existence of spirits and believed that they could also be rational actors. The Mohists believed in universal love, in which people treated others with equal regard without concern for family relationships. They rejected the traditional hierarchies so important to Confucian thought and felt that positions of authority should be based on merit. Strikingly, they argued that human society was delivered from a primitive state of self-interest when wise rulers emerged and promoted law and stability. Because Mohists emphasized the role of reason in the establishment of an orderly society, they valued the study of logic and mathematics as examples of rational thinking. The *Mo jing*, the written basis of Mohism ascribed to Mozi and his followers, included chapters on mathematics and geometry and discussed strategies for the examination of nature. Joseph Needham, the famous historian of Chinese science, believed that Mohism had the potential to grow into a form of scientific thinking:

> With their conceptual models, their deduction and induction, the Mohists, like the Greeks, reached the very threshold of the theory of science.
>
> (Needham 1978, 119)

However, while Mohism played an important role during the Warring States period of ancient Chinese history (475–221 BCE), the movement declined after the 2nd century BCE.

Daoism elevated the concept of natural order to a mystical level. Daoist philosophy probably dates from the 4th century BCE, although some traditions claim that the founder of the school, Laozi, lived considerably earlier. He is credited with writing the *Dao De Jing*, a work of eighty-one stanzas that treated the relationships between nature, humanity, and power. While Laozi himself was a semi-mythical figure, the *Dao De Jing* expressed many of the foundational views of the Daoist movement. Daoism is named for the Dao, literally the "Way"—the relentless tide of natural existence and change. It was not a god, and it could not be defined in precise terms, for the act of defining it would be a vain human attempt to grasp its essential nature. The Dao was manifested in all of the activities of the cosmos, including human life; it could not be controlled or resisted. The fall of snow, the growth of trees, the flow of a river, the birth of a child—all were expressions of the Dao.

Daoism stressed the futility of many common human enterprises, such as the pursuit of fame or the accumulation of wealth. Such activities were ultimately temporary and transient, meaningless in the face of the ongoing change or flux essential to the Dao. They did nothing but cause unnecessary strife. Similarly, human efforts to define truth or morality in absolute terms might be interpreted as expressions of vanity or control that were ultimately useless. Daoists embraced the notion of *wuwei*, passive participation in, or surrender to, the Way of Nature. Stanza 10 of the *Dao De Jing* stated:

> The Master leads
>
> by emptying people's minds
>
> and filling their cores,
>
> by weakening their ambition
>
> and toughening their resolve.
>
> He helps people lose everything
>
> They know, everything they desire,
>
> And creates confusion
>
> In those who think they know.
>
> Practice not-doing
>
> And everything will fall into place.
>
> (*Tao Te Ching*, Mitchell 1988, 3)

The processes of life—people's "cores"—were part of the Dao, while attempts to assert some form of static truth were ultimately without value and would only lead to confusion.

Some scholars have argued that Daoism contributed to a culture of natural observation in China; they link it to the pursuit of knowledge about natural processes in such fields as chemistry or alchemy. Other historians dismiss this idea and point out that Daoists would arguably reject such activity as artificial and contrived. However, while the observation of nature may have played a role in the understanding of natural order, it is clear that the Daoist tradition also defined that order in semi-mystical terms. For Daoists, the Dao was more than just the summation of natural processes; it was an entity that encompassed all the activities of the universe. The terms "order" or "harmony," in this sense, did not refer to stability, moral propriety, or rational organization, as they did in Confucian or Mohist philosophy. Rather, "order" and "harmony" meant action in accordance with an irresistible entity that could not be fully understood, only embraced. If Daoism contributed to the study of nature in ancient China, it did so within a mystical context.

The duality of yin and yang served as the basis of another important Chinese view of the cosmos. The concept probably originated early in the first millennium BCE and focused on the relationship between paired opposites: hot and cold, light and dark, male and female—yang and yin. The terms originally referred to the dark and light sides of a hill—over the course of a day, each grows and shrinks as the sun moves across the sky. The principles of yin and yang complemented each other, defined each other, and stood in opposition to each other. This notion of constant change bounded by paired opposites grew in importance over time. By the 4th century BCE, it had a significant influence on Daoist views of nature and appeared in the verses of the *Dao De Jing*.

The *Yijing* or *Book of Changes* reveals the complex character of the yin/yang duality in ancient Chinese culture. The *Yijing* was a book of divination that contained visual patterns based on the notion of yin/yang and used those patterns as a source of omens. The body of the work probably dates from before the 7th century BCE, and a series of commentaries, the "Ten Wings," were added in the 4th or 3rd century. In the *Yijing*, the principles of yin and yang were represented as lines or bars: an unbroken line for yang (active, male) and a broken line for yin (passive, female). These bars were arranged as trigrams—the eight possible sets of three bars—and associated with different aspects of nature (Figure 5.2).

This construction combined numerology with the yin/yang duality and represented the different kinds of change inherent in the universe using spatial symbols.

The trigrams were further combined into sixty-four possible hexagrams, combinations of six bars. In the *Yijing*, each of the hexagrams was

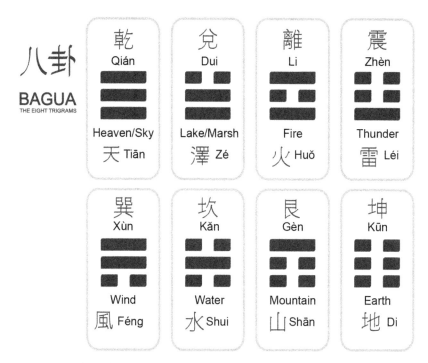

Figure 5.2 The eight trigrams of the *Yijing*.

explained, with comments on the significance of each of the component bars and a discussion of the meaning of the entire symbol. The hexagrams were placed in a specific order, the "King Wen" sequence. Scholars have advanced various schemes for how the ancient Chinese used them in the process of divination. It is possible that random numbers were generated, perhaps by throwing bones or sticks, and these numbers were matched with specific hexagrams and their associated interpretations (Figure 5.3).

The *Yijing* assumed its final form between 400 and 250 BCE with the addition of the "Ten Wings," a set of commentaries on the text attributed to Confucius (the attribution was challenged in medieval China and has been dismissed). These discussions suggested a correspondence between the trigrams and hexagrams in the *Yijing* and the order of the universe itself. The *Yijing* thus served as a guide to cosmic change. The spatial or numerical relationships between hexagrams, based on the sequencing of the bars, mirrored the relationships between the natural forces that they represented. Thus, if one understood the permutations of the hexagrams, one could grasp the secrets of nature. This version of the yin/yang principle thus assumed that the order of nature had a mystical and numerological character.

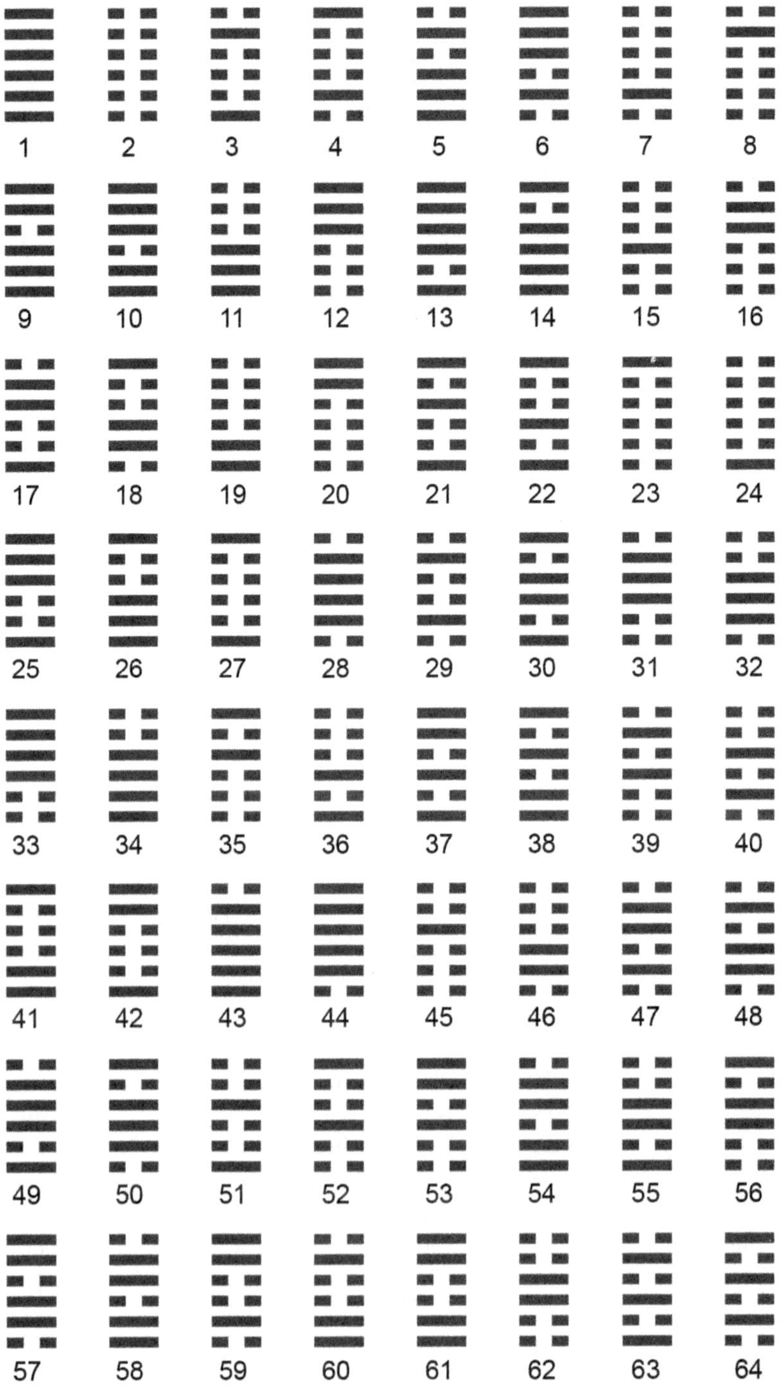

Figure 5.3 The sixty-four hexagrams of the *Yijing* in the King Wen sequence.

Numerology and correspondence also played a role in the concept of *wuxing*, another of the "hundred schools of thought" in ancient China. *Wuxing* means the "five phases" and refers to the theory of the five elements: water, fire, wood, metal, and earth. This view had roots in the Chinese Classics and was treated systematically in the 4th century BCE by the philosopher Zou Yan. The five elements were part of a larger fivefold view of the universe; they corresponded to the five colors, the five planets, the five directions, and many other aspects of nature. As in the *Yijing*, followers of *wuxing* believed that the varied transformations of the elements reflected the greater transformations of nature or history. Indeed, the *Yijing* and the idea of the five elements grew together, for both assumed a correspondence between smaller phenomena and the larger order of the cosmos and human society.

Confucianism, Mohism, Daoism, yin/yang, and *wuxing*—all originated or evolved between the 6th and 3rd centuries BCE during an age of great intellectual creativity in China. All sought to explain or embrace a sense of order that pervaded the universe and human society, and all drew on older concepts in Chinese thought or custom. These systems of ideas rivaled each other and influenced each other. Mohism, yin/yang, and *wuxing* all had explicit mathematical or numerological implications, and these affected other schools of thought.

In the late 3rd century BCE, Qin Shi Huang, the ruler of the Qin Dynasty, successfully subdued the other regions of China and established the Chinese Empire. He actively suppressed the varied schools of philosophy and mysticism in favor of Legalism, a set of values that stressed obedience to the Imperial government. According to tradition, in 213 BCE he ordered the burning of books on history, philosophy, and mysticism, although some historians dispute the story. Following the fall of the Qin in 206 BCE, the Han Dynasty came to favor Confucian thought, but some other schools revived as well. Their ideas would continue to influence Chinese culture and the Chinese view of natural order until the 20th century.

The *Jiu zhang suan shu* (*Nine Chapters on the Mathematical Art*) was produced in the 1st century BCE while the Han Dynasty was in power. A detailed commentary on the text was composed by Liu Hui in the 3rd century. That commentary began with a discussion of the roots of Chinese mathematics in the trigrams and hexagrams and the duality of yin/yang. It also mentioned the Duke of Zhou as the figure who identified the importance of the Nine Arithmetic Arts. The *Jiu zhang suan shu* was largely a book of practical problem-solving techniques, but Liu Hui apparently viewed such practical mathematics as a manifestation of the greater order in the universe, an order based on the balance between yin and yang.

Four centuries later, during the Tang Dynasty (618–906), the Imperial Chinese government instituted a set of civil service examinations for those interested in becoming government officials. While they emphasized knowledge of the Confucian classics, some of these examinations required knowledge of arithmetic. Government scholars collected ten mathematical works, including the *Jiu zhang suan shu*, and labeled them as the "Ten Classics"; these books became the basis for the mathematical component of the tests. One of the classics, *The Mathematical Classic of Master Sun*, begins with another fascinating statement of cosmic order:

> Master Sun says: Mathematics governs the length and breadth of the heavens and the earth; affects the lives of all creatures, forms the alpha and the omega of the five constant virtues (benevolence, righteousness, propriety, knowledge, and sincerity); acts as the parents for *yin* and *yang*; establishes the symbols for the stars and constellations; manifests the dimensions of the three luminous bodies (sun, moon, and stars) maintains the balance of the five phases (metal, wood, water, fire, and earth); regulates the beginning and end of the four seasons; formulates the origins of myriad things; and determines the principles of the six arts (propriety, music, archery, charioteership, calligraphy, and mathematics).
>
> The function of mathematics is to investigate the assembling and dispersing of the various orders in nature, to examine the rise and fall of the two *qi* (*yin* and *yang*), to compute the alternating movements of the seasons, to pace out the distances of the celestial bodies, to observe the intricate signs of the way of the heavens . . .
>
> (Dauben 2007, 297)

Here, the author explicitly links mathematics to cosmic order. Mathematics is the essence of the harmony of the universe, the source of yin and yang, the basis of Confucian virtue. It serves as the basis of heavenly measurement and of the balance of the five elements. It allows the investigation of different aspects of nature. The book goes on to describe the higher powers of 10 and to solve a variety of practical mathematical problems: measuring fields, dividing sums of money, calculating the number of guests in a house. For the author, the techniques of mathematical problem-solving relate directly to the greater structure of the universe; both are manifestations of the order of existence.

TRADITIONAL CHINESE NUMERATION AND CALCULATION

The ancient Chinese developed two major systems of numeration, one for writing numbers as they were expressed in spoken language and another for performing calculations. Both were based on powers of 10, and the system for calculation also incorporated the notion of place-value. While the two approaches originated in antiquity, they have both persisted to the present day, although Arabic numbers are often employed in modern Chinese calculations.

While the earliest written numbers in China occurred on oracle bones dated to around 1100 BCE, the mature system of traditional Chinese numeration arose during the Han Dynasty. It was used to represent quantity and corresponded to the spoken words used for numbers. The system employed symbols for the numbers from 1 to 9, as well as for the powers of 10 (Figure 5.4).

Figure 5.4

Numbers were written vertically, from top to bottom, and in columns from right to left. The system did not use the concept of zero and was not based on place-value. Rather, the symbols for the powers of 10 were used as multipliers. Thus, the number 65 would be written as shown in Figure 5.5.

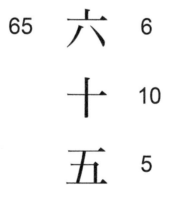

Figure 5.5

Larger numbers followed the same principle, but with additional powers of 10 acting as multipliers (Figure 5.6).

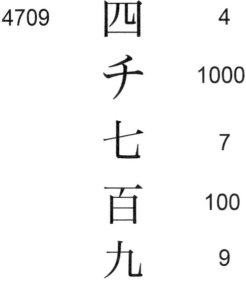

Figure 5.6

The ancient Chinese used multipliers up to 10,000. However, symbols for higher powers of 10 were added as the system persisted over time. It is still used in the modern day, in a modified form, and is often employed in financial transactions.

For actual calculations, the ancient Chinese used a counting board with straight rods to represent numbers (Figure 5.7). They developed a system to carry out mathematical operations quickly and accurately by physically manipulating these rods. This system was based on powers of 10 and the concept of place-value, but it did not have a character for zero when it was first developed in antiquity. If there was no digit at a given place-value, that place-value was just left blank. The Chinese began using a circle to represent the notion of zero later in the Middle Ages.

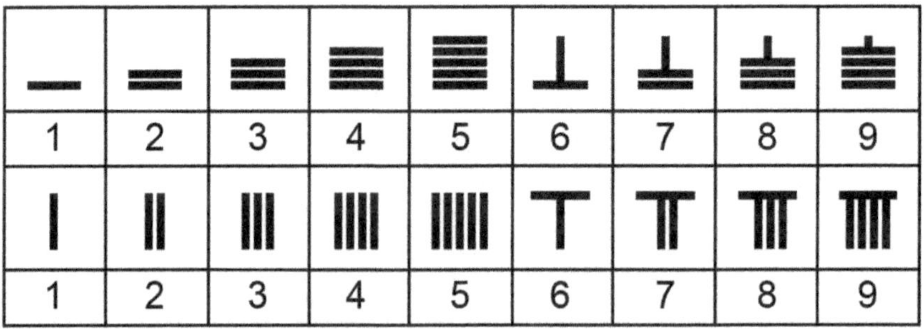

Figure 5.7 Chinese counting board numbers used horizontal or vertical strokes to represent alternating powers of 10.

Numbers were formed horizontally, from left to right. Sets of horizontal or vertical rods were used to depict the numbers from 1 to 5, and a single rod, representing 5, was combined with others to make the numbers 6 to 9. The Chinese used both horizontal and vertical rods in order to make their system of place-value easier to recognize visually. Columns representing even powers of 10—the 1s, 100s, 10,000s, etc.—used vertical rods, while columns representing odd powers of 10—the 10s, 1,000s, 100,000s, etc.—used horizontal rods. For example, the number 35,497 would look like Figure 5.8.

35,497

Figure 5.8

Counting board procedures based on the concept of place-value could be performed very rapidly. Consider the addition of two substantial numbers, 4,639 and 8,135, as shown in Figure 5.9.

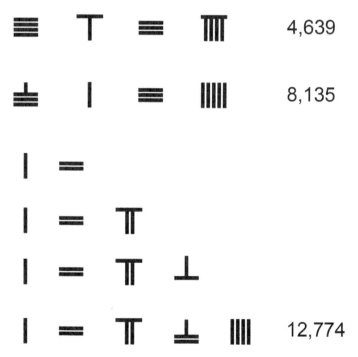

Figure 5.9 Chinese counting board addition.

The two numbers to be added would be formed in counting rods, with their corresponding place-values one above the other (1,000s, 100s, 10s, 1s). Addition begins on the left, with the highest place-value. In the first step, the figures of 4 and 8 add up to 12, so the figure for 1 is placed in the 10,000's column and the figure for 2 in the 1,000's column. Next, the figures for 6 and 1 are added to make 7, placed in the 100's column. In the third step, the figures for 3 and 3 produce a 6 in the 10's column. Finally, the figures for 9 and 5 add up to 14, so a 4 is placed in the 1's column and an additional unit is added to the 10's column, changing the 6 to a 7. The final sum is represented at the bottom (Dauben 2007, 195).

The process for subtraction resembled that of addition, while multiplication, division, roots, and problems involving fractions required more complex procedures. With practice, Chinese mathematicians could perform calculations on the counting board with great speed. Joseph Needham tells the story of a 9th-century Chinese emperor who chose his government servants based on their ability to calculate quickly (Ifrah 1994, 555). The use of counting rods persisted throughout antiquity and the Middle Ages. The well-known Chinese abacus, a set of vertical columns with moveable beads, did not emerge until after the year 1300.

WUXING: CHINESE FIVE ELEMENT THEORY

By the Han Dynasty, a concept of the cosmos based on the number 5 had crystallized in Chinese culture. This numerological picture of the universe assumed the existence of corresponding realms of existence, all with five components. Some examples included:

Planets:	Jupiter, Mars, Saturn, Venus, Mercury
Colors:	Green, Red, Yellow, White, Black
Directions:	East, South, Center, West, North
Tastes:	Sour, Bitter, Sweet, Acrid, Salt
Organs:	Spleen, Lungs, Heart, Kidney, Liver

Some of these categories proceeded from actual observation, for example, the five planets visible to the naked eye. Others appear to have been constructed to fit the system. However, the most important of the realms, the one that gave structure to the general concept of a fivefold universe, was that of the five elements: wood, fire, earth, metal, and water. The notion of the five elements, or *wuxing*, has played an important part in traditional Chinese cosmology and medicine since the Han period.

The term "elements" does not adequately describe the concept of *wuxing*, for "elements" in English suggests material substances, while the Chinese notion referred to both substances and physical processes of change. The *Shujing*, the *Book of Documents*, discussed these varied processes:

> Water is that quality in Nature which we describe as soaking and descending. Fire is that quality in Nature which we describe as blazing and uprising. Wood is that quality in Nature which permits of curved surfaces or straight edges.

Metal is that quality in Nature which can follow the form of a mold and become hard. Earth is that quality in Nature which permits of sowing, growth, and reaping.

(Needham 1978, 146)

Clearly, *wuxing* emphasized material transformation, and the *Book of Documents* defined the elements as five natural types of physical alteration. As a result, the term *wuxing* is often translated as "five phases" rather than "five elements."

The importance of ongoing change or flux in the universe is further demonstrated by the transformations of the elements themselves. The five elements constantly engaged in cyclical transmutation or conflict, giving rise to one another or overcoming one another. For example, in the cycle of generation, wood gave rise to fire, fire gave rise to earth (as ash), earth gave rise to metal, metal gave rise to water (perhaps a reference to condensation), and water gave rise to wood (presumably in terms of plant growth). The elements also interacted through a cycle of conquest or control. Thus, wood conquered earth (probably a reference to agricultural tools), metal conquered wood, fire conquered metal (through melting), water conquered fire, and earth conquered water (perhaps in terms of damming and channeling) (Figure 5.10).

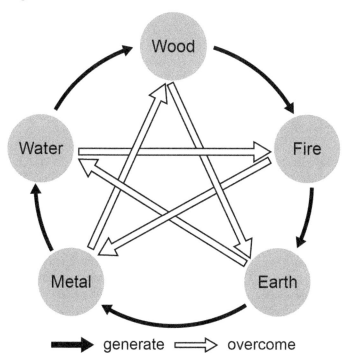

Figure 5.10 The five Chinese elements exhibited cycles of generation and conquest.

The exact origins of the five element theory, or of the correspondence of fives in nature, is unclear. References to the significance of the number 5 occur in early Chinese writings on divination and in the Five Classics as well. The discussion of *wuxing* in the *Book of Documents*, quoted earlier, illustrates the importance of the five element concept, although these sections of the book were probably not written until the later centuries of the Zhou Dynasty. However, the figure most responsible for the systematic treatment of *wuxing* was the scholar Zuo Yan. Zuo Yan lived in the late 4th and early 3rd centuries BCE, during the time of the "Hundred Schools of Thought." While none of his writings survive, his work was described by other authors. According to the *Shih Chi* (Historical Record) of Sima Qian, he had a wide range of interests:

> He began by classifying China's notable mountains, great rivers and connecting valleys; its birds and beasts; the fruitfulness of its water and soils, and its rare products; . . . Then starting from the time of the separation of the Heavens and the Earth, and coming down, he made citations of the revolutions and transmutations of the Five Powers (Virtues), arranging them until each found its proper place. . . .
>
> (Needham 1978, 142–143)

This text indicates that Zou Yan described the transmutations of the five elements, the various cycles of elemental change discussed earlier. He also sought to understand the correspondence of the elements to other aspects of nature and history. He even suggested that the dynasties of China were linked to the elements and that over time they rose and fell in accordance with the cycle of elemental conquest. Thus, the semi-mythical founder of the Chinese Empire, the Yellow Emperor, was linked to the element of earth. He gave way to the Xia Dynasty, associated with wood; the Xia were replaced by the Shang Dynasty, characterized by metal; and the Shang were followed by the Zhou Dynasty, associated with fire. According to this pattern, the next dynasty would correspond to the element of water (Schwartz 1985, 362).

Zou Yan thus explored two of the major aspects of *wuxing*: the transformations of the elements and the idea that the elements corresponded to other realms of existence. According to Joseph Needham, the historian of science in traditional China, Zou Yan was also a leader of the Naturalist school of Chinese philosophy and stressed the careful observation of nature and its processes. Needham particularly emphasized Zou Yan's role in the development of Chinese chemistry and alchemy. However, other historians dispute this and suggest that Zou Yan's main intellectual loyalty was to the teachings of Confucius.

During the Han Dynasty, the concept of *wuxing* continued to grow in complexity. The idea of continuous material change was combined with similar views from the Daoist tradition and from the yin/yang school. The transmutations of the elements could be understood as an expression of the Dao, or it could be linked to the permutations of the hexagrams in the *Yijing (Book of Changes)*. The idea of corresponding realms of five components developed and grew in detail; the five elements corresponded to the five colors, the five planets, the five bodily organs, the five human relationships, the five musical notes, the five tastes, the five directions, the five star palaces, the five types of animals—and more. The five elements became the foundation of a numerological order of nature, linked by an intricate web of correspondences. This complex cosmology would become an enduring part of Chinese culture.

LIU HUI AND THE NINE CHAPTERS ON THE MATHEMATICAL ART

The *Jiu zhang suan shu* (*Nine Chapters on the Mathematical Art*) lay at the heart of mathematics in traditional China. The book consisted of 246 problems, organized by category and arranged into nine major divisions. It was created in the 1st century BCE during the Han Dynasty, but like the *Suàn shù shū* (see earlier discussion), it probably included a lot of material from the Zhou period. In the 3rd century CE, a scholar named Liu Hui wrote a detailed commentary on the *Nine Chapters*, evaluating the mathematical operations of all the component problems and adding substantial explanations. His work would prove to be extremely influential and would become the dominant form of the *Nine Chapters* in succeeding centuries.

According to tradition, Liu Hui added the preface to the work. He begins by recounting the history of mathematics in China:

> Fu Xi created the eight trigrams in remote antiquity to communicate the virtues of the gods and parallel the trend of events in earthly matters, and he invented the nine-nines algorithm to co-ordinate the variations in the hexagrams. Subsequently, the Yellow Emperor marvelously transformed and extended these so that he regulated the calendar and harmonized the music scale. These were used to investigate the cosmic principles, and thereby the subtle and exquisite power of duality and the four diagrams could be effectively harnessed.
>
> (Dauben 2007, 230)

It is important to note that before moving on to the discussion of mathematical operations and problem-solving, Liu Hui places the discipline of mathematics itself into the context of Chinese cosmology and mysticism. The figure of Fuxi is one of the legendary generators of humanity, along with his sister Nüwa. He is also credited with formulating the principles of the *Yijing*; this passage refers to his creation of the trigrams that play such an important role in that work. Liu Hui considers the invention of the trigrams and hexagrams as a crucial development in mathematics. He also asserts that the Yellow Emperor, the mythical founder of Chinese Imperial rule, linked the trigrams and hexagrams to the calendar and the musical scale, exploring the relationship between numerology and nature.

Liu Hui goes on to describe the history of written mathematics. He claims that the Duke of Zhou, a legendary figure from the *Book of Documents*, had respect for the Nine Arithmetic Arts. However, the texts associated with such pursuits were destroyed in the book burning of the Qin Emperor:

> The Qin tyrant ordered the burning of books. Classics and literature were ruined or lost. Years after that, in the Han Dynasty, the Marquis of Beiping, Zhang Cang, and the Deputy Minister of Agriculture and Finance, Geng Shouchang, are noteworthy for their proficiency in mathematics. Working on the basis of the remains of incomplete old manuscripts, each, as they saw fit, revised and supplemented the text.
>
> (Dauben 2007, 230)

This passage suggests that some old texts survived the ravages of the Qin Dynasty, remnants of the "Nine Arithmetic Arts" referred to earlier. Liu Hui names two officials of the Han Dynasty who were responsible for the organization of the *Nine Chapters*. At this point, he is relating the history of the practical problem-solving techniques that are the subject of the rest of the book. He makes it clear that some of these originated in far older texts, and he points out the individuals, including a minister of agriculture and finance, who were responsible for collecting, ordering, and supplementing this material. The *Nine Chapters* thus represents an ancient form of knowledge with great relevance in Liu Hui's own day.

Like the *Suàn shù shū*, the *Nine Chapters* deals principally with the solution of practical problems, but it employs more sophisticated mathematical techniques and addresses new topics. The early chapters deal with the measurement of fields, the value and distribution of grains, the computation of areas and volumes for the purposes of engineering, and the calculation of taxes. The text instructs the user in arithmetical operations, treatment of fractions, ratios and proportions, square and cube roots, and the formulas for areas and volumes.

Liu Hui's commentary adds a great deal to the original text, and he offers improved approaches to the material. For example, in a problem concerning the area of a circular field, he describes his approach to finding a more precise value of π, the ratio of the circumference of a circle to its side. Liu Hui inscribed polygons with an increasing number of sides inside a circle with a given diameter in an attempt to approximate its area. His approach resembled the ancient Greek "method of exhaustion." Once he had a close approximation, he used that area figure to determine the value of π and arrived at a value of 3.1416. Up to that time, Chinese mathematicians had used 3 as the figure for π, so Liu Hui's commentary represented a significant improvement.

While the *Nine Chapters* treats problems in geometry, the focus of the work is algebraic in nature, for it provides numerical solutions to specific problems. The text reveals some advanced algebraic techniques developed by the Chinese in antiquity. For example, in the chapter on squares and cubes, Lui Hui describes the process of "completing the square," an approach capable of solving quadratic equations. The eighth chapter, *Fang cheng*, focuses on the use of mathematical arrays or matrices to solve simultaneous linear equations with multiple unknowns—a technique not known in the West until the work of Carl Frederich Gauss in the early 19th century. The closing chapter of the *Nine Chapters* concentrates on the geometry of right triangles and employs the Pythagorean theorem to solve problems. It is not clear when the Chinese discovered the principle, but much of the chapter is devoted to the "Gou-gu relation"; that is, the relationship of the base of a right triangle to its height (and to the hypotenuse).

The *Nine Chapters on the Mathematical Art* demonstrates the scope of Chinese mathematics during the period of the Han Dynasty. While it illustrates the continued importance of practical problem-solving, it also reveals important developments in more abstract mathematical thought, especially in the commentary provided by Lui Hui. The version of the work with Liu's commentary would ultimately form part of the *Ten Books of Mathematics Classics*, the basis of mathematical study in China during the T'ang Dynasty.

CHINESE MATHEMATICIANS OF THE 13TH CENTURY

The pursuit of mathematics in China suffered a significant decline after the Tang Dynasty. During the period of the Northern Song rulers (960–1127), formal instruction in mathematics waned, and the Imperial government eventually disbanded its institute of mathematics. The subject was not considered to be significant for the training of public servants (Dauben 2007, 308). After 1127, the Song Dynasty lost control of northern China to the Jin Dynasty but continued to rule in the south. The Southern Song court also neglected formal training in mathematics, and many of the mathematical classics, including the *Jiu zhang suan shu* (*Nine Chapters*), began to disappear. During the 13th century, China was still divided. The Jin Dynasty had been conquered by the growing Mongol Empire in the north, while the Song Dynasty maintained its southern domains. However, at this point in time a series of four gifted scholars revived mathematical studies, and their works have survived. Two of the authors worked in the

Mongol Empire, and two in the realm of the Southern Song. However, while some may have held government posts, they appear to have done their mathematical work as independent scholars or teachers, apparently studying the subject for its own sake.

The Chinese mathematicians of the 13th century wrote significant works that revealed some powerful new techniques of problem-solving. Li Zhi (1192–1279) worked as a government official for part of his career, but ultimately withdrew and lived the life of a scholarly recluse. He refused opportunities to work for the court of Kublai Khan, the Mongol emperor. Li wrote a collection of 170 problems, *The Sea Mirror of Circle Measurement*. The work explored the geometry of right triangles inscribed in circles, and vice versa, touching on issues of trigonometry in the process (Merzbach and Boyer 2011, 183). Li Zhi also displayed significant skill in algebraic problem-solving and included a character for zero in his computations with counting rod-style numerals.

Qin Jiushao (c 1202–1261) wrote the *Shushu jiuzhang* (*Mathematical Treatise in Nine Sections*) and took care to relate his work to Chinese mathematical tradition. With a Confucian respect for convention, he linked his book to the *Nine Chapters* (now more than 1,000 years old), emulating its general structure and its emphasis on practical applications. In Confucian terms, this represented the appropriate way to address mathematical subject matter. In his preface, Qin also linked his mathematics to the constant flux of the *Dao*:

> The Six Arts of the teaching of the Chou were truly made complete by mathematics. It is something that scholars and great officers have always esteemed. Their application was based on the idea that "the Great Void generates the One, and it oscillates without ending."
>
> (Dauben 2007, 312)

For Qin, traditional esteem for mathematics proceeded from its relationship to the natural changes inherent in the universe. It is possible that he sought to restore the waning respect for the discipline.

Like the earlier Chinese mathematical classics, much of the *Mathematical Treatise* dealt with practical issues: calendars, interest calculation, area of agricultural fields, tax calculations, etc. However, the book also discussed the *ta-yen* rule, a sophisticated method for the solution of indeterminate equations—equations with more than one solution. The author pointed out that this rule is not part of the *Nine Chapters*, perhaps a point of pride. At several points in the text, he solved particularly difficult problems to show the power of his methodology. Qin also treated higher-order algebraic equations and used an approach to the solution of polynomials that would appear in the work of Lagrange and Horner in Europe five centuries later.

Yang Hui, a contemporary of Qin Jiushao, also used this method to solve polynomials. In his *Detailed Analysis of the Mathematical Methods in the Nine Chapters*, he explored the use of mathematical matrices and wrote about the properties of the "arithmetical triangle," a triangular matrix with important applications for the treatment of binomials. The triangle was well-known to Indian mathematicians, as well as to the Persian scholar Omar Khayyam. Yang Hui also wrote a manual on computation that included problems with multiple unknowns.

Zhu Shijie was perhaps the most accomplished of the 13th-century Chinese mathematicians. In 1303 he wrote the *Siyuan yujian* (*Jade Mirror of the Four Origins*), the most advanced example of medieval Chinese algebra. The four "origins" in the title refer to four unknowns in a single equation. Zhu discussed elaborate ways to solve such equations using complex configurations of characters on the Chinese counting board. He also treated very high-order equations—as high as the fourteenth power. He opened the work with his own discussion of the arithmetical triangle, referring to it as "an old method for finding eight and lower powers." His terms suggest that this triangular matrix had been known to Chinese mathematicians for some time.

The mathematicians of the 13th century demonstrate the complexity and flexibility of Chinese algebra during the late Song and early Mongol Dynasties. While the solution of practical questions still occupied an important place in their work, they developed effective techniques for the treatment of difficult theoretical problems. However, the accomplishments of these figures were elusive at best. Interest in mathematics continued to decline in China as the Middle Ages proceeded, and much of the work of the 13th century faded into obscurity, only to be rediscovered in the modern period.

CONCLUSION

The Chinese authors of the 13th century represented the high point of medieval Chinese mathematics. Woodblock printing was invented in China in the 9th century CE, and printing with moveable ceramic type appeared in the 11th century. During the Song Dynasty, many of the classic Chinese mathematical texts were printed. However, despite the presence of these books, interest in complex mathematics declined in China in the Late Middle Ages, and much of the Chinese mathematical tradition was lost for centuries. However, the practical mathematics inherent in Chinese engineering and the elements of number embedded in various forms of Chinese philosophy endured throughout the Yuan, Ming, and Qing periods (14th to 19th centuries CE). Number remained an important component of the Chinese concept of order.

PRIMARY TEXT: *MATHEMATICAL CLASSIC OF MASTER SUN*

Master Sun says: Mathematics governs the length and breadth of the heavens and the earth; affects the lives of all creatures, forms the alpha and the omega of the five constant virtues (benevolence, righteousness, propriety, knowledge, and sincerity); acts as the parents for *yin* and *yang;* establishes the symbols for the stars and constellations; manifests the dimensions of the three luminous bodies (sun, moon, and stars) maintains the balance of the five phases (metal, wood, water, fire, and earth); regulates the beginning and end of the four seasons; formulates the origins of myriad things; and determines the principles of the six arts (propriety, music, archery, charioteership, calligraphy, and mathematics).

The function of mathematics is to investigate the assembling and dispersing of the various orders in nature, to examine the rise and fall of the two *qi* (*yin* and *yang*), to compute the alternating movements of the seasons, to pace out the distances of the celestial bodies, to observe the intricate signs of the way of the heavens, to perceive the physical features of the earth, to locate the positions of the celestial and terrestrial spirits, to verify the causes of success and failure, to exhaust the principles of morality, and to study the temperament of life. The field of mathematics covers the use of the compass and the carpenter's square to regulate squares and circles, the fixing of standard measures to estimate lengths, and the establishment of measures to determine weights. These measures are split to the accuracies of *hao* and *li* for lengths and *shu* and *lei* for weights.

Mathematics has prevailed for thousands of years and has been used extensively without limitations. If one neglects its study, one will not be able to achieve excellence and thoroughness. There is indeed a great deal to master when one views mathematics in perspective. When one becomes interested in mathematics, one will be fully enriched; on the one hand, when one keeps away from the subject, one finds oneself lacking intellectually. When one studies mathematics readily like a youth with an open mind, one is instantly enlightened. However, if one approaches mathematics like an old man with an obstinate attitude, one will not be skillful in it. Therefore, if one wants to learn mathematics fruitfully, one must discipline oneself and aim for perfect concentration; it is through this way that success in learning is assured.

(Dauben 2007, 297)

FURTHER READING

Dauben, Joseph. 2007. *Chinese Mathematics*. In *The Mathematics of Egypt, Mesopotamia, China, India, and Islam—A Sourcebook*, edited by Victor Katz. Princeton: Princeton University Press.

Ebrey, Patricia Buckley. 1993. *Chinese Civilizations, A Sourcebook*. New York: Simon and Schuster Free Press.

Ifrah, Georges. 1994. *The Universal History of Numbers,* vol. 1. London: Harvill Press.
Libbrecht, Ulrich. 1973. *Chinese Mathematics in the 13th Century.* Cambridge: MIT Press.
Merzbach, Uta and Boyer, Carl. 2011. *A History of Mathematics.* Hoboken: John Wiley and Sons.
Needham, Joseph. 1978. *The Shorter Science and Civilization in China.* Cambridge: Cambridge University Press.
Needham, Joseph. 1981. *Science in Traditional China.* Cambridge: Harvard University Press.
Schwartz, Benjamin. 1985. *The World of Thought in Ancient China.* Cambridge: Harvard University Press.

6

Numbers and the Classical Maya

Chronology of the Maya

Preclassic Period	2000 BCE–250 CE
Classic Period	250–950 CE
Postclassic Period	950–1539 CE

CULTURAL ICON: DRESDEN CODEX

For more than seven centuries during the first millennium CE, Maya civilization thrived in what is now southern Mexico and Guatemala. The Maya also developed the most elaborate written language in Mesoamerica, and they amassed substantial mathematical and astronomical knowledge. Maya civilization began to decline in the 10th century CE, and by the end of the 12th century, most city sites had been abandoned. However, the Maya people lived on in smaller communities, and they preserved some elements of their culture in written form. During the conquest of Mexico in the early 16th century, the Spanish destroyed much of this remaining literary heritage. Only four Maya manuscripts survive from the Pre-Columbian period of Mesoamerica, and the Dresden Codex is the oldest of these (Figure 6.1).

The manuscript was produced in the 13th or 14th century CE, during the Postclassic period of Maya history. However, scholars believe that the codex is a copy of an older text. The book was probably sent to Europe in the

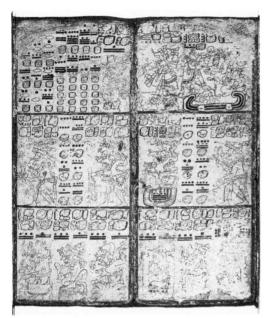

Figure 6.1 The Dresden Codex is the most well-preserved of the surviving manuscripts of the classical Maya. It records a variety of precise astronomical data, probably for ritual purposes. (Jay I. Kislak Collection, Rare Book and Special Collections Division, Library of Congress)

16th century; it thus survived the destruction of Maya manuscripts wrought by the Spanish at that time. It was acquired in the 18th century by the Royal Library of Dresden, in what was then the Kingdom of Saxony, and it was identified as Mayan in the early 19th century. The Dresden Codex continues to reside at the Royal Library, although it was damaged during the bombing of the city in the Second World War.

A codex may be defined as a book written on separate leaves of material (as opposed to a scroll, for example); most codices are bound along one edge. The pages of the Dresden Codex are bound to each other in such a way that it opens like a folding screen or a paper fan. The seventy-eight pages of the manuscript are composed of amate, a very durable paper-like substance made from the bark of tropical *Ficus* trees. The manufacture of amate dates back to the pre-Maya period in southern Mexico, and most Maya books were probably made of the material.

The Dresden Codex deals largely with mathematical and astronomical subject matter, linked to Maya religion, divination, and ritual timekeeping. Calculational tables, tables of eclipses, and tables of the positions of the planet Venus make up a substantial part of the book. For example, pages 51 to 58 of the manuscript contain an extended lunar eclipse table, recording such events over an interval of more than three decades. The Maya did not view eclipses in terms of the orbital paths of the sun, moon, and earth; rather, they saw them as events in time—events that occurred in complex mathematical cycles. As a result, they compiled an impressive collection of eclipse observations and attached great significance to them as religious events or omens. The table in the Dresden Codex indicates that lunar eclipses took place every 148 or 177 days (that is, every five or six cycles of the moon's phases). These figures correspond closely to modern values for the periodic occurrence of lunar eclipses (Aveni 2001, 176–179). The table covers a span of 11,959 days—a significant figure, for 11,960 = 260 × 46. Thus, the Maya scribes and scholars who composed this section of the codex linked the regular phenomena of lunar eclipses to the 260-day length of their ritual year, the most important element in their cyclical view of time (Aveni 2001, 176).

The Venus tables on pages 46 to 50 of the Dresden Codex exhibit a similar combination of astronomical observation and mathematical computation. Venus held great significance for the Maya, and they devoted great care to the study of its motions in the sky. Venus appears as a morning star, close to the sun at sunrise, for 263 days. It then disappears from the sky and returns as an evening star, close to the sun at sunset, for another 263 days. Following another disappearance of 8 days, the planet emerges again as a morning star. The entire cycle—the *synodic period* of Venus—takes 584 days. This figure is prominent in the Dresden Codex Venus table, although the Maya had different values for the individual motions of the planet (236 days as a morning star and 250 as an evening star). Furthermore, the Maya calculated that five Venus cycles of 584 days corresponded to 8 years of 365 days.

While impressive in itself, the blend of astronomy and mathematics featured in the codex must be understood as an expression of Maya religion and spirituality. The Maya interpreted eclipses as important cyclical omens, especially with respect to power and sacral kingship. They associated the planet Venus with warfare and with the god Kukulkan; in fact, images of Kukulkan adorn the Venus table in the Dresden Codex. For the Maya, time itself was cyclical, and the gods maintained the cycles of time. The complex cyclical movements of the sun, moon, planets, and stars reflected the activities of the gods. By charting these movements in time and relating them to their calendars of 260 and 365 days, the Maya sought to understand the actions of their gods and to participate in those actions through religious ritual and sacrifice. The Dresden Codex represents part of a greater quest to understand a divine universe and to play a proper role in its preservation.

CULTURAL COMMENTARY: TIME AND THE GODS

In our technological world, we chart the passing of time with devices—clocks, cell phones, computers. It seems that every appliance in a modern house has a digital readout, and they blink constantly at us as we go about our lives. Modern technology sometimes obscures far older ways to experience time, but these still remain with us. Our most fundamental units of time are based on astronomical phenomena: the rising and setting of the sun (or, more precisely, the turning of the earth), the phases of the moon, the passing of the seasons, the changing of the stars in the night sky as the earth moves in its orbit. Thus, the days, months, and years that punctuate our lives and our history are all derived from the motions of the earth and the heavens (so are hours, minutes, and seconds, for that matter).

People who lived before the onset of modern technology experienced these celestial events vividly. This was especially true of cultures that identified their gods with the heavenly bodies; for them, the motions of the sun, moon, and planets reflected divine activity as well as temporal regularity. An understanding of the cyclical movements of the cosmos brought with it an understanding of the gods themselves. The Babylonians, the Egyptians, the Greeks—all these groups had polytheistic religions with an astral component. Yet perhaps the most complex combination of astronomy, mathematics, timekeeping, and religious practice may be found in the thought of the Maya.

Maya civilization was centered in what is now southern Mexico and Guatemala. At its height, it grew to more than forty city-states, as well as hundreds of smaller subsidiary communities. The Maya never formed a unified state or empire; rather, Maya cities were united by common linguistic and cultural factors. Historians have traditionally estimated the population of the Maya during their Classic period at around 2 million. However, recent radar surveys of the rainforest of the Yucatán Peninsula have revealed the remains of dozens of towns and settlements obscured by tropical vegetation, and this has resulted in substantially larger population estimates.

The Preclassic period of the Maya lasted from around 1500 BCE to 100 CE. At the start of this era, the Maya were an agricultural people, but they were influenced by the Olmecs, the earliest urban civilization of Mesoamerica. Substantial Maya cities grew by the 8th century BCE. By the end of the Preclassic period, the Maya had developed many of the cultural elements that would define their society: substantial religious architecture, a system of writing based on glyphs or images, and institutions of kingship.

From the 2nd to the 9th century CE, Maya civilization reached its apex. Classic Maya culture was characterized by complex political and commercial relations among city-states. The stability of these city-states depended upon institutions of astral religion, divine kingship, and social hierarchy, institutions celebrated by the massive temples and ritual structures built during these centuries. Such structures bore detailed glyphic inscriptions with dates expressed in the elaborate Maya system of interlocking calendar cycles. Kingship lay at the center of Maya society, and scholars believe that it first appeared in the city of Cerros in the 1st century CE (Schiele and Freidel 1990, 98–103). During the Classic period, kingship became dominant in the principal city-states of the Maya, including Tikal, Uaxactun, Caracol, Calakmul, Palenque, and Copán.

For the Maya, the king served as more than just a political figure; he was a shaman, a chief priest, the bridge for his people between this world and the realm of the gods (Schiele and Freidel 1990, 65, 85–90). He was also viewed as the living embodiment of the Hero Twins, central figures from Maya mythology who had outsmarted the lords of the underworld (Schiele and Freidel 1990, 76). Maya kings fulfilled these varied roles by performing the rituals necessary to gain divine favor and to assist the gods in maintaining the universe and assuring the cyclical passage of time itself. The most important of these rituals involved blood, the vital substance required by Maya divinities. Blood offerings took two forms, human sacrifice and bloodletting—that is, the king piercing his own body as an act of devotion (Schiele and Freidel 1990, 87). Leadership in warfare also formed an important component of kingship, for war served as a crucial source of captives for sacrificial purposes.

As sacred figures and agents of sacrifice, Maya kings acted in cooperation with the gods. They worked in accordance with the divine order of the universe and the cyclical progression of time, both expressed through the motions of the heavenly bodies. By coordinating their religious practices and their political decisions with the cycles of the heavens, kings linked their realms and their people to the ways of the gods and the ongoing rhythms of the cosmos (Schiele and Freidel 1990, 215). Rituals, royal ceremonies, agricultural practices—even wars—were timed to correspond to particular astronomical events, celestial omens, calendar dates, or some combination of these. It appears that the main purpose of Maya astronomy and mathematics was to track these significant points in time, to illustrate the numerical connections that existed between different cosmic cycles— the regular motions of the heavenly bodies, the spans of the different Mesoamerican calendars, and perhaps the length of a human pregnancy. Maya priests, astronomers, and scribes—overlapping titles, to be sure— charted the positions of the sun, moon, stars, and planets, with particular attention to their periodic appearances and disappearances. Solar and lunar eclipses and the movements of Venus were of great importance. They accumulated observations over centuries and scrutinized them for mathematical patterns, developing a remarkable understanding of astronomical cycles. For example, the Maya set the synodic period of Venus—the time it takes for the planet to return to a given position in the sky with respect to the sun—at 584 days, quite close to the modern figure of 583.92. The Maya also worked to relate these astronomical intervals to their various calendar cycles: the *tzolk'in* of 260 days, the *haab'* of 365 days, and the Long Count of 2,880,000 days. Thus, they knew that five Venus cycles of 584 days were the equivalent of eight solar years of 365 days. The renewal of this eight-year cycle would be a day of great ritual significance, a day for importance religious or royal events to take place.

To illustrate the relationship between astronomical cycles and royal or religious rituals, one could consider how the official rise to power of the king of Copan was celebrated on the day that the planet Venus reappeared on the horizon (Aveni 2001, 255). Venus was an entity of great power for the Maya; its appearance was often linked to successful warfare as well. By the end of the Classic period of Maya civilization, Venus was associated with the god Kukulkan. El Caracol, a ritual observatory built in the city of Chichen Itza at this time, was oriented to a number of different celestial events, and the windows of the upper tower were arranged to admit the light of the newly re-emerged Venus into a central chamber. No one knows why.

Maya civilization began to decline in the 9th century BCE, perhaps due to prolonged drought or exhaustion of the soil due to overfarming. Cities in the southern parts of the Maya region were abandoned as political institutions collapsed. However, the city of Chichen Itza in the northern Yucatán Peninsula rose to prominence in the middle of the 9th century and remained as an important center of power in the early centuries of the Postclassic period. As the urban world of the Maya largely disappeared, the Maya people lived on in smaller communities. They had experienced a cultural calamity, and their knowledge of the past slowly declined, along with literacy itself (Schiele and Freidel 1990, 379). Yet they preserved religious objects and written manuscripts, and many of their rituals and oral traditions persisted.

The arrival of the Spanish in southern Mexico during the 16th century would have a disastrous effect on the surviving Maya culture, as Indigenous populations were ravaged by European diseases and by the violence of the Spanish conquistadors. The Spanish gained control of the region and sought to spread Christianity and to suppress indigenous religions. For example, Diego de Landa, a Franciscan priest and the Provincial of the Franciscan Order in the Yucatán, worked among the Maya since 1549 and learned a great deal about their customs. Determined to end Maya religious practices, he led an inquisition in the city of Mani in 1562, tortured inhabitants, and burned manuscripts and ritual objects. According to de Landa, twenty-seven books were destroyed, but other sources challenge that figure. De Landa was disciplined, sent to Spain for trial, and exonerated; he later returned to Yucatán as bishop. Ironically, modern knowledge of Maya culture depends significantly upon de Landa's work, for he wrote the *Relación de las Cosas de Yucatán*, a description of Maya customs, language, and writing, in 1566. The book was edited and published in 1864, and it played an important role in the deciphering of Maya writing in the 20th century.

Only four Maya manuscripts survived the destruction of the 16th century: the Paris Codex, the Madrid Codex, the Dresden Codex, and the Maya

Codex of Mexico (known as the Grolier Codex until recently). All date from the Postclassic period, but it is likely that they are copies made from earlier texts. The Madrid Codex is the longest of the four, but the Dresden Codex has proven especially important to the modern understanding of Maya culture. The four codices deal mainly with religion, astronomy, astrology, mathematics, and calendars. Because so few Maya books have survived, the study of these four may have created a distorted understanding of the Maya, with an overemphasis on these aspects of their thought. On the other hand, it is possible that most Maya manuscripts were dedicated to such priestly concerns. The recent deciphering of inscriptions on Maya buildings and monuments, many from the Classical period, has addressed this situation to some extent. These sources tell of a world of mystical royal power, a world where kings knew the mathematical cycles of the heavens and strode hand in hand with their gods.

MAYA NUMERATION

Maya manuscripts and carved inscriptions reveal the sophistication of their system of numeration. The Maya largely based their numbers on powers of 20, like many other peoples of Mesoamerica. The use of a base 20, or *vigesimal*, system probably originated from the practice of counting on one's fingers and toes. Like the Indo-Arabic numbers used in the modern world, Maya numeration employed the concepts of zero and place-value, and this sets the system apart from most other ancient approaches to quantification. In modern numeration, the placement of a digit in a number determines its value. Beginning on the right, digits represent increasing powers of the number 10. Thus, in the number 73, the characters are interpreted as follows:

3×1 $\qquad$ $3 \times 1 = 3$

7×10 to the first power $\qquad$ $7 \times 10 = 70$

Similarly, 68,412 may be read as:

2×1 $\qquad$ $2 \times 1 = 2$

1×10 to the first power $\qquad$ $1 \times 10 = 10$

4×10 to the second power $\qquad$ $4 \times 100 = 400$

8×10 to the third power $\qquad$ $8 \times 1000 = 8,000$

6×10 to the fourth power $\qquad$ $6 \times 10,000 = 60,000$

Where Indo-Arabic numeration employs ten symbols (1, 2, 3, 4, 5, 6, 7, 8, 9, 0) to indicate value, the Maya used just three: a dot to represent 1, a bar to represent 5, and a shell-like symbol for zero. In writing the numbers from 1 to 19, they combined the symbols for 1 and 5 (Figure 6.2).

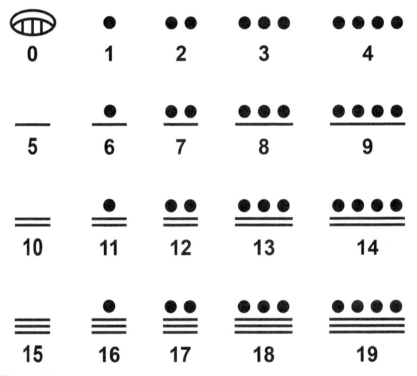

Figure 6.2

For larger numbers, the Maya used a system of place-value based on the powers on the number 20. They wrote their numbers vertically: beginning from the bottom, symbols represented increasing powers of 20 (20, 400, 8,000, etc.). The number 60 would be written as follows:

•••	3×20 to the first power	$3 \times 20 =$	60
🐚	0×1	$0 \times 1 =$	0
			60

235 would be represented as follows:

•̄	11 × 20 to the first power	11 × 20 = 220
≡	15 × 1	15 × 1 = 15
		235

Just as in modern numbers, the use of a zero indicated that there was no multiplier for a given place-value; in such cases, the zero served as a placeholder.

Some historians write of a "trade count," a number system for purposes of commerce that retained a purely vigesimal character (Aveni 2001, 145). In this system, larger numbers would continue the pattern shown earlier. Moving upwards, the progressive place-values would represent increasing powers of 20: 400, 8,000, 160,000. The number 27,812 would thus have four place-values when written:

•••	3 × 20 to the third power	3 × 8,000 = 24,000
••••	9 × 20 to the second power	9 × 400 = 3,600
≡	10 × 20 to the first power	10 × 20 = 200
•• ≡	12 × 1	12 × 1 = 12
		27,812

However, most of the Maya mathematical texts that do remain have a religious character and deal with issues of astronomy, omens, and ritual calendar-keeping. These sources indicate that the Maya departed from a purely vigesimal approach to number when dealing with calendars and religious matters.

The Maya had a series of different, interrelated calendars, but the most familiar to a modern reader would be the *Haab* of 365 days, essentially a solar year. The *Haab* had 18 months of 20 days each—360 days—and a short, irregular month of 5 days. In their number system, the Maya replaced the third place-value—that of 20 to the second power, or 400— with 360, or 18 × 20. This reflected the number of days in the 18 major months of the *Haab* and may have simplified calendar calculations. It also substantially changed the character of the number system. While the higher place-values were still generated by multiplying by 20, the use of 360 in the third place meant that subsequent place-values no longer represented actual powers of 20. For example, the fourth place-value was occupied by multiples of 360 × 20, or 7,200 (as opposed to 20 to the third power, 8,000). Thus, the number 24,516 would be written as follows:

•••	3 × (360 × 20)	3 × 7200	= 21,600
≡••	8 × 360	8 × 360	= 2,880
•	1 × 20 to the first power	1 × 20 =	20
≡•≡	16 × 1	16 × 1 =	16
			24,516

Undoubtedly, the choice of 360 as the basis of the third place-value had great ritual significance. Yet it made the system irregular, for all larger place-values were multiples of 360 rather than powers of 20. This did not matter to the priests and astronomers of the Maya, for their concern lay with the cycles of time and the motions of the stars.

THE LONG COUNT

The Mayan Long Count calendar reflected the cyclical understanding of time inherent in the culture of the Maya. The Maya believed that the universe renewed itself periodically, and the Long Count tracked the passage of time from one creation event to the next. The calendar was closely linked to Mayan religion and astronomy and to the system of numeration used by priests and officials to describe important spiritual and astral events.

It is highly likely that Mayan numeration and Mayan calendars developed together. As discussed earlier, the Maya adopted a vigesimal, or base 20, system of numbers, complete with a notion of place-value and a concept of zero. In a purely vigesimal system, each place in a numerical expression would represent a power of 20. So, numbers in the first placement would represent multiples of 20 to the zero power—1's. Numbers in the second placement would represent multiples of 20 to the first power—20's. Numbers in the third placement would represent multiples of 20 to the second power—400's. However, in the Mayan system, figures in the third place denoted groups of 360 rather than 400, probably because the number of 360 is close to the number of days in a solar year. Hence, the Mayan sense of time had a defining effect on the Mayan approach to number.

A date in the Mayan Long Count depicted the number of days that had passed since the last creation event or time of universal renewal. Long Count dates had five places, each representing a unit of time—comparable to days, weeks, months, and years. The Mayan units were based on the Mayan number system; they contained multiples of 20 days, with an exception in the third unit, the *tun* of 360 days:

1 day   = 1 *kin*
1 *uinal* = 20 *kin* (20 days)
1 *tun*   = 18 *uinal* (18 × 20 = 360 days)
1 *k'atun* = 20 *tun* (360 × 20 = 7,200 days)
1 *baktun* = 20 *k'atun* (7,200 × 20 = 144,000 days)

(Aveni 2001, 136)

One can see that the figure of 360 days for the *tun* moved the calendar away from a purely vigesimal or base 20 form, as higher units incorporate the number 360 as a multiple. The Long Count included 13 *baktun*; at the end of the last one, a new cycle would begin. Like Mayan numbers, Long Count dates recorded in manuscripts or on monuments appeared vertically, but they may be written horizontally in modern numbers for the purposes of illustration. Thus, the date 11.6.4.14.5 may be interpreted as 11 *baktun*, 6 *k'atun*, 4 *tun*, 14 *uinal*, and 5 *kin*, or 1,628,925 days (4,459 years, 9 months) since the date of creation. An entire cycle of the Long Count, 13 *baktun*, included 1,872,000 days, or 5,125 years (Aveni 2001, 137).

Modern scholars disagreed for a long time on the relationship between Mayan dates and the Gregorian calendar used in the Western world, but the Goodman-Martinez-Thompson (GMT) correlation between the systems has been broadly accepted. That places the beginning of the classic Mayan Long Count at August 12, 3114 BCE. According to the creation story recounted in the *Popol Vuh*, the Classic Maya lived during the third cycle of creation; this period would also encompass most of recorded human history. This cycle ended on December 20, 2012. Popular notions that the latter date would be associated with a cataclysm proceeded from a misunderstanding of the Mayan view of cyclical time; the end of one Long Count cycle would be followed by the start of a new one. The society of the classical Maya collapsed in the 9th and 10th centuries CE, well before the end of the Long Count cycle. One may only wonder how they would have acknowledged the renewal of creation.

THE *TZOLK'IN, HAAB'*, AND CALENDAR ROUND

In addition to the Long Count, the Maya used two other calendar cycles, both probably rooted in astronomical phenomena to some extent. The *tzolk'in* was a 260-day calendar important to Mayan religious rituals, while the *haab'* was a 365-day calendar originally based on the length of a solar year. While the two ran independently of each other, the Maya took great interest in the time of convergence between the two calendars, that is, the period required for the two cycles to come together on a given date. Modern scholars refer to this period as the Calendar Round. It is important to point out that the *tzolk'in* and the *haab'* were not unique to Mayan culture; both arose earlier in Mesoamerican history and were used by other peoples of the region for centuries

The *tzolk'in* proceeded from the interaction of two smaller cycles, a sequence of 20 days with different names and a count of 13 days. The two ran together. For example, two counts of 13 (26 days) would fill one 20-day sequence and the first 6 days of the next sequence. After 260 days (20 × 13), the two cycles would converge where they began, that is, the first day of the 13-day count would fall on the first day of the 20-day sequence. At this point, the *tzolk'in* would be complete and a new one would begin. The *tzolk'in* illustrates the importance of ritual numbers and cyclic convergences in the culture of the Maya. The number 20 was the basis of the Mayan system of numeration, while 13 represented the number of levels in the Mayan upper cosmos (in addition to nine levels of the underworld) (Aveni 2001, 145). The product of these two numbers, 260, might have had additional significance on its own. The planet Venus, crucial to Mayan religion, is visible in the sky for periods of 263 days—close to the 260-day figure. The planet Mars, as viewed from the earth, returns to the same position in the sky every 780 days, that is, 3 × 260. The average period of gestation for a human fetus is 266 days— again, close to 260, and very interesting as a metaphor for the birth of a new time cycle (Aveni 2001, 144–145). All of these correlations are speculative, for the actual origins of the ritual numbers are uncertain.

The *haab'* of 365 days reflected the length of a solar year, the time it takes for the sun to return to the same position in the sky. The Maya divided the *haab'* into 18 months of 20 days each, with an additional month of 5 days. These days, the nameless days of the month of Uayeb, were considered unlucky. The Maya did not use any form of leap year, so the *haab'* moved out of synchrony with astronomical phenomena and with the progress of the seasons.

The Maya believed that cosmic order resided in repeating patterns of time. The reconciliation or convergence of the *tzolk'in* and the *haab'* was a temporal ideal of great importance to them. They knew that the 260-day and 365-day cycles would converge every 18,980 days, that is, the least common multiple of 260 and 365. In other words, the correspondence between a particular date in the *tzolk'in* and a particular date in the *haab'* would take place again in 18,980 days, or 52 *haab'*—a Calendar Round in the language of modern historians and anthropologists. The end of one 52-year cycle and the beginning of the next probably held great significance for the Maya as a time of renewal. The 16th-century Spanish missionary Bernardino da Sahagun worked among the Nahuatl-speaking peoples of the central Mexican plateau for decades, and he described their customs at the end of a Calendar Round:

> . . . When it was evident that the years lay ready to burst into life, everyone took hold of them, so that once more would start forth—once again—another

period of fifty-two years . . . Behold what was done when the years were bound—when was reached the time when they were to draw the new fire, when now its count was accomplished. First they put out fires everywhere in the country round. And the statues, hewn in either wood or stone, kept in each man's home and regarded as gods, were all cast into the water. Also these—the pestles and the hearth stones; and everywhere there was much sweeping—there was sweeping very clear. Rubbish was thrown out; none lay in any of the houses.

(Aveni 2001, 148)

Perhaps the Maya of the Classical period greeted a new 52-year cycle with similar rituals of restoration and rebirth.

EL CASTILLO

Located in the northern part of the Yucatán Peninsula, the city of Chichen Itza was an important population center during the Classic and early Postclassic periods of Maya civilization. More than 50,000 people lived in the city at the time of its ascendancy, and the surrounding region may have supported thousands more. The ruins of Chichen Itza serve as a potent reminder of the grandeur of Mayan culture. The major buildings cover more than five square kilometers, and the massive structure of El Castillo dominates the site. While the Spanish name El Castillo means literally "the castle," the great stepped pyramid was actually a temple to the god Kukulkan (Figure 6.3).

Figure 6.3 El Castillo is a Maya stepped pyramid built in the city of Chichen Itza during the Postclassic period. Its design incorporates many elements of Maya calendars and religious ritual. (Petraarkenian/iStockphoto.com)

Kukulkan, the feathered serpent, played a crucial role in Mayan religion. Known as Gucumatz to the Quiche Maya of Guatemala and as Quetzalcoatl to the Nahuatl-speaking peoples of central Mexico, Kukulkan helped to shape the world and brought law and knowledge to human beings. The temple of Kukulkan in Chichen Itza features large carved serpent heads at the base of its northern stairway. Near sunset on the spring and autumn equinoxes, the steps of the northwest corner of the pyramid cast their pointed shadows onto the side of this stairway, creating the impression of a large serpent made of alternating light and dark triangles, stretching down the side of the entire temple and terminating at the carved head at the bottom. The spectacle may have been an important spiritual experience for the residents of the city, and it has become a major attraction for modern visitors to the site.

The relationship of the temple to the sun demonstrates the ability of the Maya to orient their monuments to the heavens with great precision. The structure of El Castillo also serves as a physical expression of the numerological patterns and cycles central to the Mayan approach to time and space. The pyramid has nine levels, perhaps corresponding to the nine levels of the Mayan underworld. Each of the four faces of the temple has a large, central stairway with ninety-one steps. Combined with the large platform at the summit, these add up to 365—the number of days in the *haab'* calendar. The stairways divide the nine levels into eighteen terraces on each side—the number of months in the *haab'*. Finally, each pyramid face is decorated with fifty-two stone plaques; fifty-two *haab'* cycles corresponded to seventy-three cycles of the *tzolk'in* to make up one complete Mayan Calendar Round (Aveni 2001, 275).

El Castillo illustrates the link between time, space, and divinity that was so important to the Mayan view of the cosmos. Built as a site of religious devotion and ritual, the temple incorporated many of the elements of the Mayan calendar, elements based on the vigesimal number system and on astronomical observation. The structure of the temple expressed the notion that the gods acted to guarantee the passage of time. The mathematical cycles of the calendar and the periodic motions of the heavenly bodies reflected the regular, repeated activities of powerful divine beings. The religious ceremonies performed at the temple were understood to support the gods in their endeavors; sacrifices, possibly involving human victims, nourished the gods and ensured their continuing benevolence. The Maya viewed their religion as a way to maintain the temporal stability of the universe, the stability of time itself.

EL CARACOL

When one thinks of religious architecture, images of temples or shrines come to mind. El Caracol, a structure built in the city of Chichen Itza during the 10th century CE, had a special relevance to the Mayan view of divinity, space, and time: it was an astronomical observatory. Mayan astronomy was positional; it emphasized the motions, appearances, and disappearances of the heavenly bodies and sought to discern the regular mathematical patterns exhibited by these phenomena. El Caracol was constructed to provide sight lines to a variety of astronomical events, events with particular spiritual significance to the Maya.

At first glance, the ruins of El Caracol actually look like a modern observatory; the roof seems to have the domed shape of a structure designed to house a large telescope. The resemblance is coincidental, for the building had no dome—the roofs and walls of the upper levels have collapsed over time to produce the current appearance (Figure 6.4).

Figure 6.4 El Caracol was built in the Maya city of Chichen Itza during the 10th century CE. The orientation of the structure and the placement of its windows correspond to the movements of the heavenly bodies, particularly the planet Venus. (Jannis Werner/Dreamstime.com)

The intact building consisted of two parts: a rectangular platform below and a second platform crowned by a cylindrical tower above. The lower platform was constructed around the year 800; the tower was added later (Aveni 2001, 274). A spiral staircase led to the upper level of the tower and gave the tower its current name in Spanish—*El Caracol* means "the snail." The height of the compound structure raised the observer above the level of the vegetation surrounding the city and allowed for an unobstructed view of the horizon. This was crucial, for the Maya recorded the motions of the stars and planets in terms of their positions on the horizon when they rose or set. The structure was deliberately oriented to a number of astronomical phenomena considered significant by the Maya. For example, the diagonal connecting the northeast and southwest corners of the building lie on a line that points to the sunrise on the summer solstice, the day of the sun's maximum elevation in the Northern Hemisphere. Architectural elements of El Caracol correspond to more than twenty different celestial events, but the observatory seems to be dedicated to the movements of the planet Venus in particular.

The modern understanding of the motions of Venus in the night sky is based on the arrangement of the solar system. The orbit of Venus is closer to the sun than that of the earth; as a result, Venus is only visible near the sun from the perspective of a viewer on earth. That is, it may only be seen near sunrise as a bright "morning star" or at sunset as an "evening star." Specifically, Venus is visible as a morning star for 263 days. It then disappears behind the sun for 50 days and reappears as an evening star for another 263 days. It then disappears for eight days and re-emerges as a morning star again. The entire cycle takes 584 days; this period—the synodic period of Venus—results from the combined motions of Venus and the earth around the sun.

Mayan astronomy was not directed towards the plotting of planetary orbits; rather, the Maya were interested in the planets' behavior with respect to time. They recorded the days when heavenly bodies appeared and disappeared in the sky and looked for mathematical relationships between the cycles of different planets, the sun, and the moon. The Maya knew of the 584-day cycle for Venus, although their values for its appearances as a morning or evening star differ significantly from modern ones. This could be because they sought to relate the coming and going of Venus to the phases of the moon (Aveni 2001, 84). The Maya also realized that five Venus cycles of 584 days (2,920 days) were equal to eight earth years of 365 days and placed great importance on this relationship.

The entire lower platform of El Caracol is oriented to the northernmost point on the horizon where Venus sets as it moves in its 584-day cycle; a line perpendicular to the front face of the building runs to this point. The placement of some of the upper-story tower windows are also based on the movements of Venus. Only three such windows survive, but two provide sightlines from the inner chamber of the tower to the northernmost and southernmost points on the horizon where Venus sets over the course of its synodic period. In other words, one window is precisely oriented to the setting position of Venus at the start of its cycle, the day it first appears in the sky. The other is based on the setting position of the planet on the day it disappears from view at the end of its cycle. Other windows on the collapsed part of the tower may had similar sightlines to the points where Venus rises (Aveni 2001, 273–276). Indeed, the building seems to be based largely on the positional astronomy of the planet.

One can only speculate about the purpose of El Caracol. It may have been built as a physical representation of Mayan astronomical knowledge—a record in stone of the movements of the heavens. However, it is important to point out the spiritual importance of the heavenly bodies to the Maya. Venus was associated with war, and the motions of the planet may have regulated Mayan military campaigns to some degree (Aveni 2001, 167). Venus was also associated with the god Kukulkan, a figure of great significance to the Maya in general and to Chichen Itza in particular. El Caracol may have been a shrine, a temple oriented precisely to the motions of the sun, moon, stars, and planets. The windows in the tower allowed the viewer to look out on the light of Venus as it appeared and disappeared; they also admitted that same light into the inner chamber. Did the light play a role in rituals, ceremonies, or sacrifices conducted there, literally in the sight of Kukulkan? There is no way to know.

CONCLUSION

The Maya understanding of mathematics was intricately interwoven with their conception of nature and of time. The twenty digits of the human body and the motions of the sun were reflected in their view of number. They observed the heavens with great astronomical skill, precisely recording the positions of the sun, stars, and planets. They conceived of the heavenly bodies as divine agents of time and sought the numerical patterns in their movements, appearances, and disappearances. The Maya believed that their religious practices had to be coordinated with these cycles of divine time and heavenly change. In particular, Maya rulers, as shamans or priest-kings, carried out their ritual functions and royal duties in accord with such cycles. Maya astronomy, Maya mathematics, Maya calendars, Maya religion, and Maya rulership all formed part of a single worldview, a coordinated picture of the cosmos.

FURTHER READING

Aveni, Anthony. 2001. *Skywatchers*. Austin: University of Texas Press.
Culbert, Patrick. 1993. *Maya Civilization*. Montreal: St Remy Press.
Demaraest, Arthur. 2004. *Ancient Maya*. Cambridge: Cambridge University Press.
Ifrah, Georges. 1998. *The Universal History of Numbers*, vol I. London: Haverill Press.
Knowlton, Timothy. 2010. *Maya Creation Myths*. Boulder: University of Colorado Press.
Milbrath, Susan. 1999. *Star Gods of the Maya*. Austin: University of Texas Press.
Schiele, Linda and Freidel, David. 1990. *A Forest of Kings*. New York: Harper Collins.
Schiele, Linda and Freidel, David. 1993. *Maya Cosmos*. New York: Harper Collins.

7

Numbers in Ancient and Medieval India

Chronology of Sanskrit Works/Authors

Rig Veda	(1500–1200 BCE)
Upanishads	(800–300 BCE)
Pāṇini	(4th c BCE)
Āryabhaṭa	(476–550 CE)
Bhāskara I	(7th c CE)
Brahmagupta	(7th c CE)
Bhāskara II	(12th c CE)
Kerala School	(14th to 18th c CE)

CULTURAL ICON: THE BAKHSHALI MANUSCRIPT

The Bakhshali Manuscript is the oldest surviving mathematical document in Sanskrit (Figure 7.1). Discovered in 1881 near the town of Bakhshali in present-day Pakistan, the document is written on birch bark in a form of Sanskrit influenced by local dialects. Historians believe it is a copy of an older text and that it was produced some time between the 7th and 12th centuries CE, based on its language and

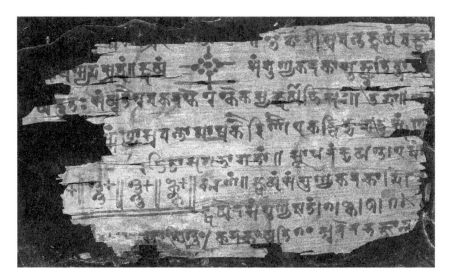

Figure 7.1 The Bakhshali Manuscript is the oldest example of a mathematical text in Sanskrit. It includes calculations based on a decimal place-value system. (Bodleian Library, Oxford)

physical characteristics (although some scholars place it at an earlier date). However, recent radiocarbon dating has assigned different ages to different parts of the manuscript, and this has resulted in some controversy.

The text consists of a series of mathematical rules composed in verse, complete with sample problems and commentary. Significantly, the Bakhshali Manuscript contains numbers made up of ten different characters, including a zero. Further, the zero is used as an actual number, not merely a placeholder. While the Indian numeration system emerged centuries before the likely date of the manuscript, these are among the earliest examples of handwritten Indian numbers. Some of the symbols bear a striking resemblance to their modern equivalents (Figure 7.2).

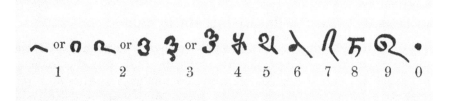

Figure 7.2 The numbers in the Bakhshali Manuscript are composed of ten symbols, including a zero. (Hoernle, Dr. R. *On the Bakhshali Manuscript*, 1887)

The characters are organized in a place-value system based on powers of 10, and this allows for effective calculation. The Bakhshali Manuscript thus embodies the Indian approach to numeration that would ultimately spread around the world.

The manuscript is written in a compact manner, and boxes are drawn to separate the numbers from the textual commentary. The mathematical operations themselves are indicated by abbreviated words: "added," "subtracted," etc. The text includes negative numbers as well. The mathematical principles represented in the document are linked to practical problems of trade and administration; it appears to be a type of mathematical manual. However, the fact that the rules are written in *sutras*, or verses, links the manuscript to the formal tradition of Sanskrit mathematics.

The twenty-seventh verse or sutra serves as a good representation of the text. The rule concerns the computation of the impurity in gold:

> Having multiplied the weights of the gold pieces by their own impurities, one should then divide their sum by the weights of the gold pieces added together. This result is indeed the loss of gold per unit weight of the alloy.

(Plofker 2007, 438)

A specific problem serves as an example of the rule:

> Four gold pieces, the qualities of which are one, two, three, and four *suvarnas* respectively, are inferior by one, two, three, and four negative *masas* per *suvarna* in order. They are melted into one having a single luster. What is the impurity of the alloy?

(Plofker 2007, 438)

The commentary then guides the reader through the solution of the problem.

The Bakhshali Manuscript is part of the collection of the Bodleian Library at Oxford University. In 2017 samples from the manuscript were subjected to radiocarbon dating. The results surprised historians: the different samples yielded different ages. One was dated to the 3rd to 4th centuries CE, one to the 7th to 8th centuries, and one to the 9th to 10th centuries. The Bodleian Library released its results to the public with an emphasis on the earliest figure; the article stressed that the manuscript contained the first known use of a symbol for zero, dating perhaps to the 3rd century CE. A number of historians of Indian mathematics responded with an article pointing out that the Bakhshali Manuscript is a unified mathematical treatise, not a series of fragments; that the author may have written it on sheets of birch bark of different ages; and that the latest radiocarbon date therefore had significance. The age of the document remains in dispute.

CULTURAL COMMENTARY: HEAVEN AND EARTH

In the year 662 CE, a bishop in Syria wrote of Indian mathematics and commented on their "valuable methods of calculation which surpass description" (Plofker 2009, 255). He was referring to the Indian approach to numeration—ten symbols, including a zero, employed in a place-value system based on the powers of 10. It was probably conceived in India during the first millennium BCE and spread to the Arab world in the 8th century CE. The system passed on to Western Europe in the High Middle Ages and ultimately became the standard way of representing number in the modern world.

Mathematics occupied an important place in Indian intellectual life in antiquity and the Middle Ages. It was linked closely to the study of positional astronomy, and Indian knowledge of algebra, geometry, and trigonometry grew in conjunction with a complex spherical view of the cosmos. Indeed, Indian mathematicians developed the trigonometry of sines as a tool in the calculation of celestial motions. In the 15th and 16th centuries, the scholars of the Kerala School in western India would use the sine function as the basis for their work on infinite series, a precursor of calculus. The classic texts of Indian mathematics and astronomy were written in Sanskrit verse and represent a fascinating fusion of literature, scholarship, and spirituality.

Sanskrit is one of the oldest languages of the Indo-European language group, the set of languages that includes English and other Western languages. Languages develop from other languages over time, as groups of people separate, migrate, or mix with other groups. The three oldest known Indo-European languages that have endured in textual form are Sanskrit, Hittite, and Mycenaean Greek. The Linear B tablets of Bronze Age Greece represent a form of Mycenaean, and Hittite survives in the written archives of the Hittite Empire in Anatolia (present-day Turkey). The classical texts of Indian religion, philosophy, mathematics, and astronomy are written in Sanskrit, for it served as the language of scholars in ancient and medieval India for more than 2,000 years. For part of that time, it was a spoken language as well. Sanskrit, Hittite, and Mycenaean Greek probably descended from an original Proto-Indo-European language spoken in the region of the Black Sea and the Caspian Sea in southwest Asia during the Neolithic Period.

The divergence of Indo-European languages and their spread to the Indian subcontinent may be explained by the migration of people during the Bronze Age (roughly 2500–1100 BCE). The ancestors of the Mycenaean Greeks moved into southeastern Europe in the middle of the third millennium BCE, while the Hittites entered Anatolia around 2000 BCE and established a great empire 400 years later. Another group, the Indo-Aryans, migrated from the area northeast of the Black Sea, through modern-day

Iran, to the Indian subcontinent in the years between 1800 and 1500 BCE. Some stayed in Iran, while others moved farther east. Sanskrit resulted from that migration; it combined the Indo-European language of the Indo-Aryans with linguistic elements from the older inhabitants of India.

The *Rig-veda* is the oldest surviving text written in Sanskrit. It dates from between 1500 and 1200 BCE and consists of over 900 prayers and hymns to the Vedic gods, a set of polytheistic entities derived from both Indo-Aryan and local Indian traditions. Along with the other Vedas, the *Yajur-veda*, the *Sāma-veda*, and the *Atharva-veda*, it continues to be revered in Hindu spirituality. All of the Vedas, and some of the early *Upanishads*—religious texts written between 800 and 300 BCE—were composed in Vedic Sanskrit, the oldest form of the language. Classical Sanskrit emerged between 600 and 400 BCE and was formalized by the grammatical text of Pāṇini, a scholar who lived during the 4th century BCE. Philosophical, astronomical, and mathematical works would be written in Classical Sanskrit until the 17th century CE. The heart of mathematical study in Classical Sanskrit lies in the various *siddhānta* texts composed between 500 and 1000 CE. These were essentially treatises of astronomy, but they also contained discussions of the mathematics necessary for astronomical work. Several mathematical surveys were also written in Sanskrit in the 12th and 13th centuries, essentially compiling the accumulated knowledge of the *siddhāntas*. By that time, Sanskrit had started to decline, as Muslim influence grew in northern India. By the time of the Mughal Empire of the 16th century, it had been largely supplanted as the language of scholarship. However, perhaps the most innovative of Indian mathematicians, the scholars of the Kerala School, wrote their work in Sanskrit verse between the 15th and 17th centuries BCE.

Hindu spirituality provided cultural support for the study of mathematics. Hinduism embraces a broad range of beliefs, and many varieties of Hindu thought accept intellectual pursuits as part of one's journey to greater enlightenment. A common theme in Hinduism is the notion of transcending the self, of moving beyond individual desire towards union with *Brahman*—the divine or the absolute. Intellect can play an important part in this process; while all aspects of the universe have a divine essence, the intellect exists on a higher plane than the material world or the physical body. The *Chandayoga Upanishad* discusses intellect as a means of spiritual enlightenment:

> One who worships intelligence as Brahman attains worlds of intelligence [i.e., things he regards as important]. He becomes true and attains the world of truth. He is firmly established and also attains a world which is firmly established. He is free from pain and also attains a world free from pain. One who worships intelligence as Brahman can do what he pleases within the limits of intelligence. (*Chandayoga Upanishad* Chapter 7 verse 5)

To "worship intelligence as *Brahman*" suggests that the life of the mind may be a path to the divine. A similar theme appears in the *Bhagavad Gītā*, a segment of the Hindu epic the *Mahabharata*, written between 200 BCE and 200 CE. The *Bhagavad Gītā* tells of an exchange between the god Krishna and a young prince, Arjuna, who has lost his resolve before a battle. Krishna points out that performance of duty is an important part of self-denial and goes on to state:

> They say that the senses are superior to the body, higher than the senses is the mind, yet higher than the mind is the intellect, but what is greater than the intellect is "He" (the Supreme). (*Bhagavad Gītā* 3.42)

Here, Krishna states that intellect occupies a place higher than body and mind, but lower than the soul itself.

The *Bhagavad Gītā* holds an important place in *Vedanta* Hinduism, a version of Hindu thought that is central to much of modern Hindu spirituality. However, other schools of Hinduism also developed in the first millennium BCE. Hindu philosophy lists six *pramāṇas*, or sources of understanding: perception, inference, analogy, circumstance, nonperception, and testimony. Different schools of Hinduism accepted different combinations of the *pramāṇas*, different versions of the nature of truth. However, perception and inference played a role in many Hindu ideologies. As an expression of intellect and as an embodiment of perception and inference, mathematics would be accepted as a form of spiritual enlightenment by most forms of ancient and medieval Hindu thought.

Significant references to mathematics appear in the Vedic and early Hindu traditions. The *Rig-veda* reveals signs of a decimal number system, and the *Yajur-veda* contains a striking statement of devotion to number itself as an object of praise:

> Hail to one, hail to two.... Hail to a hundred, hail to a thousand, hail to ten thousand, hail to hundred thousand, hail to million, hail to ten million, hail to hundred million, hail to billion, hail to ten billion, hail to hundred billion, hail to trillion, hail to the dawn, hail to the daybreak...
>
> (Plofker 2009, 14)

A more complex discussion of mathematical material in Sanskrit occurs in the *Śulba-sūtras*, written between 800 and 300 BCE. These texts spoke of the proper shapes and sizes for the ritual fire altars used in religious sacrifice. Such altars took the form of different geometrical figures: rectangles, triangles, circles. Further, they needed to have specific dimensions and areas. The altars were laid out with the use of measuring chords, and the *Śulba-sūtras* contained detailed instructions on the formation of figures and their conversion into other figures with particular areas and proportions. This form of sacred practice revealed a command of practical geometry.

Language occupied a special place in Hindu thought, and some of the classical works on the Sanskrit language itself probably had a mathematical character. Pāṇini's treatment of Sanskrit grammar in the *Eight Chapters* (5th–4th c BCE) used symbols to represent different kinds of phonetic values; this may have had an influence on the emergence of Indian numeration and algebra (Plofker 2009, 54). In addition, the *Chandah-sūtra*, a 3rd-century BCE text on poetic meter by Pingala, contained detailed analyses of the patterns of accented and unaccented syllables in Sanskrit verse. In considering the number of possible metrical arrangements, Pingala constructed a version of the arithmetical triangle, a triangular mathematical matrix also discovered by the ancient Chinese and later described by Omar Khayyam and Blaise Pascal.

The link between mathematics and astronomy so crucial to the development of Indian mathematical ideas first appeared in the *Jyotiṣa-Vedāṅga*, a text on lunar and solar calendars probably written in the 5th or 4th century BCE. However, as stated earlier, the most significant expressions of mathematical knowledge in classical Indian thought may be found in the astronomical *siddhāntas*, or treatises, written in Sanskrit verse during the first millennium CE. A collection of five early treatises, the *Pañca-siddhāntikā*, was compiled in the 6th century CE; it probably contained significant material from earlier periods. The texts of the *Pañca-siddhāntikā* treated issues of calendar construction and the movements of the planets, as well as the mathematics of spherical astronomy, including the earliest known table of sines (Plofker 2011, 50–52).

More elaborate astronomical *siddhāntas* appeared in the 6th century and afterwards. They included detailed calculations of celestial movements, based on the assumption of a central spherical earth. They also contained chapters on mathematical problems and instruction, intended to foster the skills necessary for positional astronomy. Different schools of astronomical thought developed. The author Āryabhaṭa composed one of the most influential *siddhāntas*, the *Aryabhatiya*, and this work served as the basis of an important commentary written by Bhāskara I in the early 7th century. At the same time, in the region of Gujarat, Brahmagupta authored the *Brahma-sphuta-siddhānta*, in which he challenged some of the astronomical assumptions of Āryabhaṭa. In addition, he made use of the ten-symbol Indian number system and treated zero as an actual number, as opposed to a mere placeholder. Yet another school was based on the *Sūrya-siddhānta*, an expansion of one of the five treatises contained in the *Pañca-siddhāntikā*.

While the various *siddhāntas* focused on astronomy and mathematics, they also embodied the values of Sanskrit literature; they were part of a larger culture of Sanskrit verse. The authors of the *siddhāntas* used a variety of techniques to express numbers in literary terms; they used code words or phonetic syllables to stand for the digits of numbers, while retaining a decimal place-value system. These techniques allowed for a poetic description of mathematical operations and processes. Further, while the *siddhānta* tradition emphasized the mathematical description of celestial movements, its adherents continued to revere the scriptures of Hinduism. This produced some conflicts between mathematically based astronomy and the influence of religious texts and truths.

The varied astronomical *siddhāntas* all featured calculations of planetary position based on intricate circular geometry. They also used the trigonometry of sines in the process of describing the angles and arcs of celestial motions. It is likely that Indian scholars were influenced by Hellenistic mathematics, for the Hellenistic Greeks had used similar geometrical techniques in their descriptions of planetary movements, and they had pioneered the use of trigonometry as an astronomical tool. However, no direct evidence exists for the transmission of Greek texts to India in antiquity or the Middle Ages.

On the other hand, the influence of Indian astronomy and mathematics on the Arabic intellectual tradition has been well-documented. Indian travelers brought copies of astronomical *siddhāntas* to Baghdad in the 8th century CE, and those texts played a role in the adoption of Indian numerals by Arab and Persian mathematicians in succeeding centuries. The Arabs also embraced Indian knowledge of trigonometry and combined it with their growing knowledge of Hellenistic Greek astronomy and geometry.

During the 12th century, Bhāskara II wrote comprehensive treatises of the mathematical knowledge expressed in the various *siddhāntas* and added his own innovations and commentaries. His treatises were copied widely and served as standard texts for centuries. While Sanskrit started to decline in the later medieval period, the Kerala School of the 15th and 16th centuries represented the apex of classical Sanskrit mathematics. The scholars of Kerala continued to do astronomical study, and they built on the mathematical knowledge of their predecessors. Madhava, the founding scholar of the community, developed infinite series to find the values of π, sine, and cosine, and his successors continued his work. The concept of an infinite series is an important component of modern calculus; however, the Kerala mathematicians did not take the step to the functions of calculus itself.

As the Islamic Mughal Empire came to dominate India, Arabic and Persian mathematics would mix with Indian intellectual traditions. Later, with the onset of European imperialism, Western mathematics and science would enter the subcontinent and have a major influence on Indian culture. Yet the Sanskrit texts of classical Indian mathematics would have an enduring legacy. They played an important part in the cultural exchanges of the Middle Ages, and they remain as an icon of Hindu culture.

INDIAN NUMERATION

The modern number system used around the world originated in India in the early centuries of the first millennium BCE. The features of the system are familiar to any reader: ten symbols from 0 to 9, arranged in place-values based on the powers of 10. When a given symbol is in a given place-value, it acts as a multiplier of the power of 10 corresponding to that place-value. Thus, in the number 714, the symbol "7" is multiplied by 10 to the second power, 100, to represent a value of 700.

While tantalizing signs of the emergence of this system have survived, there is no clear evidence of exactly when and where it arose. The actual symbols for Indian numbers came from *Brâhmî*, a written script used in the Indian subcontinent in the first millennium BCE. *Brâhmî* was the ancestor of scripts throughout south and southeast Asia. However, numbers in *Brâhmî* were not based on place-value; they had an additive quality common to many numeration systems.

No clear evidence exists of the origins of place-value notation in India, but it probably arose during the period of the Gupta Empire (3rd–6th c CE). The *Lokavibhāga*, a Jainist cosmological text dated 458 CE, contained references to place-value and the use of zero. Many Sanskrit texts of the first millennium CE combined the notion of place-value with the use of verbal terms for individual digits of numbers. Some historians claim that the concept was known in India as early as the 3rd century CE and suggest that it came from China (Plofker 2009, 48). Chinese counting board notation had been based on decimal place-value since the middle of the first millennium BCE.

Modern numbers combine ten symbols and place-value notation. The oldest known example of such numbers is a copper plate from Gujarat engraved with the date 346 in the local calendar, corresponding to the year 594 or 595 CE. This indicates that the numbers were in use in the region before that time. The 7th-century Indian mathematicians Bhāskara I and Brahmagupta both used them in their work. The system was transmitted to the Arab world in the 8th century and to Western Europe in the 12th. The symbols varied with time and place, but the basic principles of the system remained unchanged, a creation of classical Indian culture.

INDIAN ASTRONOMY IN THE FIRST MILENNIUM CE

Indian mathematics in late antiquity and the Middle Ages was closely tied to the study of positional astronomy. During the first millennium CE, a series of authors composed *siddhāntas* on the motions of the heavens, treatises on astronomy in Sanskrit verse. These focused on the mathematical analysis and prediction of celestial movements—those of the sun, moon, stars, and planets. They also placed great emphasis on the prediction of solar and lunar eclipses. The *siddhāntas* often contained chapters on the mathematical principles (*gaṇita*) necessary for astronomical calculations, as well as discussions of *Gola*, the model of the universe based on circular and spherical geometry.

Any treatment of the astronomy of the *siddhāntas* must first address the far older cosmology of traditional Hinduism. The *Puranas*, Sanskrit texts from the period between 300 and 100 BCE, describe a universe made up of circular disks, stacked vertically in space. The earth is the largest disk, located in the center of the stack, while the heavenly bodies are situated in the upper disks, turning in circles above the earth. Below the earth are the disks of the underworld. The entire structure is contained in an egg-like shell and supported from below by a giant tortoise on snake (Plofker 2011, 52).

The mathematical astronomy of the *siddhāntas* largely rejected this traditional cosmology. Indian astronomers based their predictive calculations on the concept of a spherical earth in the center of the universe. The pattern of stars and constellations surrounded the earth and moved in a circle around it; the astronomer Āryabhaṭa referred to "the cage of constellations formed of circles" (Plofker 2011, 112). The sun, moon, and planets occupied the space between the earth and the stars. The authors of the *siddhāntas* sought to account for the positions of the heavenly bodies and to predict their future motions through the use of complex circular geometry.

It is important to note that because the planets actually move around the sun in elliptical orbits, any attempt to support an earth-centered system will involve significant mathematical difficulty. The astronomers of the *siddhāntas* used a variety of circular geometrical techniques to make their predictions conform to observed positions. They assumed that the heavenly bodies did not move around the earth in simple circular paths. Rather, a planet moved around a point in space that in turn moved around the earth. This combination of a small circle—the epicycle—and a larger circle—the deferent—gave astronomers a number of different parameters that could be adjusted to produce accurate predictions of planetary positions. The Indians also used the geometry of eccentrics, in which the earth was not placed in the exact center of the circular course of a planet. In the *Sūrya-siddhānta*, the author suggested that a planet's deviation from a pure circular path resulted from a divine being regularly pulling the planet farther from the earth, an interesting mix of observational astronomy and religious explanation (Plofker 2011, 84).

The process of deriving the circular movements of the sun, moon, and planets from observed positions involved complex calculations, the essence of Indian mathematics. Trigonometry played an important part in spherical astronomy. It was used to link angular to linear values, to determine distances, angles, and arcs of movement within the sphere of the heavens. Hellenistic Greek astronomers had used the trigonometry of chords to find these values, based on the geometry of a triangle within a circle (representing the circle of the heavens). The authors of the Hindu *siddhāntas* employed the trigonometry of sines, a far more efficient method for solving the same problems. The purely mathematical sections of the *siddhāntas* focused on the computational skills (*gaṇita*) necessary for such challenging analysis and prediction.

While the authors of the *siddhāntas* demonstrated great mathematical sophistication in their calculations, they do not discuss the origins of their methods, data, or models of the universe. No substantial tables of astronomical observations are present in the texts (Plofker, 2011, 114). It is possible that the Indians did accumulate astronomical observations over time; it is also possible that they were influenced by Hellenistic Greek mathematics, data, and cosmological models. The Greeks had developed a detailed astronomy based on epicycles and eccentrics in the Hellenistic period, and they had gathered substantial observations of celestial movements. In any case, the astronomical calculations of the *siddhāntas* assumed a central earth and accounted for celestial motions in complex geometrical ways. This resulted in some conflict between the spherical earth and cosmos of the Indian mathematicians and the flat, disk-shaped earth of Hindu tradition. The authors of the various *siddhāntas* steadfastly maintained the accuracy of their models. Because of the great variety inherent in Hindu religious belief, they apparently had the ability to reject traditional ideas. Indeed, in a sense, *gaṇita* and *Gola* themselves could viewed as intellectually rigorous forms of devotion.

THE *ĀRYABHATĪYA*

The mature *siddhānta* tradition that took shape in the 6th century CE represented a period of growth in Indian astronomy and mathematics. Written in Sanskrit verse, the *siddhānta* treatises had three major components: calculation of the positions of heavenly bodies, discussion of the spherical geometry of the universe—*Gola*, and a treatment of mathematical operations themselves—*gaṇita*. The *siddhāntas* also linked their mathematical treatment of astronomy with Hindu scriptures. Various schools of thought developed, based on geography and on different ways of computing time and position.

Āryabhaṭa, an astronomer born in 476 CE, wrote one of the most important *siddhāntas*, the *Āryabhatīya*. The work was quite concise and was made up of 123 verses. The mathematician Bhāskara composed a commentary on the book in the year 629 and explained the author's mathematics in considerably more detail.

The *Āryabhatīya* treated positional astronomy based on a spherical view of the earth and the universe:

> The sphere of the earth made of earth, water, fire, and air, in the middle of the cage of the constellations formed of circles, surrounded by the planetary orbits, in the center of the heavens, is everywhere circular.
>
> (Plofker 2009, 112)

Āryabhaṭa's celestial calculations made use of the circular geometry central to the *siddhānta* tradition; he assumed that the planets moved uniformly in compound circular paths. He employed the trigonometry of sines to calculate these motions and provided a list or table of sine values, one of the earliest ever recorded. However, while Āryabhaṭa explained the positions of celestial bodies in terms of complex circular geometry, he also made the significant argument that some heavenly motions could be explained by a central rotating earth:

> In the same way that someone in a boat going forward sees an unmoving object going backward, so someone on the equator sees the unmoving stars going uniformly westward. The cause of rising and setting is that the sphere of the stars together with the planets apparently turns west at the equator, constantly pushed by the cosmic wind.
>
> (Plofker 2009, 111; Plofker 2007, 417)

Here, the author equates the earth to a moving boat and the stars to an "unmoving object" that only appears to move because of the earth's rotational motion. However, in the second part of the passage, he refers to the sphere of the stars as moving, pushed by a "cosmic wind." It appears that he wishes to point out that the phenomena of the stars rising and setting may be explained in two different ways, both dependent on circular motion (Plofker 2009, 111).

The second chapter of the *Āryabhatīya* discussed mathematical operations, or *gaṇita*; it served as a guide to the tools necessary for astronomical calculation. It is the oldest surviving example of mathematical instruction in Sanskrit. The discussions were quite concise; readers were expected to grapple with the complicated Sanskrit wording as part of the intellectual exercise of confronting the subject matter (Plofker 2007, 400). Āryabhaṭa opened his chapter on *gaṇita* with a list of powers of 10, important in a decimal-based system of numeration. He went on to relate processes for finding square and cube roots and rules for finding the areas and volumes of geometrical figures. Some of these were invalid; for example, he provided incorrect formulas for the volume of a pyramid and a sphere.

In the *Āryabhatīya*, the author used a circle with a circumference of 62,832 and a diameter of 20,000 as the basis of his value for π. The ratio of the circumference to the diameter yielded the following figure, an important departure from earlier Sanskrit sources:

$$\frac{62832}{20000} = 3.1416$$

Āryabhaṭa went on to discuss briefly how to divide a circle into arcs and to calculate sines from those arcs. The author Bhāskara would add substantially to this treatment in his commentary on the *Āryabhatiya*, for these operations were crucial to the practice of positional astronomy.

Āryabhaṭa treated right triangles in some detail and discussed the relationship between the "arm," the "upright," and the hypotenuse—the Pythagorean theorem. His work addressed the solution of proportion problems (the rule of three quantities), operations with fractions, the finding of unknowns, and the calculation of interest. He closed the chapter with a discussion of his own method for solving linear indeterminate equations, that is, equations of the type $ax + by = c$. Here, x and y are unknowns and a, b, and c are integers. Āryabhaṭa's approach involved a successive series of divisions that reduced the magnitude of the figures in the equation and produce a simpler one. Again, Bhāskara would elaborate on this in his commentary and give the process a name: *Kuṭṭaka*, "the pulverizer."

The commentary of Bhāskara (or Bhāskara I—the number distinguishes him from a later author of the same name) was a significant work on its own. Written in prose, it provided extensive explanations of the verses in the *Āryabhatiya*. It also made use of Indian place-value numeration in its discussions, as opposed to the verbal numbers used by Āryabhaṭa. Bhāskara was a follower of Āryabhaṭa's school of astronomical thought, and would produce two *siddhāntas* of his own in addition to his famous commentary.

INDIAN MATHEMATICS AND THE HELLENISTIC GREEK TRADITION

In the contemporary world, information moves almost instantaneously. Computer technology, fiber-optic transmission, and wireless communication allow the easy transfer of text, images, sound, and video. However, in a premodern world, the movement of ideas required physical travel; books had to be carried from one place to the next. New concepts moved along trade routes like merchandise; they moved with migrating people and conquering armies; they moved with diplomats and traveling scholars. Just as the migration of the Indo-Aryans had a profound effect on Indian culture, other influences may have shaped Indian mathematical and astronomical thought in antiquity and the Middle Ages. Similarly, Indian ideas had a profound effect on their geographical neighbors and trading partners, particularly Iran and the Arab world.

In the late 4th century BCE, the Macedonian armies of Alexander the Great conquered Egypt and the Persian Empire, including most of the modern Middle East and Iran. He also invaded India and conquered territory there before returning to Babylon with his army. After Alexander's death in 323 BCE, his empire fractured into smaller domains ruled by his generals. The larger realms were controlled by the Ptolemid rulers of Egypt and the Seleucid emperors of Mesopotamia. However, small kingdoms in western India were also ruled by Macedonian Greeks, and these maintained contact with the Greek Hellenistic world. These "Indo-Greek" states came under control of Scythian rulers by the 1st century CE, but their Greek inhabitants remained significant (Plofker 2011, 48). This Greek presence led to the transmission of Greek astrology and its absorption into the Indian calendrical tradition of *Jyotiṣa*. The twelve constellations of the Zodiac and its division into 360 degrees of arc (both originally Babylonian concepts) became part of the Indian picture of the universe (Plofker 2011, 49).

Hellenistic Greek positional astronomy flourished between the 4th century BCE and the 2nd century CE. It was based on the concept of a spherical earth at the center of the universe and on the idea of circular celestial motions. The observed positions of the sun, moon, stars, and planets were explained in terms of circular geometry; explanatory techniques included epicyclic motion, eccentric planetary courses, and the concept of the equant (see Chapter 4). Hipparchus of Nicaea developed the trigonometry of chords as a means of determining celestial positions. The Hellenistic astronomical tradition reached its apex in the *Almagest* of Claudius Ptolemy, written around 150 CE. Ptolemy included tables of observed planetary positions and improved upon the circular geometry and trigonometry of his predecessors in the process of accounting for his positional data.

Some historians have suggested that Hellenistic astronomy had a significant influence on the authors of the astronomical *siddhāntas* of the first millennium CE, perhaps through the agency of the Indo-Greek populations of western India. Indian mathematical astronomy also assumed a central, spherical earth, and it accounted for celestial positions with a similar set of circular geometrical approaches, most particularly, epicycles and eccentrics. The Hellenistic Greeks had employed a form of trigonometry in the process of interpreting their positional data (based on the interception of a chord of a circle by an angle of a triangle). Indian astronomers used the more sophisticated trigonometry of sines, and it is possible that one evolved from the other. The calculations of the *siddhāntas* differ substantially from those in Ptolemy's *Almagest*, but it is possible that Indian authors drew on scholars from earlier in the Hellenistic mathematical tradition, the same scholars who had informed Ptolemy's work (Plofker 2011, 113–115 and 118–119).

However, it is important to point out that no direct evidence exists for a Hellenistic influence on Indian astronomy. The mathematics of the *siddhāntas* do diverge from Ptolemy's book, and this could indicate an independent origin for Indian conceptions of the universe. The Indian trigonometry of sines represents a significant distinction between the two bodies of work. Finally, some historians have argued that an analysis of the calculations in the *siddhāntas* points to the existence of a corpus of Indian astronomical observations made in the first millennium CE, observations that formed the foundations of the *siddhānta* texts. Yet direct evidence of such observation does not exist (Plofker 2011, 113–118).

In short, it cannot be established with certainty whether Indian astronomy reflected Hellenistic influence or represented an independent strain of thought. The authors of the *siddhāntas* displayed a more pragmatic frame of mind than the Greeks; they did not appear bound to particular cosmological models or philosophical assumptions about the character of circular motions (Plofker 2011, 120). Yet they explained the motions of the heavens in a similar fashion. In her survey of Indian mathematics, Kim Plofker examines the question and suggests a compromise: the Indians adopted some Hellenistic concepts, but added to them and altered them based on their own mathematical traditions and views of validity. However, she points out that the issue is certainly open to further research (Plofker 2011, 120). It remains a fascinating inquiry in the history of mathematics.

INDIAN MATHEMATICS AND THE ISLAMIC WORLD

The rapid growth of the Arab Empire into Persia (Iran) during the 7th and 8th centuries CE brought it to the borders of the Indian subcontinent. This resulted in substantial contact between Arab, Persian, and Indian cultures and set the stage for significant intellectual exchanges. The Indian system of numeration would have a profound effect on scholarship in the Arabic world, that is, the region dominated by Islam in which Arabic became the primary language of scholarship. Some small works of astronomy written in Persia in the 8th century reflect the influence of Indian numerals. Perhaps most significantly, Arab sources tell of a visit to Baghdad in the 770s by Indian travelers bearing one of the astronomical *siddhāntas* in Sanskrit. The Arabic translation of the work was called the *Sindhind*, and it served as a vehicle for the transmission of Indian astronomy and mathematics, as well as Indian numbers based on place-value. Considering the proximity of India and Persia, other transfers probably took place around the same period of time. In 820, the Persian scholar al-Khwārizmī composed a well-known treatise on the use of Indian numbers, and other authors would write on the same topic in Arabic in succeeding centuries. The use of Indian numbers would spread throughout the Islamic world, from the borders of India in the east to Spain in the west. The 10th-century European author Gerbert de Aurillac learned of the system during his travels, and Indian (or Indo-Arabic) numbers would eventually enter Western Europe through Arab Spain in the 12th century.

Other aspects of Indian mathematics spread throughout the Arabic cultural sphere. The use of sines in astronomical calculations would inspire significant Arabic work in trigonometry; the Persian mathematician Abū al-Wafā would describe all six common trigonometric functions in the 11th century. Through the *Sindhind*, the Arabs were also exposed to Indian problem-solving techniques and earth-centered astronomy. However, it appears that medieval Arabic astronomy was primarily inspired by translations of Hellenistic Greek texts, particularly Ptolemy's *Almagest*. Further, the development of Arab algebra in the work of al-Khwārizmī and other authors had a character distinct from Indian methods (Plofker 2009, 258–259). Arab scholars probably drew on ideas from the Babylonian and Hellenistic traditions (the work of Diophantus, for example) and added their own insights and innovations.

With the dawn of the second millennium CE, Islamic rulers began to assert their authority in northern India. A series of invasions of the Indian subcontinent took place, culminating in the establishment of the Mughal Empire in 1526. These political developments resulted in increased Islamic influence on Indian culture, including the field of mathematics.

Abu Rayhan Muhammad ibn Ahmad al-Biruni traveled extensively in northern India during the early 11th century. He was an accomplished mathematician, astronomer, and natural philosopher, and he authored treatises on a wide variety of subjects. His *Book of Inquiry Concerning India* described Indian culture and scholarship, as well as his personal reflections on the region and its people. Al-Biruni's background included extensive command of Euclidean geometry and Aristotelian logic, and he pointed out the lack of formal geometrical proof in Indian mathematics. He also criticized the role of religious belief in Indian astronomy. However, he recognized Indian accomplishments in mathematics and commented on the Indian approach to mathematical proportions (Plofker 2009, 265–266).

As the Islamic presence grew in India, Arabic and Persian mathematical texts, translated into Sanskrit, had an influence on Indian scholarship. Between the 14th and the 18th centuries, numerous works on astronomy, astrology, and the use of the astrolabe were transmitted in this fashion. In the early 18th century, Jai Singh II, a Hindu prince, assembled a group of astronomers and encouraged the study and criticism of both Islamic and

Indian views of the heavens. He constructed a series of observatories, sponsored the translation of numerous Arabic and ancient Greek texts, and ultimately included European astronomical ideas as well (Plofker 2009, 268–269).

At some points in its history, the Mughal Empire provided a forum for the mixing of Islamic and Indian ideas on mathematics and astronomy. Some Mughal rulers employed both Islamic and Hindu astronomers. Akbar the Great reigned from 1556 to 1605 and was known for his views on religious tolerance and his enthusiasm for different forms of spirituality. He and his court showed interest in the mathematical surveys of Bhāskara II, the Līlāvatī, and the *Bīja-gaṇita* (Plofker 2009, 270). However, despite these examples of intellectual exchange, Hindu scholars never really accepted Islamic or Greek styles of geometrical proof. Whatever ideas they adopted from the Islamic or Persian traditions were placed in the context of existing Indian approaches.

BHĀSKARA II

While the medieval *siddhāntas* formed the core of the Sanskrit mathematical tradition, they focused primarily on astronomy; mathematical knowledge, or *gaṇita*, was presented as a necessary component of astronomical calculation. The first treatment of mathematics as a discipline in its own right was written by the Jainist scholar Mahāvīra in the 9th century CE. During the 12th century, Bhāskara II wrote a series of mathematical works that organized the mathematical knowledge of the *siddhāntas* and expanded upon that knowledge. His books would serve as standard mathematical texts in India for several centuries and were reproduced in thousands of manuscript copies.

Bhāskara II (so called to distinguish him from the 7th-century figure of the same name) was born in 1114 in western India. He came from a scholarly family, and his father was probably his most important teacher. Like the medieval authors of the astronomical *siddhāntas*, Bhāskara wrote his mathematical treatises in the form of Sanskrit verse. However, he also provided his own prose commentary, and this contributed significantly to the impact of his work (Plofker 2009, 182). His most well-known book, the *Līlāvatī*, may have been named for his daughter. According to tradition, Bhāskara used his astrological knowledge to determine the date and time most appropriate for his daughter's marriage and constructed a simple water clock to indicate the moment. When his daughter examined the clock, she unknowingly allowed a pearl from her wedding dress to fall into it, blocking the flow of water. The appointed time passed unnoticed, and as a result the wedding could not take place. As a consolation for his daughter, Bhāskara named his book after her so that she would be remembered forever. Whether the story is true or not, several of the verses in the *Līlāvatī* refer to an "intelligent girl" and ask her to perform mathematical operations. Perhaps the daughter of Bhāskara lives on in the text.

The *Līlāvatī* begins with an invocation to the Hindu god Ganesh or Gaṇeśa, a deity associated with intellect, learning, and the removal of obstacles—appropriate to a treatise on the solution of problems:

> Having bowed to Gaṇeśa who causes the joy of those who worship him, who, when thought of, removes obstacles, the elephant-headed one whose feet are honored by multitudes of gods. I state the arithmetical rules of true computation, the beautiful Līlāvatī, clear and providing enjoyment to the wise by its concise, charming, and pure quarter verses.
>
> (Plofker 2007, 448)

Significantly, Bhāskara mentions mathematics as a source of enjoyment, as pure intellectual pleasure. The work proceeds with numbered verses on units of measurement and basic mathematical operations, asking questions and offering solutions. A problem in multiplication could be addressed to his daughter:

> Fawn-eyed child Līlāvatī, let it be said, how much is the number resulting from one-three-five multiplied by twelve, if you understand multiplication by separate parts and by separate digits. And tell me, beautiful one, how much is that product divided by the same multiplier?
>
> (Plofker 2007, 449)

Bhāskara discusses square and cube roots, as well as the treatment of fractions. He moves on to discuss more complicated problem-solving and the finding of unknown quantities, touching on the procedures of algebra. Like his predecessors, he includes problems of a practical character, like this one on the computation of interest:

> If the interest on one hundred for a month is five, tell me what is the interest on sixteen when a year has passed? Also, tell me, mathematician, the time from the principal and the interest, and the amount of the principal when the time and the result are known.
>
> (Plofker 2007, 456)

The text continues with more challenging material on algebraic operations and mathematical series. Bhāskara devotes more verses to geometry than to any other topic in the *Līlāvatī*. He discusses right triangles and the Pythagorean theorem at length and applies those principles to complex problems:

> There is a hole at the foot of a pillar nine hastas high, and a pet peacock standing on top of it. Seeing a snake returning to the hole at a distance from

the pillar equal to three times its height, the peacock descends upon it slantwise. Say quickly, at how many karas from the hole does the meeting of their two paths occur?

(Plofker 2007, 459)

The mathematics of proportions, indeterminate equations, and combinatorics occupy the final part of the text.

Bhāskara II wrote a series of other works. In his second major treatise, the *Bīja-ganita*, he focused on the problem-solving techniques of algebra; the text is also noted for its treatment of operations involving zero. He recognized that division by zero has some association with infinity, but his treatment of the topic was unclear (Plofker 2009, 192–193). In addition to his purely mathematical writing, Bhāskara authored a *siddhānta* of his own, the *Siddhānta-Śiromaṇi*, devoted to positional astronomy. Interestingly, he discussed the mathematics of sines in his *siddhānta*, but not in his earlier treatises. He may have considered that particular subject matter to be exclusively astronomical in nature, essential to the calculation of planetary positions. Some historians view the *Līlāvatī* and the *Bīja-ganita* as part of an expanded *Siddhānta-Śiromaṇi*; if that is the case, then the author apparently chose to place the trigonometric material in the culminating part of the work where it was most applicable.

Bhāskara II took a systematic approach to the subject matter of mathematics, and his treatises were particularly effective as a form of instruction. The frequency with which they were copied makes him the most influential of the medieval Indian mathematicians.

THE KERALA SCHOOL

By the late Middle Ages, Sanskrit was declining as a literary language in India. However, on the western coast of the subcontinent, in the region of Kerala, a community of scholars continued to study mathematics and astronomy and to express their ideas in Sanskrit verse from the 14th to the 18th centuries. The members of the Kerala School maintained some of the traditions of the medieval *siddhāntas* and wrote commentaries on those classic texts. However, they also broke exciting new ground in mathematics through the study of infinite series—a topic recognized as a component of modern calculus. While they did not take the step to the derivative and integral operations of calculus itself, the scholars of the Kerala School arguably represented the zenith of mathematical study in Sanskrit.

The Kerala School grew in an unusual political and social atmosphere. The area around Kerala was dominated by groups from different *varnas*, major divisions of Indian society. Foreign merchants also played a role. As a result, the Kerala mathematicians did not come exclusively from the

Brahmin castes of the Indian populace. Some historians have suggested that this contributed to the creativity of the scholarly community there (Plofker 2009, 218).

Mādhava, the founding member of the Kerala School, was born in the late 14th century. A follower of Aryabhata's school of thought, he wrote several texts on astronomy. Yet his most important efforts lay in the field of pure mathematics. Mādhava pursued the study of infinite series—summations of a group of quantities based on a given formula or rule. Such quantities get smaller and smaller as the series progresses, so the summation of an infinite number of terms can have a finite result. Mādhava developed these techniques to find the value of π and the measurements of sines and cosines. While only fragments of his writings survive, his successors commented on his views in detail. Mādhava worked in the late 14th and early 15th centuries, and his ideas shaped the thought of the scholars of the Kerala School for more than 200 years. One of the most important of these was Nilakantha, born in 1444. Nilakantha, in turn, was a teacher of Sankara, who worked in the 16th century. His writings explored the ideas of Madhava and explained their significance; much of what we know of Madhava's mathematics comes from the work of Sankara.

While the writings of Mādhava and his successors on infinite series represent the most important work of the Kerala School, its scholars pursued other topics as well. For example, Nilakantha composed a commentary on the *Aryabhatiya*, as well as a separate treatment of astronomy, the *Tantrasangraha*. Sankara later wrote a commentary on that. Some of Nilakantha's views on astronomy demonstrated great intellectual independence. He challenged the continuing influence of Hindu scripture on positional astronomy and made a case for constructing models of the universe based on observations and analysis. In doing so, he argued that the intellectual ability of scholars was a form of divine enlightenment itself: "The favor of the deity is just the cause of mental clarity" (Plofker 2009, 250). Nilakantha suggested a view of the cosmos in which the epicycles of the planets were based on the position of the sun (Plofker 2009, 250–251). Essentially, he maintained that the planets circled the sun while the sun circled the earth, an arrangement similar to the model of Tycho Brahe, a Danish astronomer who worked a century later. Clearly, the vibrant intellectual atmosphere of the Kerala School inspired creative work in numerous disciplines.

Some historians have suggested that the ideas of Mādhava and his followers on infinite series may have been transmitted to Europe through the activities of Jesuit missionaries. They maintain that such ideas could have inspired the work of Newton and Leibniz on calculus in the late 17th century. However, there is no evidence of such contact, and the intellectual roots of the European mathematics of the period are well known. The innovations of the Kerala School probably never circulated much further than the immediate circle of scholars who produced them for their own sake. However, these mathematical ideas embodied a compelling spirit of pure enquiry, and they were among the most creative ideas in the long tradition of Sanskrit mathematics.

CONCLUSION

The body of mathematical work in Sanskrit had many cultural dimensions: an exercise of pure intellect, an expression of Hindu devotion, a tool for the study of astronomy, an embodiment of Sanskrit literary tradition, a set of practical administrative skills, a vehicle for great creativity, a statement of joy. It was a product of individual insight and cultural exchange, and it reached a high level of complexity. Through its influence on the Islamic world, it had a transforming effect on global mathematics, an effect that is still felt in the present day.

FURTHER READING

Clark, Walter. 2006. *The Aryabhatiya of Aryabhata*. Whitefish: Kessinger Publishing.
Ifrah, Georges. 1994. *The Universal History of Numbers II—The Modern Number-System*. London: Harvill Press.
Joseph, George Gheverghese. 2016. *Indian Mathematics*. London: World Scientific Publishers.
Merzbach, Uta and Boyer, Carl. 2011. *A History of Mathematics*. Hoboken: John Wiley and Sons.
Plofker, Kim. 2007. *Mathematics in India*. In *The Mathematics of Egypt, Mesopotamia, China, India, and Islam*, edited by Victor Katz. Princeton: Princeton University Press.
Plofker, Kim. 2009. *Mathematics in India*. Princeton: Princeton University Press.
Rao, S. Balachandra. 1994. *Indian Mathematics and Astronomy: Some Landmarks*. Bangalore: Jnana Deep Publications.

8

Numbers in the Medieval Arabic World

Chronology of Events

Life of Muhammad	570–632
Umayyad Caliphate	661–750
Conquest of Spain	711–718
Abbasid Caliphate	750–1258
Reign of al-Mansur	754–775
Founding of Baghdad	762
Reign of al-Ma'mun	813–833
Seljuq Control of Baghdad	1055
Mongol Conquest of Baghdad	1258

CULTURAL ICON: MEDIEVAL ARABIC ASTROLABE

The astrolabe is an ancient instrument with many functions (Figure 8.1). On the one hand, it may be used to measure the positions of the sun, moon, planets, and stars; on the other, it acts like a moving two-dimensional model of the night sky. One may determine one's latitude with a simple version; with an advanced astrolabe, one may predict the movements of the stars, find the local time, or determine direction. The scholars, mariners, and engineers of the Arab Empire employed the astrolabe in navigation,

Figure 8.1 The astrolabe is a complex instrument used to make astronomical observations and to model the movements of the heavens. Medieval astrolabes illustrate the mathematical sophistication of the Arabic world. (Germanisches Nationalmuseum, Nuremberg, Germany/The Bridgeman Art Library)

positional astronomy, and construction. It was also used to find the *Qibla*, the direction of Mecca, and to indicate the proper times for daily prayer.

The astrolabe was invented by the Hellenistic Greeks, perhaps by Hipparchus of Nicaea himself. Hypatia of Alexandria authored a treatise on its principles in the 5th century CE. The medieval Arabs adopted the astrolabe along with a wide range of Hellenistic mathematical and astronomical texts. They combined this technology with knowledge of Indian numbers and astronomy, as well as their own mathematical innovations, to produce a more versatile instrument. Al-Khwārizmī, perhaps the most important mathematician of the medieval Middle East, wrote a discussion of its operation in the 9th century, and many other authors followed over the years. In a sense, the astrolabe serves as an image of the complex character of Arabic intellectual life during the Middle Ages.

An astrolabe consists of a circle of bronze or brass about a foot in diameter with a central axis; the parts of the instrument attach to the axis so they can turn freely. It has two working faces. One side of the bronze circle, or mater, is engraved along its circumference with a 90-degree scale—90 degrees is located at the top and 0 degrees at the horizontal. A long pointer with attached sights—the alidade—turns on the axis and points at the scale. The astrolabe must hang vertically if it is to be used for measurements (there is usually a ring at the top for the purpose). Holding it at eye level, the user looks through the sights on the alidade at a given object: a star, the sun, or the top of a mountain.

The altitude of the sighted object is indicated on the engraved scale as an angle. For example, if one sights the sun and reads a value of 60 degrees on the scale, then the sun is located at an angle of 60 degrees above the horizon at that point in time (Figure 8.2).

The ability to measure the angular altitude of an object with accuracy has many applications. If one has access to trigonometric tables, one may determine the height of the object—a mountain, for example. In the Northern Hemisphere, the altitude of the North Star (Polaris) over the horizon is equivalent to the latitude of the observer. Thus, if one stood at the North Pole with an astrolabe and sighted the North Star, the alidade would point straight up at the number 90 on the scale, for the North Pole is at 90 degrees north latitude. It is also possible to find one's latitude using the sun. At a given latitude, the sun will be at a particular elevation on a particular date at noon. The position changes from day to day. If an observer has a chart of these different values, they can take a sighting of the sun at noon and determine their latitude. The medieval Arabs began the work of compiling such charts—solar declination tables.

Figure 8.2 The alidade is used to find the angle of elevation of a given star or planet. The ring at the top of the astrolabe allows the instrument to hang freely so that it will be vertical with respect to the horizon. (Album/Alamy Stock Photo)

These navigational uses of the astrolabe were of great importance to long-distance commerce. Arab trade routes stretched across the Indian Ocean to the east and the Sahara Desert to the west. However, the astrolabe also had great importance for the study of astronomy. One side of the instrument is for sighting, but the other side has a set of stellar coordinates engraved on its face—curves for altitude and azimuth (see Figure 8.1). Azimuth comes from the Arab term *al-samt*, "the direction," and refers to the direction of a star, measured in degrees, with respect to true north. The North Star has an azimuth of zero, for it lies directly to the north. An object located at true east would be at a 90-degree angle to the North Star and have an azimuth of 90. Calendar coordinates are also engraved around the circumference of the astrolabe.

Rotating freely over the engraved coordinates is a device called a rete. Essentially, it is a metal star map, with a number of pointers corresponding to specific stars. From the perspective of the earth, all stars move together; they stay in place with respect to each other as they rise and set or as they change with the seasons. As a result, the position of one star can be used to determine the positions of the others. A user can find the altitude of a star using the alidade, flip the astrolabe over, and turn the rete until the pointer for that star rests on the proper altitude reading. The other pointers on the rete will indicate the positions of all the other stars, including those that are not yet visible. Similarly, by rotating the rete and reading it in conjunction with a given calendar date, one can find the time and azimuth of a star's rising and setting on that date. Essentially, the movement of the rete over the engraved coordinates corresponds to the motions of the stars in the sky over the course of a year.

Astrolabes were used throughout the Middle Ages in the Middle East and in Europe, but more accurate and less cumbersome instruments rendered them obsolete in the 17th century. Yet their intricate construction makes them objects of genuine beauty, as well as potent symbols of the sophistication of Arabic mathematics and astronomy in the medieval period.

CULTURAL COMMENTARY: SYNTHESIS AND INNOVATION

At its height in the 9th century CE, the Arab Empire stretched from Spain in the west to the boundaries of India in the East, a distance of over 5,000 miles. At a time when the Byzantine Empire was in decline and Western Europe was fragmented, the Arabs dominated the Mediterranean, North Africa, and the Middle East. As a complement to their military and political power, they grew to be an important cultural force in world history. During the period from 700 to 1250, Middle Eastern thinkers created an energetic intellectual climate and adopted components of ancient Greek and Asian knowledge. They expanded on those traditions in creative and original ways and laid many of the foundations of modern mathematics, astronomy, and geography. A thriving scholarly community emerged in the Islamic world, bound together by the Arabic language.

The prophet Muhammad was born in the city of Mecca (Makkah) around the year 570 CE. According to Islamic tradition, he had a mystical experience at the age of forty, a visit from the angel Gabriel. The revelations that he received became the basis of the Koran (Qur'an), the foundation of the Islamic faith. Muhammad sought to spread his teaching in Mecca for twelve years, but he met with significant opposition, including threats to his life. In the year 622, he and his followers left Mecca for the nearby city of Medina; his journey, the *Hijrah*, serves as the starting point for the Muslim calendar. Once in Medina, the number of his followers increased, and he conquered the city of Mecca over the course of eight years. Muhammad was a military and political leader as well as a religious figure; by the time of his death in 632, he had united the entire Arabian Peninsula under the banner of Islam.

Muhammad was succeeded by a series of four rulers, or *caliphs*, related to him by marriage. From 632 to 661, they conquered vast areas at the expense of the Sassanid and Eastern Roman (Byzantine) empires. The Byzantine provinces of Egypt, Syria, and Palestine all fell to the Arab armies, and most of the Sassanid domains in Mesopotamia and Persia were vanquished in two years. In 661, the Umayyad dynasty came to power in Arabia and continued to add territory to the realm. The Umayyad rulers subdued Afghanistan and parts of the Indus River Valley in India, completed the conquest of North Africa, and in the year 711, crossed into Europe and took all of Spain. By 750, when the Abbasid caliphs rose to power, the Arab Empire stretched from the Atlantic Ocean, across Africa and the Middle East, to the Indian Ocean (Figure 8.3).

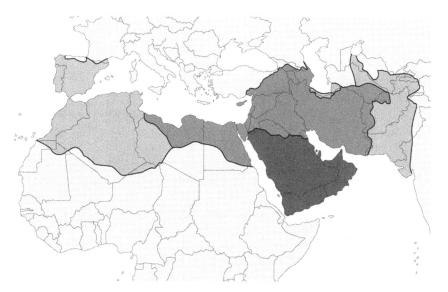

Figure 8.3 The vast extent of their empire led the Arabs to draw influences from Persia, the Byzantine Empire, and India.

This explosive growth had important cultural implications, for the Arabs now controlled eastern Mediterranean cities steeped in the Greco-Roman past: Alexandria, Damascus, Caesarea. Beginning in the late 7th century, they began to translate works of Greek scholarship into Arabic, with a special emphasis on philosophy, mathematics, astronomy, and medicine. The Umayyad caliphs established a royal library in their capital at Damascus to house such books, as well as works of Islamic spirituality and Arab literature. The Arabs also sought out the written knowledge of the ancient cultures of Mesopotamia and Persia. In addition, as the empire grew, significant trade links developed with India, Ceylon, southeast Asia, and China.

A great deal has been written about the Silk Road, the overland caravan route that linked the medieval Middle East with eastern Asia. However, the Arabs also maintained a flourishing seaborne commerce across the Indian Ocean. Sailing in small, nimble ships of Indian teak, they traveled from the ports of the Persian Gulf and the Red Sea eastward to the island of Ceylon (now the nation of Sri Lanka). Located about halfway between the Arabian Peninsula and southeast Asia, Ceylon served as a meeting place for merchants from Arabia, India, China, and the islands of what is now Indonesia. Arab mariners also ventured farther into Asia; they traded extensively in India, and a community of Arab merchants existed in the Chinese city of Guangzhou (Canton) in the 9th century. Trade routes often serve to transport cultural influences as well as merchandise; the spread of Islam throughout the coastal regions of the Indian Ocean illustrates this. Similarly, Asian ideas and scholars could easily enter the Arab world.

In short, by the 8th century, Middle Eastern scholars had access to intellectual resources from ancient Greece, Hellenistic Alexandria, Mesopotamia, Persia, and India, and they eagerly seized the opportunity to translate books into Arabic. Because of its importance in Islam as the language of the Koran, Arabic became an international language of scholarship. Indeed, one may speak of an Arabic intellectual culture that united the literate people of the Islamic world—Arabs, Persians, Berbers, and others—into a community of written discourse. Arabic thought was closely tied to Islamic spirituality; indeed, many of the intellectual institutions, schools, and libraries in the Arab Empire were associated with mosques. Yet scholars, and the rulers who supported their work, came to see the pursuit of non-religious knowledge as complementary to Islam.

A significant factor in the growth of this intellectual culture was the adoption of papermaking technology from the Chinese. Commercial and political contact between the Islamic Middle East and China grew over the course of the 8th century, and the Arabs adopted the Chinese technology for making paper from plant fibers, particularly flax (the source of linen) and hemp. Paper was an ideal material for writing and book production, and its manufacture spread throughout the Arabic world (Lyons 2009, 57–58). Trade in paper and in books flourished, and the existence of the book trade fueled a growing Arabic literary and scholarly mindset that would set the pattern for the establishment of the House of Wisdom by the Abbasid dynasty.

The Abbasids came to power in the year 750. Caliph Abu Jafar al-Mansur began the construction of the new capital of Baghdad in 762, at a site located near the center of his vast domains. Completed in 765, Baghdad rapidly became the one of the most important hubs of administration, trade, and culture in the Middle Ages. While the city was well-defended, its location on the River Tigris ensured its commercial, diplomatic, and cultural links to the rest of the empire, and the world. These links played a crucial role in the gathering of knowledge and the patronage of scholarship, initiatives that were very important to al-Mansur. For example, Indian visitors to the court of Baghdad in the early 770s brought one or more of the Hindu astronomical *siddhāntas*, treatises of astronomy and mathematics in Sanskrit verse. Translated into Arabic and called the *Sindhind*, these texts influenced Middle Eastern astronomy and contributed significantly to the spread of the Indian system of numeration throughout the Arab Empire. This system, based on decimal place-value and the use of ten basic symbols, would ultimately become the basis of modern mathematical expression. In addition, al-Mansur sponsored the acquisition and translation of other works in Greek, Persian, and Sanskrit and founded a royal library, the likely predecessor of the *Bayt al-Hikma*, the House of Wisdom.

The fifth caliph of the Abbasid dynasty, Al-Rashid, was particularly interested in literature and poetry, but he also promoted the collection of philosophical and mathematical works, including part of Euclid's *Elements*. The Abbasid caliphate had developed a consistent policy of literary patronage. With such royal support, Arab intellectuals eagerly continued to seek ancient Greek texts for translation. The Byzantine Empire served as one fertile source of such texts, and the newly translated works inspired novel scholarship in a variety of fields. This intellectual exuberance would only increase during the reign of Caliph al-Ma'mun.

Al-Ma'mun rose to power in 813, and for the twenty years of his reign he was devoted to the growth of scholarship and to the study of astrology, astronomy, and mathematics. Sources refer to the House of Wisdom that thrived under his patronage—a repository of written work and a community of scholars actively pursuing new knowledge. Historians disagree about the nature of this institution: some identify it with the libraries of the earlier caliphs and the intellectual community of Baghdad itself, while others maintain that al-Ma'mun actually founded a new library and center of study. Clearly, though, al-Ma'mun enthusiastically embraced a role as a benefactor of scholars. Like the Ptolemaic rulers of Alexandria in the Hellenistic period, he made intellectual life a royal priority and gave it substantial institutional sustenance. He prized the thought of the Greek philosopher Aristotle and encouraged the translation of many of his works; one Arab author claimed that Aristotle had come to the caliph in a dream and inspired him to pursue philosophy (Lyons 2009, 77). Al-Ma'mun's patronage drew scholars to Baghdad; intellectuals from all over the Abbasid Empire came to work at the House of Wisdom, with its burgeoning library. The prosperous city became a great center for the pursuit of philosophy, mathematics, medicine, astronomy, and optics for much of the Middle Ages. The caliph took a special interest in the stars, partly because of the allure of astrology. He actively negotiated with the Byzantine Empire for manuscripts of Euclid and Ptolemy, and he had a new translation of Ptolemy's *Almagest* made by the scholar al-Hajjaj. In an attempt to clarify the dimensions of the earth as described by Ptolemy, al-Ma'mun sent teams of astronomers to the desert north of Baghdad to measure the length of a single degree of the diameter of the earth by noting the change in the altitude of the sun as they walked (Lyons 2009, 69).

The resident intellectuals of the House of Wisdom emphasized the translation and study of ancient Greek, Persian, and Indian manuscripts and used them as an inspiration for original and innovative thought. Arab and Persian scholars combined Indian numeration and problem-solving techniques with Greek geometry and logical analysis to produce a powerful synthesis that formed the foundation for later mathematical work. Arabic mathematicians developed the techniques of algebra and did pioneering work in trigonometry. They made important advances in fields involving applied mathematics, particularly astronomy and geography. Mohammed ibn Mūsā al-Khwārizmī, a Persian thinker who came to Baghdad and served as a leader of the House of Wisdom, embraced the use of Indian numerals and wrote an important book on their use. He produced a series of star tables and a major work of geography. In addition, he composed one of the pioneering works on algebra and dedicated it to al-Ma'mun. Al-Kindi, a contemporary of al-Khwārizmī, also wrote a major work on Indian numerals. A true polymath, he produced dozens of treatises on geometry, optics, medicine, and philosophy. He was fascinated with the ideas of Aristotle and insisted that Greek philosophy was compatible with the theology of Islam. Thābit ibn-Qurra played a crucial role in the ongoing acquisition of classical Greek mathematics. Born in 826, he came to Baghdad for his education and was eventually appointed as court astronomer. He translated works by Ptolemy, Archimedes, Euclid, and Apollonius of Perga into Arabic and contributed a great deal to the Arab mastery of Hellenistic geometry. He also did geometrical work of his own and made original contributions in number theory and the geometry of triangles (Merzbach and Boyer 2011, 213).

By the end of the 9th century, the ardor of the Abbasid caliphs for supporting intellectual pursuits had cooled, and the empire was breaking up into smaller autonomous regions. However, Baghdad had become a focal point of scholarship throughout the Islamic world, the center of an intellectual community based on the Arabic language. Cairo would also grow to be an important cultural hub, and a flowering of Persian culture would take place in such cities as Nishapur, Bukhara, and Samarkand. Yet despite regional differences and identities, scholars across the Middle East, Persia, North Africa, and Spain were linked by a common language and by the movement of books and ideas. The continuing development of mathematics and astronomy in these areas illustrates some of these intellectual connections.

Abū al-Wafā was born in Persia but worked in Baghdad during the late 10th century. He advanced the study of trigonometry, building on the knowledge of the sine function in the Indian tradition. He ultimately explored all six of the basic trigonometric relationships: sine, cosine, tangent, cosecant, secant, and cotangent (Merzbach and Boyer 2011, 216). Al-Wafā also did important work in algebra, translated the ancient Greek author Diophantus's book on problem-solving, and wrote a commentary on al-Khwārizmī's book on algebra. Finally, like many mathematicians in the Arabic world, he pursued interests in astronomy. He engaged in a correspondence with Abu Rayhan al-Biruni in Persia, in which they tried to establish the difference in time between Baghdad and al-Biruni's home city of Kath, over 1,500 miles away. Al-Biruni was a Persian scholar with wide interests: he wrote treatises on mathematics, astronomy, geography, philosophy, history, and religion, as well as a detailed account of the culture of India. He in turn carried out an energetic correspondence with the great Persian physician and philosopher Ibn Sina ("Avicenna" in the Latin tradition). Ibn Sina, a great admirer of Aristotle, worked to reconcile Aristotelian logic with Islamic belief. In their written discussions, Al-Biruni attacked different aspects of Aristotle's cosmology and physics.

Al-Biruni and Ibn Sina spent their lives in Persia during the 10th and 11th centuries, a time when the region was no longer ruled by the Arabs in Baghdad. Yet they were in contact with each other, with scholars in Baghdad, and probably with the rest of the Arabic intellectual community. Books and ideas traveled across the Arabic world; for example, the works of Ibn Sina were available in Islamic Spain in the 12th century. Many leading Islamic thinkers, including al-Khwārizmī, al Biruni, and ibn Sina, were Persian by birth, but their ideas circulated in Arabic throughout the Middle East. The Arabic intellectual community, bound by a common language and a common religious heritage, provided a fertile climate for intellectual growth, a climate that fostered the development of mathematics and its related disciplines for much of the Middle Ages.

The Abbasid Empire had begun to fragment in the late 9th century as rival dynasties took power in parts of its domain—the Umayyads in Spain, the Fatimids in Egypt and North Africa, a variety of dynasties in Persia. In the 11th century, the Seljuq Turks conquered much of the Middle East and took control of Baghdad, but the intellectual life of the region remained vibrant. However, the Mongol invasion that began in 1219 had a devastating effect on culture and scholarship in the Islamic world. Under Genghis Khan, the Mongol armies vanquished the Khwarazmian Empire of eastern Persia and destroyed its major cities by 1221. His grandson, Hulagu Khan, struck to the east in 1255 and conquered southern Persia. In 1258 Hulagu Khan besieged Baghdad; he took the city after just twelve days and executed the last Abbasid caliph. The resulting ruin of medieval Baghdad was terrible. Tens of thousands of people were killed or enslaved, and buildings were plundered and burned. Sources tell of the destruction of the city's libraries; one account claims that so many books were thrown into the Tigris River that the waters ran black with the dissolved ink. While such stories may be exaggerated, it is clear that a great tragedy occurred, a loss of life, knowledge, and culture.

After conquering Syria, Hulagu Khan was defeated in battle by the Mameluke army of Egypt in 1260. The Mongols were stopped, and many of those who remained in the Middle East ultimately converted to Islam. However, the cultural damage wrought by the Mongol conquest was immense. No one knows how many scholars or how many books were lost when Baghdad and the cities of Persia went up in flames. The great intellectual energy that had characterized the Golden Age of the Arabic world had waned—but it had not disappeared. The career of Nasir al-Din al-Tusi demonstrates the perseverance of Arabic scholarship in the years after the Mongol conquest.

Nasir al-Din al-Tusi was a Persian mathematician, astronomer, and polymath born in 1201. He was educated in Nishapur and worked for the rulers of one of the small states in Persia. When Hulagu Khan attacked southern Persia in 1255, al-Tusi was captured, and Hulagu became his new patron. Despite the fractious time in which he lived, al-Tusi did important work and ranks among the foremost mathematicians of the Islamic Middle Ages. He continued the study of Euclid's postulates and laid some of the groundwork for the non-Euclidean geometry of the 19th century. He also pursued the study of trigonometry as an independent discipline, not merely as a tool in astronomical calculations. However, he was best known for his astronomy, and after 1259 he led the scholars at the Maragha observatory under the patronage of Hulagu Khan (Merzbach and Boyer 2011, 220–221).

And so, as empires fell and cities crumbled, the scholars of the medieval Middle East still pursued the mysteries of mathematics and the secrets of the stars.

INTRODUCTION: MATHEMATICS AND ISLAM

Mathematics played an interesting role in medieval Islamic spirituality. Like Judaism, Islam did not permit any representations of the divine form. As a result, Islamic religious decoration was often geometrical in nature. Intricate tracery or elaborate patterns of repeating polygons were used to adorn sacred spaces and to suggest the transcendence of divinity. Such motifs were often combined with Arabic calligraphy, spelling out holy verses. The exterior and interior decoration of the Dome of the Rock, built in the later 7th century and rebuilt in the 11th century, serves as an excellent example. Mathematical knowledge also aided in the determination of the *qibla*, the direction of Mecca necessary for Islamic daily prayer. The significance of the *qibla* would inspire important work in geography during the Middle Ages.

INDO-ARABIC NUMBERS

The adoption of Hindu or Indian numbers by the Arabic world had great implications for the history of world mathematics. Because of the broad reach of Arabic language and culture, the Indian number system spread throughout the Middle East and the Mediterranean and then to medieval Europe. The fundamentals of the Indian system laid the foundations for modern numbers; it employed nine number symbols and a zero to represent all quantities, and these symbols were arranged in place-values based on the powers of the number 10.

It is impossible to know the exact time and place of the transfer of Indian numbers to the Arabs; indeed, there may have been multiple incidents of transmission. However, a likely exchange took place in the late 8th century, when Indian visitors to the court of the Abbasid Caliph al-Mansur brought one of the Sanskrit astronomical *siddhāntas* (or the *Sindhind*, as the Arabs came to refer to it). The 10th-century astronomer Ibn al-Ādamī recorded the incident, and Hasan al Qifti, a court author in the 13th century, summarized his account:

> Al-Husayn Ben Muhammed Ben Hamid, known as Ibn al-Ādamī, tells in his Great Table, entitled Necklace of Pearls, that a person from India presented himself before the Caliph al-Mansur in the year 156 (of the Hegira = 776 CE) who was well-versed in the sindhind method of calculation related to the movement of the heavenly bodies and having ways of calculating equations based on kardaja calculated in half-degrees, and what is more various techniques to determine solar and lunar eclipses, co-ascendants of ecliptic signs and other similar things.
>
> (Ifrah 1998, 1045)

Al-Ādamī describes the transfer of Indian astronomy, and he states that this knowledge involved substantial mathematical material. It is quite likely that some transmission of Indian numbers took place at this point in time (Ifrah 1998, 1046). In any case, Indian numbers must have been known to the Arabic scholarly world by the late 8th century, for the great Persian mathematician al-Khwārizmī, leader of the House of Wisdom in Baghdad, wrote a treatise on their use around the year 820.

The Indian or Hindu numbers that entered the Arabic world were adopted by some of the scholars in the eastern part of the Abbasid Empire. However, the form of the numbers was altered by the Arabs and Persians, possibly due to copying techniques. The symbols continued to evolve as they spread westward over time, and different characters were ultimately used in the eastern and western parts of the Arabic literary world. European numbers evolved from the characters employed in the West. However, regardless of the symbols used, the system remained based on the powers of the number 10, arranged by place-value. This is the approach still used in modern mathematics (Figure 8.4).

	0	1	2	3	4	5	6	7	8	9
European (descended from West Arabic)	0	1	2	3	4	5	6	7	8	9
Arabic-Indic	٠	١	٢	٣	٤	٥	٦	٧	٨	٩
Eastern Arabic-Indic (Persian and Urdu)	٠	١	٢	٣	۳	۵	۶	٧	٨	٩
Devanagari (Hindi)	०	१	२	३	४	५	६	७	८	९
Tamil	0	க	௨	௩	௪	௫	௬	எ	அ	௯

Figure 8.4 The Arabs adopted Indian numbers and ultimately transmitted them to Western Europe.

It is important to point out that Indian numbers were not universally accepted by all Islamic scholars once they were transmitted. For centuries, many Arabic authors, including some renowned mathematicians, used a system of numeration based on the Arabic alphabet, similar to the ancient Greek approach. Letters of the alphabet were used to represent numbers. This system was also based on decimal place-value and was preferred by many authors of note: Abū al-Wafā and al-Karajī, for example. In addition, because the alphabetic numbers were accepted as essentially Arabic and not foreign, they were often linked to astrology and omens—matters of great importance. They played a role in Islamic mysticism as well: because the Koran was written in Arabic, it was believed by some that any mystical truths about the nature of God that were revealed through number must also be written in Arabic (Ifrah 1998, 1093). Arabic words could have numerical equivalents, with special revelations contained therein.

Arabic mathematicians adopted not only Indian numbers but Indian modes of calculation. The ancient Indians performed mathematical operations on a dust board; the board could be used for figuring and then erased and easily reused. While early dust board methods resembled the operations of an abacus, the Indians eventually abandoned such procedures and adopted dust board calculations that took advantage of the power of written symbols arranged in columns of decimal place-value (Ifrah 1998, 1106). Arabic mathematicians imitated these practices and continued to develop them over the centuries, experimenting with different methods. Modern modes of written calculation crystallized in early modern Europe, but they find their roots in these medieval Arabic techniques.

AL-KHWĀRIZMĪ

Muhammad ibn Mūsā al-Khwārizmī occupies a place among the most influential mathematicians of history. Born in Persia in the late 8th century, he worked as an early member of the House of Wisdom in Baghdad, under the patronage of Caliph al-Ma'mun. During his career there, he composed a series of vital works that advanced the study of mathematics, astronomy, and geography in the Islamic Middle East. Al-Khwārizmī's book on the use of Indian numerals contributed to the adoption of this system in the Arab Empire, a development that would have profound implications for later Middle Eastern mathematical advances. His more substantial work on algebra and problem-solving became one of the foundations of the Arabic mathematical tradition. In addition, al-Khwārizmī shaped the character of mathematical thought in Europe; his work had a tremendous effect on the West after it was first translated in the 12th century. The adoption of Indo-Arabic numerals and algebraic problem-solving techniques by European mathematicians in the late Middle Ages may be attributed to his influence.

Al-Khwārizmī's treatment of Indian numerals no longer exists in an Arabic version, but a Latin translation survives in Europe under the title *De Numero Indorum (On Indian Numbers)*. In it, the author demonstrated the power of Indian numbers as a means of representing quantity and as a basis of calculation. The combination of nine number symbols, the concept of place-value, and the notion of zero made the system an efficient means of carrying out mathematical operations. The procedures familiar to modern readers would develop later, but al-Khwārizmī recognized the power of the system, and his approaches to addition and subtraction would be used for centuries in the Middle East and Europe. His influence on European mathematics was so profound that the word for figuring with the new numbers—*algorism* or *algorithm*—was based on the Latin version of his name. The term *algorithm* now refers to a specific set of mathematical operations dedicated to a specific problem.

Kitab al-jabr wa'l muqabala (The Book of Restoring and Balancing), al-Khwārizmī's book on problem-solving, represented the growing capability of Middle Eastern mathematics in the 9th century. While the ancient author Diophantus had written the *Arithmetica*, a work on the solution of problems, it had an idiosyncratic character; that is, the discussion was tailored to individual problems with no real sense of system. Al-Khwārizmī, on the other hand, sought to establish general principles of solving different classes of problems, and he organized his book on that basis. He was probably influenced by the works of Diophantus and the Indian author Brahmagupta, but unlike them he did not employ syncopation—the use of symbols in writing mathematical statements or operations. Consequently, his book did not contain recognizable algebraic equations. Rather, he described problem-solving verbally, leading the reader through the steps of his operations.

The terminology involved seems familiar and foreign at the same time to a modern reader: al-Khwārizmī referred to unknowns as *roots*, a word derived from Greek geometry. Yet the steps he took in the solution of problems bore a significant resemblance to modern elementary algebra (Merzbach and Boyer 2011, 206). Consider his solutions to problems that we would represent as:

$$\frac{x^2}{3} = 4x \text{ and } 5x^2 = 10x$$

using "x" as a variable or an unknown (or a root, as he would say):

> Another example: the third part of a square equals 4 roots. Then the root of the square is 12 and 144 designates the square. And similarly, five squares equal 10 roots. Therefore, one square equals two roots and the root of the square it 2. Four represents the square.
>
> (al-Khwārizmī 1974, 22)

In the first problem, al-Khwārizmī essentially multiplied both sides of the equation by 3 and found the root or unknown: 12. In the second case, he divided both sides of the equation by 5 and discovered the root to be 2.

The book divided problems into six major categories, including more complicated quadratic equations. Al-Khwārizmī solved these through the process of "completing the square":

> Another possible example: half a square and five roots are equal to 28 units. The import of this problem is something like this: what is the square which is such that when to its half you add five of its roots the sum total amounts to

28? Now however, it is necessary that the square, which here is given as less than a whole square, should be completed. Therefore the half of this square together with the roots which accompany it must be doubled. We have then a square and 10 roots equal to 56 units. Therefore take one-half of the roots, giving 5, which multiplied by itself produces 25. Add this to 56, making 81. Extract the square root of this total, which gives 9, and from this subtract half of the roots, 5, leaving 4 as the roots of the square roots of t the square.

(al-Khwārizmī 1974, 24)

This procedure of "completing the square" originated in the geometric problem-solving techniques of the ancient Greeks. In *The Book of Restoring and Balancing*, al-Khwārizmī used Indo-Arabic numbers and computational techniques, as well as ancient Babylonian mathematical procedures. He combined them with elements of Greek geometry; the later parts of the book focus on Greek geometrical proofs (Merzbach and Boyer 2011, 208–210). His work on algebra represents a powerful problem-solving synthesis, produced in the vibrant intellectual atmosphere provided by al-Ma'mun and the House of Wisdom. As a fitting tribute, al-Khwārizmī' dedicated his book to the Arab ruler.

Other treatments of algebra appeared at almost the same time, particularly the work of ʿAbd al-Hamīd ibn Turk. This suggests that knowledge of these problem-solving techniques was common in the scholarly community of the Arab Empire. Taken in this context, al-Khwārizmī's book appears to have been a well-received treatment of an important subject. This is supported by the fact that it was often commented upon and quoted by later authors in the Arabic mathematical tradition (Lyons 2009, 74). Al-Khwārizmī's treatment was translated into Latin in the 12th century and became the basis for the European study of algebra; the term "algebra" comes from the word *al-jabr* (restoration) in his title.

As a member of the House of Wisdom, al-Khwārizmī did not limit his interests to mathematics; he did extensive work in astronomy and geography as well. Just as in his treatment of mathematics, he embraced knowledge from India and used it as a foundation for further investigation. The *Sindhind* was a collection of astronomical, geographical, and mathematical knowledge brought to the court of the Abbasid caliph by Indian sages in the late 8th century. Al-Khwārizmī wrote commentary on its contents and used it to construct two star tables, the *Zīj al-Sindhind* (Lyons 2009, 73). Such knowledge of stellar positions and movements was essential to the evolution of the astrolabe, and al-Khwārizmī wrote an early treatise on that instrument as well.

The body of al-Khwārizmī's scholarship demonstrates the scholarly energy of the House of Wisdom in the 9th century. The potent combination of royal support and access to the works of India, Greece, and Persia helped him to produce work that would influence mathematical thought on three continents. His insights and his ability to communicate effectively made him one of the most important mathematicians of the Middle Ages.

AL-HAYTHAM (ALHAZEN) AND OPTICS

Many Arab scholars of the Golden Age pursued the study of optics, for it featured prominently in the work of the ancients: Aristotle, Euclid, and Ptolemy. Hasan Ibn al-Haytham, or Alhazen as he became known to the West, made the most important contributions to this emerging discipline; he also did important work in pure mathematics and astronomy. While his approach to the exploration of nature had its roots in Aristotelian observation and logic, he applied mathematics to the study of optics, discussing the behavior of light in terms of Euclidean geometry. He also added a new level of rigor to the examination of existing theories and employed experimental thinking in his work.

Alhazen lived in the late 10th and 11th centuries and had access to the ancient texts that had been amassed by Arab copyists since the 8th century. In his study of optical phenomena, he evaluated the ideas in those texts with a critical eye and freely rejected ideas that he felt were unconvincing. For example, Euclid and Ptolemy both supported the idea that vision worked through emission, that is, through the eye giving off rays that made objects visible. Aristotle, on the other hand, discussed the idea that the eye admitted immaterial influences or qualities that originated in objects. Alhazen rejected the idea that the eye could function by emitting rays:

> ... I inquire whether those rays return something to the eye. If perception occurs only through such rays, and they do not return anything to the eye, then the eye does not perceive. But the eye does perceive the object of sight, and we have supposed that it perceives only by the mediation of rays. Therefore, those rays that perceive the visible object must transmit something to the eye, by means of which the eye perceives the object.
>
> (Alhazen 1974, 403)

He adopted a version of Aristotle's view, but argued that the conduct of the rays that entered the eye should be described geometrically. In a short work on parabolic mirrors, Alhazen clearly linked the study of light to the work of ancient geometers:

> One of the most exquisite things invented by geometers, about which the ancients were concerned, in which the most excellent properties of geometrical figures are apparent, and which follows (according to hem) from natural things, is the construction of mirrors that produce combustion through the reflection of a solar ray.
>
> (Alhazen 1974, 413)

In his *Book of Optics*, written in the early 11th century, Alhazen discussed his ideas on the functioning of the eye and the characteristics of light in great detail. He studied the phenomena of refraction and reflection, applying geometrical analysis to the phenomena:

> The reason why light is reflected along a line having the same slope as the line by which the light approaches the mirror is that light is moved very swiftly, and when it falls upon a mirror it is not allowed to penetrate but is denied entrance into that body.
>
> (Alhazen 1974, 418)

As a more significant example, he used the geometry of conics to solve the problem, first posed by Ptolemy in the 2nd century CE, of finding the point on a curved mirror that would reflect light from a given source to a given target.

Alhazen's methods went beyond the mere observation of nature; he constructed instruments to aid him in his study of optical phenomena. Here, he discusses a device for investigating the refraction of light by the surface of water:

> That light indeed passes through the air and is extended along straight lines was declared in the first book of this work; however, air is but one of the transparent substances. Light also passes through water, glass, and transparent stones, and in them it proceeds along straight lines. But this is learned by experience. If one would like to experience it, let him take a round plate of copper . . .
>
> (Alhazen 1974, 420)

He examined the behavior of light systematically, designed apparatus to help him in his endeavors, and varied the conditions of his observations. Taken together, these intellectual activities constitute an experimental approach to nature. Alhazen had great respect for the authors of antiquity, but he recognized that they contradicted each other and that their ideas could be flawed. He turned to nature itself for answers to his questions and devised methodical means to find those answers.

Like many of his scholarly contemporaries in the Arab world, Alhazen was a polymath—he had interests in different intellectual disciplines. In addition to his work on optics, he wrote on pure mathematics, medicine, religion, and astronomy. He criticized Ptolemy's *Almagest*, pointing out contradictions in the text and arguing that parts of the author's system of the planets were materially impossible, even if they were mathematically elegant. In his *Doubts on Ptolemy*, he wrote:

> Truth is sought for its own sake. . . . It is not the person who studies the books of his predecessors and gives a free rein to his natural disposition to regard them favorably, who is the seeker after truth. But rather the person who is thinking about them [and] is filled with doubts . . . who follows proof and demonstration rather than the assertion of a man whose natural

disposition is characterized by all kind of defects and shortcoming. . . . A person who studies scientific books with a view to knowing the truth, ought to turn himself into a hostile critic of everything that he studies . . . if he takes this course, the truth will be revealed to him and the flaws . . . in the writings of his predecessors will stand out clearly.

Alhazen made it clear that he did not accept the authority of authors alone; he would subject their ideas to intense scrutiny—informed by "proof and demonstration." Some historians argue that he anticipated the modern notion of falsifiability in scientific statements. This seems anachronistic, but it is clear that he believed that one could only learn the truths of nature through rigorous questioning. Alhazen's methods remained his own; he did not inspire an extended tradition of experiment in the Arab world. Yet his work represents an extraordinary view of natural inquiry: rigorous, geometrical, and experimental.

OMAR KHAYYAM

The work of Omar Khayyam ranged from algebra and geometry to astronomy and poetry. Born in 1048 in the Persian city of Nishapur, he traveled substantially during his long life and enjoyed the support of the rulers of Persia and of the sultans of the Seljuq Turkish Empire. His career thus demonstrates the ongoing tradition of royal patronage of scholarship in the medieval Middle East, a tradition with its origins in the House of Wisdom in Baghdad.

Khayyam took up the study of mathematics when he was a young man; he composed his great works on mathematics when he was in his twenties. For example, he was immersed in the study of Euclidean geometry and wrote a treatise on Euclid's fifth postulate. Euclid stated that if two lines crossed a line segment and formed internal angles that added up to less than two right angles, those two lines must eventually meet (Figure 8.5).

The postulate leads to the conclusion that if the

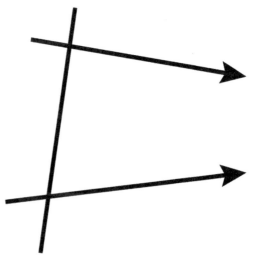

Figure 8.5 Omar Khayyam explored Euclid's fifth postulate, a topic that would have important implications for modern geometry.

internal angles are both right angles, the two lines will never meet; they will be parallel lines. Throughout the Hellenistic period, mathematicians had worked to prove the postulate without success, and the problem particularly fascinated scholars in the Islamic world. Omar Khayyam attempted to prove the postulate based on the construction of a quadrilateral with two equal sides that were both perpendicular to the base; he then attempted to prove that the upper angles had to be right angles. He did not produce a definitive proof, but his approach to the problem may have influenced the development of non-Euclidean geometry centuries later.

In addition to his commentary on Euclid, Omar Khayyam did crucial work in algebra. His most important insights involved the treatment of cubic equations, that is, problems involving a root or variable raised to the third power. Khayyam did not believe that these problems could be solved using the techniques of arithmetical algebra; he resorted to a geometrical approach based on conic sections, in the tradition of Apollonius of Perga. His elaborate, painstaking geometry succeeded, but unlike the Greeks, he expressed his solutions as numbers rather than as geometrical statements (Merzbach and Boyer 2009, 219). His mathematics embodied the eclectic character of Islamic intellectual life: a Persian author combining Hellenistic Greek geometry with Indian arithmetical techniques and Arab algebra. One could argue that his combination of these elements anticipated the thought of the French mathematician Rene Descartes in some ways (Merzbach and Boyer 2009, 219). Omar Khayyam also wrote on the characteristics of the triangular array of numbers known as the "Arithmetical Triangle" or "Pascal's triangle" and pointed out its importance in algebra.

Pure mathematics made up just one part of Omar Khayyam's work. At the request of the Seljuq emperor, he collaborated with others to produce a highly accurate calendar based on solar motion—the Jalali Calendar. He may also have been a poet of some renown. Shortly after his death, contemporary authors began to attribute verse to him, written in both Persian and Arabic. Over the centuries, his literary reputation grew in both the Middle East and in Europe. In 1859 Edward Fitzgerald published an English translation of some of his work, the *Rubaiyat of Omar Khayyam*, that resulted in great popular acclaim. Scholars disagree about the actual authorship of the poetry, but it appears likely that Omar Khayyam did write some of the material. The fame of Fitzgerald's translation and the controversy over the origins of the original poetry overshadow the greater significance of Omar Khayyam's intellectual career: he was part of a vibrant network of mathematicians and philosophers drawing on the ideas of Greeks, Indians, Arabs, and Persians and often supported in their work by royal patronage.

ARABIC ASTRONOMY

The pursuit of astronomical knowledge in the medieval Islamic world reflects the importance of mathematics in Arab intellectual culture, as well as the influence of Greek, Persian, and Indian ideas. Any discussion of astronomy in the Arabic tradition must begin with the astronomical *siddhāntas*, a collection of Sanskrit texts on astronomy and mathematics brought to Baghdad by a group of Indian scholars around the year 770. Translated into Arabic by the astronomer Muhummad Al-Fazārī and referred to as the *Sindhind*, this Indian knowledge had a profound influence on the intellectuals of the Abbasid Empire. The Persian scholar Muhammad ibn Mūsā al-Khwārizmī (late 8th to early 9th century) produced two star tables based on the *Sindhind*, and these inspired many more by later authors, adapted to specific latitudes and locales. Al-Khwārizmī clearly prized Indian learning, for he also wrote an extensive treatment of the Indian number system and pointed out its advantages.

Al-Khwārizmī enjoyed the patronage of the Abbasid Caliph al-Ma'mun, a sophisticated ruler with a passion for scholarship. In addition to Persian and Indian books, al-Ma'mun actively sought works of Classical Greek philosophy, astrology, and astronomy so that they might be translated into Arabic. He eagerly pursued the works of Aristotle, and he negotiated with the Byzantine Empire to acquire a copy of Ptolemy's *Almagest* as well. Ancient Greek astronomers believed that the earth lay at the center of the universe and that the observed movements of the heavenly bodies had to be explained in terms of uniform circular motion. Aristotle wrote that the heavens were composed of concentric spheres that carried the heavenly bodies in circular paths around the earth. Ptolemy, on the other hand, was a positional astronomer who wished to account for observed planetary positions through different applications of complex circular geometry. Both ancient figures inspired the astronomers of the Arab world; indeed, those astronomers dedicated a great deal of effort to commenting on Greek systems of the universe and striving to improve upon them. They also made extensive observations to refine their understanding of the movements of the sun, moon, stars, and planets, and these observations led to improved star charts and instruments. Numerous Middle Eastern astronomers wrote treatises on the astrolabe, and their work resulted in the steady evolution of that instrument.

Many of the scholars who studied the heavens were polymaths; they were interested in a variety of related subjects. Astronomy, astrology, geography, and optics were viewed as complements to mathematics, and vice versa. Al-Khwārizmī serves as an excellent example, for he did work on mathematics, astronomy, and geography. Alhazen (late 10th to early 11th century) is best known for his work on optics and mathematics, but he also authored over twenty treatises on astronomy, including a number of works on observation. In his *Doubts Concerning Ptolemy*, Alhazen criticized the inconsistencies in Ptolemy's works, as well as the character of his celestial geometry. Alhazen continued to embrace the essential assumption of ancient Greek astronomy: that the heavenly bodies engaged in uniform circular motion. However, he believed that Ptolemy's geometrical models could not represent physical reality, and he attempted to construct an improved system. Al-Biruni (late 10th to early 11th century) wrote on a wide variety of topics ranging from history to languages, but the bulk of his work focused on mathematics and astronomy. He commented on Indian astronomy, criticized elements of the Aristotelian and Ptolemaic systems, and produced a treatise on the astrolabe.

Al-Biruni challenged aspects of Aristotle's cosmology in an exchange of letters with Ibn Sina, the great Persian physician and philosopher. Their correspondence illustrates the existence of networks of scholars working on astronomical and mathematical questions in the Arabic world—sharing information, collaborating, disagreeing. Al-Biruni also worked with Abū al-Wafā on charting the time difference between Baghdad and the city of Kath. A group of thinkers in Islamic Spain challenged the system of Ptolemy over the course of the 12th century and sought to replace it with a modified Aristotelian approach; it included ibn Tufayl, al-Bitruji, ibn Rushd, and the Jewish scholar Moses Maimonides (Lyons 2009, 198–199).

Perhaps the most significant collaboration in Islamic astronomy took place at the Marāgha observatory in northwestern Persia (now Iran) during the 13th century. Built in 1259 with the support of the Mongol ruler Hulagu Khan, a grandson of Genghis Khan, the observatory was intended to collect the data needed for a new set of star tables. Led by the polymath Nasir al-Din al-Tusi, a group of Islamic astronomers worked to produce the *Zij Ilkhani*, star tables named for Khan Hulagu. Al-Tusi made other important astronomical discoveries; most notably, he found a way to revise Ptolemy's system to make the motions of the planets more uniform. Ptolemy's astronomy made use of the epicycle, a form of compound circular motion in which a planet circled an imaginary point that in turn circled the earth. Al Tusi noted that a particular combination of circular motions could produce a repeating motion in a straight line. Known in modern terminology as the "Tusi Couple," this geometrical technique addressed some of the objections to Ptolemy. In addition, Nicolas Copernicus used it in his discussion of heliocentric (sun-centered) astronomy, published in 1543. It is not clear if Copernicus knew of the work of al-Tusi and the Marāgha school, but it is quite possible.

In many ways, the observatory at Marāgha serves as a symbol of the astronomical endeavors of the Arabic world. It represented a form of government support for scholarship, and it provided an opportunity for astronomers and mathematicians to work together and share ideas. That intellectual community drew on Greek, Persian, and Indian knowledge, as well as the accumulated work of four centuries of Arabic astronomy and mathematics. The observatory produced a set of star tables with astrological as well as astronomical significance. Finally, al-Tusi's work represented a crucial development in the ongoing Arabic fascination with the Ptolemaic system. The Marāgha observatory disappeared by the early 14th century as its financial support waned. However, its brief existence testifies to the vibrant character of Middle Eastern astronomy in the Middle Ages.

AL-ANDALUS—ISLAMIC SPAIN

Over the course of the late 7th century, the Arab Empire expanded rapidly across North Africa. In 711 Tariq ibn-Ziyad led an Islamic army of Moors and Berbers across the Mediterranean to the Iberian Peninsula, that is, modern-day Spain and Portugal. A larger Arab army followed, and the Islamic forces conquered the entire kingdom of the Visigoths in Spain by the year 720. They continued northwards and invaded southern France, but they were defeated by the Frankish king Charles Martel at the Battle of Tours in 732. However, the Iberian Peninsula—al-Andalus—remained firmly under the control of Islamic forces, a major province of the Arab Empire on the continent of Europe.

In the years after the Islamic conquest, al-Andalus suffered from internal conflict between Arabs and Berbers, even though it was part of the greater Arab domains ruled by the Umayyad dynasty in Damascus. When the Umayyads were driven from power by the Abbasid dynasty in 750, an Umayyad prince, Abd al-Raman, left Damascus and traveled west. In 756 he gained control of al-Andalus and named himself Emir of Cordoba. His descendant Abd al-Rahman III claimed the title of Caliph of Cordoba in 929 and ushered in a golden age of Islamic culture in the Iberian Peninsula.

The Visigothic capital had been located at Toledo in the center of Iberia, but the city of Cordoba in the south was the focus of Islamic power and culture. In the period of the caliphate, it grew to become the largest city in Europe. Aqueducts brought fresh water into the city, and outdoor lighting graced its streets. The Great Mosque of Cordoba was a wonder of Islamic architecture and a significant center of scholarship. While al-Andalus was politically separate from the Abbasid realm of the Middle East and the Fatimid Caliphate of Egypt, it was still part of a great sphere of Arabic learning that stretched from the Atlantic Ocean to the borders of India. Scholars traveled freely from Baghdad to Cordoba, and the Arabic language provided a common medium of written communication. Some sources claim that the library of the caliphs of Cordoba included 400,000 books. While such a figure is unlikely, the collection was clearly impressive. Many of the works amassed in the House of Wisdom in Baghdad, books by Greek, Persian, and Arab authors, would doubtless have been available in Cordoba. For example, the star tables of al-Khwārizmī played a major role in Andalusian astronomical study (Lyons 2009, 72).

The Caliphate of Cordoba fell in the early 11th century, and al-Andalus broke up into a number of smaller states. Other dynasties would reunite the Islamic regions of Iberia in the 12th century: the Almoravids and the Almohads. Yet while the great period of Cordoba had passed, it left an enduring intellectual legacy. Al-Andalus continued as a center of study and scholarship. The astronomer and mathematician al-Zarqālī compiled star tables and created instruments in the 11th century. A group of 12th-century scholars from al-Andalus, including ibn Tufayl, al-Bitruji, Moses Maimonides, and ibn Rushd, wished to replace the Ptolemaic system of astronomy with a modified version of Aristotle's cosmology, largely because they objected to the philosophical implications of Ptolemy's geometry (Lyons 2009, 198–199). Ibn Rushd was one of the most influential of medieval Islamic scholars. Born in Córdoba in 1126 and trained in philosophy, law, and medicine, he wrote dozens of works on a variety of subjects, including mathematics and astronomy. However, his commentaries on Aristotle were his most significant achievements, and he would have a profound influence on medieval European philosophy.

Since the late 10th century, the Christian kingdoms of northern Iberia had expanded at the expense of the Islamic dominions. This process accelerated after the decline of the caliphate of Córdoba; for example, King Alfonso VI of Castile took the city of Toledo in 1085. As territory changed hands, a profound cultural shift took place as well. Alfonso's control of a sophisticated repository of Arabic scholarship in the Iberian Peninsula had great implications for European intellectual life. Some few Europeans had come to al-Andalus to study earlier in the 11th century, but after the conquest of Toledo, the city developed into a center for the translation of books from Arabic into Latin, the European language of scholarship. Arabic works, and ancient Greek works translated into Arabic, were now more available to Western intellectuals. Al-Andalus, the far western edge of the Islamic world, became a bridge between the Arabic world and European culture. The Christian reconquest of Spain would continue; Córdoba would fall in 1236. However, the real significance of the period lay in the movement of ideas. Greek philosophy and geometry, Arabic astronomy and algebra—all would transform the European mind in the High Middle Ages.

THE MANUSCRIPTS OF TIMBUKTU

During the Middle Ages, Islamic religion and Arabic literary culture spread with the growing boundaries of the Arab Empire. Spain became part of the Arabic world in this way. However, Arabic cultural influence was also transmitted along trade routes; for example, Islam traveled across the Indian Ocean to southeast Asia. The Arabs also developed commercial connections through the Sahara Desert to western Africa. The Empire of Ghana, located in the present-day nations of Mali and Mauretania, served as an important trading partner, and by the 11th century Islam had clearly taken root there. As the Empire of Mali began to grow in the 13th century, it incorporated much of the territory of Ghana and became a focal point of West African trade. Islam played an important part in medieval Mali; many of its rulers embraced the religion. The most famous of these, Mansa Musa, ruled in the early 14th century and went on a pilgrimage to Mecca in 1324–1325.

The city of Timbuktu grew into a major trading center during the reign of Mansa Musa, as well as a center of culture and scholarship. For 300 years, the city thrived on the exchange of merchandise and ideas with the rest of the Arabic world. Mosques graced the city, and the scholars in its schools, most notably the Sankoré Madrasah, developed a thriving Arabic literary culture. They had access to much of the collected knowledge of the Middle East and the Mediterranean and produced thousands of manuscripts of Arabic books, as well as original works.

Timbuktu began to decline in the 17th century; by the 19th century it was no longer a significant urban center. It is part of the modern nation of Mali, and its monuments have been designated as a World Heritage Site. The city has retained important elements of its past, including a huge collection of medieval and early modern manuscripts in Arabic and a variety of West African languages. These include Islamic religious texts, as well as works of law, astronomy, mathematics, literature, history, and other subjects. Many of the manuscripts have not been catalogued; conservationists are working to preserve them and to record them digitally. This archive illustrates the complex links between Western Africa and the Arabic cultural sphere.

CONCLUSION

The spread of Islam and the rise of the Arab caliphate led to the growth of an Arabic intellectual culture that had a profound effect on global intellectual history. Supported by royal patronage, Islamic scholars adopted texts and ideas from ancient Greece, India, and Persia and used them to foster an atmosphere of inquiry and innovation, particularly in the fields of mathematics, astronomy, geography, philosophy, and medicine. A community of scholars formed across the Islamic world, united by the Arabic language. Arabic scholarship began to decline in the 13th century, and it was severely affected by the Mongol conquest of much of the Middle East. However, the intellectual accomplishments of the Arabic world would inspire the development of medieval European philosophy and mathematics.

PRIMARY TEXT: AL-KHWĀRIZMĪ, *BOOK OF ALGEBRA*

Al-Khwārizmī wrote one of the foundational texts on algebraic problem-solving. In this selection, he briefly states the practical importance of mathematics and discusses the nature of the decimal number system adopted by the Arabs from India. He closes with a brief description of the terms that he will use in the following math problems: "roots" are unknowns or variables in modern terminology.

The Book of Algebra and Almucabola
Containing Demonstrations of the Rules of the Equations of Algebra

The Book of Algebra and Almucabola, concerning arithmetical and geometrical problems.

In the name of God, tender and compassionate, begins the book of Restoration and Opposition of number put forth by Mohammed al-Khwarizmi, the son of Moses. Mohammed said, Praise God the Creator who has bestowed upon man the power to discover the significance of numbers. Indeed, reflecting that all things which men need require computation, I discovered that all things involve number and I discovered that number is nothing other than that which is composed of units. Unity therefore is implied in every number. Moreover, I discovered all numbers need to be arranged that they proceed from unity up to ten. The number ten is treated the same manner as the unit, and for this reason doubled and tripled just as is the case of unity. Out of its duplication arises 20, and from its triplication 30. And so, multiplying the number ten you arrive at one-hundred. Again, the number one-hundred is doubled and tripled like the number ten. So, by doubling and tripling etc. the number one hundred grows to one thousand. In this way multiplying the number one thousand according to the various denominations of numbers you come even to the investigation of number to infinity.

Furthermore, I discovered that the numbers of restoration and opposition are composed of these three kinds: namely, roots, squares, and numbers. However, number alone is connected neither with roots nor with squares by any ratio. Of these then the root is anything composed of units which can be multiplied by itself, or any number greater than unity multiplied by itself: or that which is found to be diminished below unity when multiplied by itself. The square is that which results from the multiplication of a root by itself.

al-Khwārizmī. "Book of Algebra." In *Robert of Chester's Latin Translation of the Algebra of al-Khwārizmī*. New York: The Macmillan Company, 1915, 66.

FURTHER READING

Alhazen. 1974. *Book of Optics*. In *A Source Book in Medieval Science*, edited by Edward Grant. Cambridge: Harvard University Press.

Al-Khwārizmī. 1974. *The Book of Restoring and Balancing*. In *A Source Book in Medieval Science*, edited by Edward Grant. Cambridge: Harvard University Press.

Belting, Hans. 2011. *Florence and Baghdad*. Cambridge: Belknap Press.

Hourani, Albert. 1991. *A History of the Arab Peoples*. Cambridge: Belknap Press.

Ifrah, Georges. 1998. *Universal History of Numbers*, vol. 2. London: The Harvill Press.

Lindberg, David. 1976. *Theories of Vision from al-Kindi to Kepler*. Chicago: University of Chicago Press.

Lyons, Jonathan. 2009. *The House of Wisdom*. New York: Bloomsbury.

Merzbach, Uta and Boyer, Carl. 2011. *A History of Mathematics*. Hoboken: John Wiley and Sons.

O'Leary, Delacy. 2001. *How Greek Science Passed to the Arabs*. Abingdon: Routledge Press.

Smith, Tim Mackintosh. 2020. *Arabs: A 3000-Year History of Peoples, Tribes, and Empires*. New Haven, Yale University Press.

9

Numbers in Medieval Europe

Chronology of Major Figures

Martianus Capella	360–428
Boethius	c. 477–524
Isidore of Seville	560–636
Gerbert of Aurillac	946–1003
Adelard of Bath	1080–1152
Gerard de Cremona	1114–1187
Leonardo of Pisa	1170–1245 (approximate)
William of Moerbeke	1215–1286
Roger Bacon	1220–1292
Thomas Bradwardine	1300–1349
Nicole Oresme	1320–1382
Leonardo da Vinci	1452–1519
Raphael	1483–1520

CULTURAL ICON: THE TRANSLATIONS OF GERARD DE CREMONA

Mathematical books translated by Gerard:

Euclid, *The Elements*
Theodosius, *Three Books on the Sphere*
Archimedes, *On the Measurement of the Circle*
Ahmad ibn Yusuf, *On Similar Arcs*
Mileus, *Three Books on Spherical Figures*
Thābit ibn Qurra, *On the Divided Figure*
Banu Musa (Three Brothers), *On Geometry*
Ahmad ibn Yusuf, *On Ratio and Proportion*
Abu Uthman, *On the Tenth Book of Euclid*
al-Khwārizmī *On Algebra and Almucabala*
    Book of Applied Geometry
Anaritius, *Commentary on Euclid*
Euclid, *Data*
Tideus, *On Mirrors*
al-Kindi, *On Optics*
    Book of Divisions
Thābit ibn Qurra, *Book of the Roman Balance*

(Grant 1974, 36)

> "... for love of the Almagest, which he could not find at all among the Latins, he went to Toledo; there, seeing the abundance of books in Arabic on every subject, and regretting the poverty of the Latins, he learned the Arabic language, in order to be able to translate."
>
> (Grant 1974, 35)

So writes the anonymous author of a testimonial to Gerard de Cremona, a tribute written after his death in 1187. The short passage reveals a great deal about European intellectual history in the 12th century: that Western Europe suffered from a dearth of scholarly material ("the poverty of the Latins"); that this lack included Ptolemy's *Almagest* (Figure 9.1); that Spain possessed a wealth of books written in Arabic; and that European thinkers, including Gerard, traveled there (specifically, to Toledo) in order to translate these books into Latin and bring them back to Europe.

Western Europe had lost most of the literary, philosophical, and mathematical heritage of the ancient world when the Roman Empire collapsed in the 5th century. For the next 500 years, European scholarship was

Numbers in Medieval Europe 195

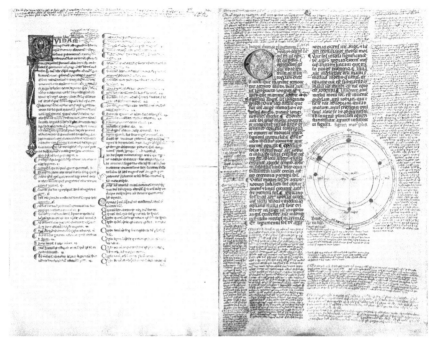

Figure 9.1 Gerard of Cremona translated many texts from Arabic to Latin in the late 12th century CE, including the *Almagest* of Ptolemy. His work played an important part in the transmission of Greek and Arabic knowledge to medieval Europe. (State Library of Victoria, Australia)

severely limited and focused largely on religious thought. At the same time, the Arab world enjoyed an intellectual golden age as Arab scholars studied the work of ancient Greek, Persian, and Indian authors and built upon that work. In particular, the Arabs developed sophisticated traditions in medicine, astronomy, and mathematics. A great wealth of books by Greek and Arab authors was available in Arabic in the libraries of al-Andalus—Muslim Spain.

By the beginning of the 12th century, a growing culture of intellectual inquiry in Europe, combined with a dramatic increase in travel and trade, led European intellectuals to travel to Spain in search of this knowledge. Over the next 200 years, a burst of scholarly activity resulted in the translation of hundreds of books from Arabic to Latin. Some Arabic texts were transported back to Europe for translation. The Englishman Adelard of Bath traveled to the Middle East and returned with works that he rendered into Latin, including the astronomical tables of al-Khwārizmī and Euclid's *Elements* (Lindberg 1978, 62–63). Most translations, however, were done in Spain. Some works were translated from Arabic to Spanish and then from Spanish to Latin. Spanish Mozarabs—Christians who had lived under Muslim rule and whose primary written language was Arabic—were instrumental in the process, as were multilingual Jewish scholars. In addition, many Europeans learned Arabic for the purposes of the work; for example, Robert of Chester translated al-Khwārizmī's work on algebra.

The most productive of all the translators of this period, Gerard of Cremona was born in Italy, but worked in Toledo for much of his adult life. According to the testimonial written in his honor and appended to one of his medical translations, he had mastered the knowledge available in Christian Europe and traveled to Toledo in search of Ptolemy's compendium of ancient astronomy, the *Almagest*. Toledo had been conquered by the Christian kingdom of Castile in the late 11th century, but it retained Muslim, Mozarab, and Jewish communities; one source relates that Gerard worked with a Mozarab scholar named Galippus (Lindberg 1978, 66). As the testimonial relates, Gerard learned Arabic, and he ultimately translated more than seventy books on subjects ranging from logic to alchemy. He had a particular interest in the works of Aristotle on logic and on the natural world, and he translated seventeen books on mathematics, including Euclid's *Elements*, Archimedes's *On the Measurement of the Circle*, al-Khwārizmī's *Algebra*, and Thābit ibn Qurra's *On the Divided Figure*. These texts would inform European work on algebra and geometry for three centuries, until new versions of the Greek geometrical classics emerged during the Italian Renaissance.

Spain was not the only site of translating activity during the High Middle Ages; many classical books were translated directly from Greek to Latin in southern Italy. Yet no other medieval figure had as large an effect on the movement of ideas as Gerard of Cremona. His life and work are representative of the acquisition of Greek and Arab knowledge—knowledge that transformed European intellectual life and defined the curricula of the new European universities for the rest of the Middle Ages.

CULTURAL COMMENTARY: DISASTER AND RESURGENCE

The story of mathematics in medieval Europe, and of intellectual life in general, is one of loss and recovery: the loss of most Greco-Roman knowledge when the Western Roman Empire collapsed in the 5th century and the recovery of that knowledge in a series of stages from the 11th to the 16th centuries. The growth of European university culture after 1100 was fueled by the acquisition of ancient texts from the Arab world and the Byzantine Empire, as well as more recent Arab works and commentaries. Similarly, the Italian Renaissance inspired a search for lost ancient knowledge and transformed European scholarship as a result. By 1500 Western Europeans had recovered much of the classical heritage, and this development gave rise to a period of great intellectual energy.

The Early Middle Ages (500–1000)

Over the course of the 5th century, Roman authority steadily decayed in Western Europe, and the empire was ultimately supplanted by a series of small, unstable Germanic kingdoms. Warfare and raiding were constant, while travel and trade decreased dramatically. Urban life declined; the population of the city of Rome decreased by more than 80 percent by the end of the 6th century. In this atmosphere of cultural deterioration, intellectual life suffered terribly. Books are fragile things, easily destroyed. As cities shrank in size, the neglect of libraries led to the loss of many ancient works. After all, in an age before printing, the number of copies of any given work was limited. If forty copies of a particular book existed in a region and all were destroyed or lost, then that book ceased to exist, as if it were never written. Literacy itself faded in the centuries after Rome fell, and the classical Latin language itself was mainly preserved because of its role in the Roman Catholic Church.

Indeed, the Roman Church became the dominant intellectual institution in Europe during the Early Medieval Period (500–1000) and would remain so throughout the Middle Ages. This trend actually began well before the fall of Rome in the West. The tremendous growth of Christianity in the 4th century had resulted in an emphasis on religious revelation as a source of truth; Christian theology took intellectual precedence over other forms of thought throughout the entire Roman world. In the Eastern Roman (or Byzantine) Empire, Christianity had largely eclipsed ancient polytheism, classical literature, and Greek philosophy by the 6th century. However, in the West, change was more dramatic. The fall of the Roman Empire meant that the institutions of the Church were the only intellectual establishments remaining in Western Europe, and they, too, had suffered significant decline. Latin literacy survived because the Church preserved it; books survived because Church clerics copied them. As a result, the literary heritage of Early Medieval Europe reflected the intellectual priorities of the Church, with a stress on the revealed word of Judeo-Christian scripture and church teaching. Some Classical texts were also saved, for early medieval churchmen revered the Greco-Roman tradition and sought to preserve what they could of it. Yet the limited scholarly culture that did exist in the Early Middle Ages was overwhelmingly Christian in its focus.

What books survived the fall of Rome in the West, and how was the Greek mathematical tradition represented? In addition to Christian religious works, scholars had preserved a portion of Roman history and literature. However, Greek texts did not fare as well. If Latin literacy in Western Europe had decreased significantly after the fall of Rome, Greek literacy had almost disappeared. It is important to point out that the Romans never developed significant traditions of philosophy or mathematics of their own, even though they ruled over the energetic intellectual community of the eastern Mediterranean. The Roman thinkers who wished to pursue these subjects did so in Greek, using the works of Hellenic or Hellenistic Greek authors. Very few of these texts remained in the West after the collapse of the empire and the rise of the Germanic kingdoms. For example, of the thirty dialogues of Plato, only one was available in early medieval Europe. Some of Aristotle's works on logic were translated into Latin by the 6th-century Roman scholar Boethius, but the bulk of Aristotelian thought was gone, as were the works of the Stoics, the Epicureans—indeed, most of the schools of ancient Greek philosophy. Boethius also translated Euclid's *Elements*, but the complex geometrical proofs were ultimately lost over time, and only the initial propositions survived. The books of the other great mathematicians of antiquity—Pythagoras, Eudoxus, Archimedes, Apollonius, Hipparchus, Ptolemy, Diophantus—were no longer available in Western Europe.

With the decline of cities in the Early Middle Ages, Western European literate culture became concentrated in Christian monasteries—isolated communities of men or women devoted to lives of prayer and contemplation. In addition to physical work and religious ceremonies, monasteries focused on the copying and preservation of books. For 500 years, monastic libraries dominated European intellectual life and served as repositories of written knowledge. While Christian thought clearly took precedence, monastic scholarship and education also stressed the seven liberal arts, intellectual divisions rooted in the culture of the Late Roman Empire. These subjects—grammar, rhetoric, dialectic, music, astronomy, geometry, arithmetic—played a crucial role in European thought for the rest of the Middle Ages. However, because of the sheer number of Greco-Roman texts that had been lost, deep knowledge of the liberal arts was limited before the 12th century. For example, the understanding of geometry involved little more than the basic definitions of Euclid, while the study of arithmetic focused on elementary number theory and operations. Indeed, monastic culture viewed knowledge itself as static and authoritative: one learned it, preserved it, and ensured that it was passed on. Mathematics was thus treated as a body of truth to be mastered or memorized, rather than as a dynamic process of reasoning and inquiry.

The High Middle Ages (1000–1300)

Southwest of Christian Europe lay the domains of Muslim Spain, or al-Andalus, part of the vibrant cultural community of the Arab Empire. The works of philosophy, medicine, and mathematics collected and produced in the Arab world were available in al-Andalus; in the late 10th century, the great library of Cordoba was said to contain 400,000 volumes. Gradually, the intellectual wealth of Muslim Spain came to influence the rest of medieval Europe. The life of Gerbert of Aurillac serves as an illustration of this influence at an early stage. Gerbert came from a monastery in southern France, but he traveled to Catalonia in northern Spain to study in the year 967. By this time, the northern parts of Spain had come under the control of Christian kingdoms, but there was still significant cultural contact with the Muslim south. Gerbert spent three years studying mathematics and music, and he probably encountered Arab scholarship in these areas (Mahoney 1978, 60). When he returned to France, Gerbert may have brought books with him, including treatises on the astrolabe. He continued to pursue mathematical studies for the rest of his life and introduced a new kind of abacus to Europe that employed Arabic numerals rather than counting stones (Mahoney 1978, 60–61). His reputation as a scholar and churchman continued to grow; he taught at the cathedral school at Rheims in northern France, served as abbot of the monastery at Bobbio, and was ultimately elected pope (Sylvester II) in the year 999.

Gerbert's students at Rheims were perhaps his most important legacy; many of them played a crucial role in the growing cathedral schools of the 11th century (Grant 1977, 14). In this period, travel and trade began to revive in Western Europe, towns began to grow again, and cathedrals often served as important administrative centers in the towns. A cathedral is the church of a bishop; the word comes from the Latin *cathedra*, or seat. As bishops acquired more wealth and more complex roles in local government, they needed educated clergy to manage their affairs, and they established cathedral schools to train those clergy. Rapidly, the new schools became focal points of European intellectual life. Like monastic schools, they were based on the study of the seven liberal arts, with particular emphasis on Latin literacy and the Aristotelian logic available in the translations of Boethius. Further, the influence of Gerbert and his students led to increased emphasis on music, astronomy, geometry, and arithmetic. This is illustrated by an exchange of letters in the year 1025 between two scholars, Ragimbold of Cologne and Radolf of Liege. The letters deal with mathematical questions and demonstrate the growing interest in such subject matter among European thinkers. However, the correspondence also reveals the continuing lack of real mathematical expertise; the two authors have no real sense of method or proof, and they cannot effectively address such simple problems in geometry as the relation of the side of a square to its diagonal (Grant 1977, 14–15).

The resurgence of European commerce and the emergence of cathedral schools led to the most important development in medieval European intellectual life: the large-scale translation of ancient Greek texts and medieval Arabic works into Latin, the major language of scholarship in Western Europe. Increased travel made it possible for Europeans to journey to al-Andalus and draw on its great intellectual resources, and this was augmented by the continued growth of the Christian kingdoms in the northern parts of the Iberian Peninsula. The city of Toledo fell to the king of Castile in 1085 and became the most important center of translating activity in the 12th century. European scholars worked to translate texts from Arabic, often in conjunction with Spanish Mozarabs, Muslims, or Jews who were fluent in that language. The translators sought out Greek works of philosophy, medicine, and mathematics, as well as Arab treatises and commentaries on these subjects. While they paid particular attention to the works of Aristotle on logic and natural philosophy, mathematical texts also drew great attention. For example, Gerard of Cremona translated over seventy texts, including Ptolemy's *Almagest*, Euclid's *Elements*, al-Khwārizmī's *Algebra*, and the *Geometry of the Three Brothers* (Grant 1977, 17). Substantial translating activity also took place in the states of Italy during the 12th and 13th centuries, particularly in the Norman kingdom of Sicily. Here, scholars learned Greek and began to translate ancient texts directly from Greek into Latin. Again, Aristotelian philosophy took precedence, but translations of Euclid's *Elements* and his other treatises were done in the mid-12th century. One hundred years later, William of Moerbeke translated dozens of books, including seven mathematical treatises of Archimedes, the first available to Western Europe since the fall of Rome.

In short, during the span from 1100 to 1300, Western Europe recovered a substantial portion of the ancient Greek intellectual legacy, along with a great deal of Arab knowledge. The complete philosophy of Aristotle, the astronomy of Ptolemy, the medical works of Galen and Ibn Sina, and the mathematics of Euclid and al-Khwārizmī transformed European intellectual life. European universities emerged during these same two centuries, and Aristotelian thought came to dominate university curricula in conjunction with Christian belief. The newly acquired Greek and Arab texts allowed these institutions to educate their students thoroughly in all of the seven liberal arts; for example, the works of Euclid and the Arab mathematicians made substantial study of geometry and arithmetic possible. Over the course of the 13th century, the use of Indo-Arabic numerals based on place-value and Arab approaches to basic mathematical operations grew rapidly in Europe—largely due to al-Khwārizmī's book on the subject. A short text by John of Sacrobosco (1244–1256) on the system contributed a great deal to its rapid spread. The Latin term *algorismus*, derived from al-Khwārizmī's name, came to refer to such numerals and calculating techniques in Europe.

However, because Europeans had lacked a real mathematical tradition for so long, the real assimilation of Greek geometry and Arab algebra was a more gradual process. European thinkers were confronted all at once with mature mathematical systems, and they had to develop the analytical habits of mind necessary to understand them. In addition, the central place occupied by Aristotelian philosophy in European universities affected the study of mathematics. Aristotelian methodology is logical rather than quantitative; its approach to nature does not depend on numbers or measurement. Mathematics, while valued, was secondary; European university scholars often used the proofs of Euclid as examples of Aristotelian logical thought rather than as pure geometrical exercises in their own right. A truly investigative mathematical culture would finally begin to emerge in European universities in the 14th century.

While university faculty were initially slow to grasp the intricacies of ancient mathematics, two extraordinary figures in Italy mastered Greek and Arab ideas and produced real mathematical innovations. Italian merchants had developed trade links with the Byzantines and the Arabs in the Eastern Mediterranean by the late 11th century. Such merchants might speak Greek or Arabic, they might have access to a variety of books in those languages, and they certainly used basic mathematics in their business dealings. These conditions set the stage for the work of Leonardo of Pisa (often referred to as Fibonacci) and Jordanus of Nemore, two 13th-century mathematicians who took full advantage of the intellectual resources available to them. Fibonacci helped to introduce Indo-Arabic numbers to Italy; he especially emphasized their usefulness in commercial transactions (Merzbach and Boyer 2011, 229–230). He also did original work in algebra and number theory and wrote a book on the practical use of geometry that preserved parts of an otherwise lost text of Euclid. Jordanus authored works on mechanics and algebra and introduced a technique for expressing mathematical operations by using letters of the alphabet to represent quantities. The work of both authors would influence European mathematics in the 15th century and afterwards, but they were not well-known in their own day.

The Late Middle Ages

The work of Euclid and his Arab commentators, combined with the Aristotelian character of European universities, produced a unique style of mathematical thinking in the 14th century. Thomas Bradwardine at Oxford University and Nicole Oresme at the University of Paris both explored the mathematics of ratio and proportion, building on Euclid's treatment in Books V and VII of the *Elements* and a number of Arab authors. They also applied their mathematical knowledge to the description of moving bodies, perhaps as a response to Aristotle's emphasis on the description of the natural world. The faculty of Merton College at Oxford, which included Bradwardine, formulated a theorem on the motion of a uniformly accelerating body, comparing it to that of a body moving at a constant rate. Oresme represented the acceleration of such an accelerating body graphically, essentially producing an early diagram of a mathematical function. Such unique work represented the apex of medieval university mathematics.

The translations of the 12th and 13th centuries had revived philosophical and mathematical study in Europe. A new round of translations would take place during the Italian Renaissance of the 14th and 15th centuries, and these would again have a revolutionary effect on European intellectual life. The Italian Renaissance emerged from the urban culture of Late Medieval Italy. Leading citizens of the free Italian city-states identified with the ancient Romans and embraced Roman culture and values. Italian Humanism, the signature intellectual movement of the Renaissance in Italy, grew out of this trend. Humanist scholars prized classical morals, literary styles, and standards of beauty, and this led to the recovery of many ancient texts during the 14th and 15th centuries. By exploring the monastic libraries of Italy, Renaissance intellectuals found many Roman books that had supposedly been lost for centuries—this in turn inspired a search for Greek works in the eastern Mediterranean and the crumbling Byzantine Empire. Between 1420 and 1500, much of the ancient Greek literature and philosophy known to the modern world was translated into Latin by Renaissance Humanists. Dozens of "new" lost works of philosophy and mathematics were transmitted to Western Europe during this time, and better translations of Aristotle, Ptolemy, Euclid, and Archimedes replaced those made during the High Middle Ages in Spain and Sicily. Their impact on European ideas was dramatic: where High Medieval scholastic thought had been based largely on Aristotle, European theorists could now draw on substantial texts by Plato, Epicurus, Lucretius, the Stoics, Pythagoras, and the Neo-Platonists. Informed by this broad range of ancient ideas, they began to pose skeptical challenges to Aristotelian worldviews, challenges that would contribute to the rise of modern science. The newly transmitted books also included some of the most challenging mathematical thought of the Hellenistic period: works by Apollonius of Perga and Hipparchus of Rhodes, as well as improved translations of Archimedes and Ptolemy. These would influence the progress of European mathematics and astronomy in the 16th and 17th centuries, a period of great intellectual energy.

As the term implies, Italian Humanism stressed humanity itself as worthy of study. Humanists adopted the optimistic view of the human individual characteristic of the ancient Greeks and Romans and stressed the ability of humanity to improve itself in this world. Renaissance artists adopted classical values and celebrated the nude human figure, while Renaissance philosophers defined humanity as the crowning achievement of God's creation. The *Vitruvian Man*, a drawing made by Leonardo da Vinci around the year 1490, expresses this view of humanity as the "measure of all things," as the ancient Greek scholar Protagoras once said. Leonardo drew a nude male torso with arms and legs in two different orientations: one inscribed in a circle and one inscribed in a square. His symbolism was clear—God had created humanity based on the perfect proportions of the two most perfect geometrical figures. Leonardo's vision linked geometry with Humanism and with Plato's ideas of pure intellectual beauty (Figure 9.2).

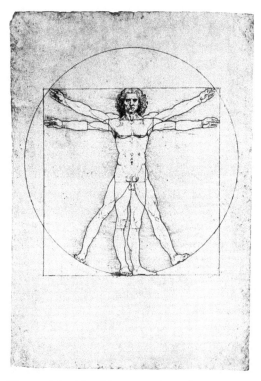

Figure 9.2 The *Vitruvian Man*, by Leonardo da Vinci, serves as a potent image of Italian Humanist thought. Leonardo relates the proportions of the male human body to the fundamental figures of geometry. (Jakub Krechowicz /Dreamstime.com)

Italian mathematics during the Renaissance itself touched upon the academic, practical, and artistic dimensions of contemporary culture, and one author in particular served to illustrate these different strains of thought. Luca Pacioli taught mathematics in late 15th-century Venice and wished to summarize his subject matter; his *Summa de Arithmetica, Geometria, Proportioni et Proportionalita* was published in 1494. Pacioli's treatment of Arabic numerals and algebra drew upon the work of Fibonacci and on the university scholarship of his day. However, as a citizen of the great trading city of Venice, Pacioli also described the techniques of the double-entry bookkeeping that powered Italian business and banking. Later in his career, he published a discussion of proportion and perspective in art, probably illustrated by Leonardo da Vinci. His mathematical works thus embody the complex society that produced the Italian Renaissance.

So, medieval European society, largely deprived of its classical intellectual heritage after the fall of Rome, regained a version of that heritage over the course of 1,000 years. Mathematical knowledge fell—and rose again—as part of a larger story. Christian monks and Arab scholars, Italian merchants and French university professors—all played a part in the emergence of a new mathematical vitality, built on the achievements of Greece, Arabia, and India and poised to confront new challenges in the modern period.

THE TRANSMITTERS: BOETHIUS, CASSIODORUS, ISIDORE

Faced with the catastrophic loss of books and the decay of literacy itself after the fall of Rome in Western Europe, a handful of intellectuals worked to preserve whatever Greco-Roman knowledge they could. While their efforts were limited in scope, they managed to save a number of significant texts and to transmit an appreciation of Classical learning to their Early Medieval successors. The most important of these figures worked in Spain and Italy in the 6th and early 7th centuries: Boethius, Cassiodorus, and Isidore of Seville.

Boethius (d 524) was born in the city of Rome in the late 5th century, at a time when Italy came under the control of the kingdom of the Ostrogoths. A member of a noble Roman family, he served the Ostrogothic king Theodoric for most of his adult life, attaining the offices of senator and consul. The Ostrogothic rulers employed Roman advisors and retained Roman governmental titles in an attempt to enhance their prestige and legitimacy. In addition to his official role, Boethius worked tirelessly as a scholar. Fluent in both Greek and Latin, he recognized that Greek literacy would disappear in Western Europe during his lifetime, and he sought to preserve as much of the Greek philosophical tradition as possible. While he planned to translate all of the books of Plato and Aristotle into Latin, he only completed work on Aristotle's treatises on logic. He also wrote commentaries on Aristotle and Porphyry and a treatment of arithmetic, *De Arithmeticae Institutiones*, based on the work of Nichomachus of Gerasa. Nichomachus had worked in Syria in the 1st century CE; his *Introduction to Arithmetic* was a summary of the subject, combined with a great deal of number mysticism. Boethius may have shared those views of number, or he may have chosen Nichomachus as a source because he had nothing else available. Boethius also apparently translated part of Euclid's *Elements*, but only the basic propositions survived. In the year 523, he got immersed in the political intrigues of the Ostrogothic court and was sentenced to death for treason. As he waited for his execution, he composed the *Consolation of Philosophy*, a mix of Plato's philosophy and Christian thought.

Boethius had a tremendous influence on medieval intellectual life; his *Consolation of Philosophy* was one of the most widely read books in medieval Europe. His translations of Aristotle's logic and his commentaries on the subject served as the basis of European logical studies for the next six centuries. Finally, his mathematical work, though rudimentary and somewhat mystical, was one of the few books available on the subject in the Early Middle Ages. Cassiodorus (485–585), another Roman, also worked for the Ostrogothic court; he may have succeeded Boethius in his role. Upon his retirement, he founded a monastery, the *Vivarium*. For the rest of his long life, he emphasized the copying of both Christian and classical books; he also wrote a systematic guide to the works available in the monastic library. His work contributed to the organization of medieval knowledge and the pursuit of the seven liberal arts in monastic schools.

Isidore (560–636) served as bishop of Seville in the Visigothic kingdom in Spain for more than thirty years. Concerned with the loss of Greco-Roman learning, he compiled a summary of knowledge in twenty books, the *Etymologies*. While Isidore saved a great deal of material that might otherwise have been lost, his discussions are often vague and incomplete; a sign of the sparseness of his sources. This becomes clear in his treatment of mathematical material; he presents definitions and propositions without proofs and often omits important detail. For example, he defines a circle as follows:

> The first of these, the circle, is a plane figure which is called a circumference, in the middle of which is a point upon which everything converges, which geometers call the center and the Latins call the point of the circle.
>
> (Isidore of Seville 1974, 8)

The works of Boethius and Isidore on mathematics illustrate the intellectual crisis that existed in Europe after the fall of Rome. As literacy waned and books were lost, both struggled to save what they could. Often, they used summaries of Greek knowledge, for the actual books were already gone—consider Boethius's use of Nichomachus's text. As a result, their treatments of the subject lack any real sense of mathematical reasoning; they are collections of facts to be memorized, shadows of a tradition that has largely disappeared. However, the "transmitters" served a vital purpose, for they preserved the memory of what once had existed; they conveyed a sense of the grandeur of the Classical past.

THE SEVEN LIBERAL ARTS

The notion of the seven liberal arts that ultimately defined medieval European education found its roots in the work of Martianus Capella, a Roman author of the early 5th century CE. A resident of Roman North

Africa, Martianus wrote *The Marriage of Mercury and Philology*, the kind of intellectual handbook or summary of knowledge that was common in the Western Roman Empire during the Late Antique period. Because his book survived the fall of Rome in the West, it had a tremendous influence on Early Medieval culture; monastic schools valued it as a guide to the knowledge of the Classical world. As a result, Martianus's classification of human intellectual pursuits would influence European scholarship for more than 1,000 years.

The Marriage of Mercury and Philology begins with a story of the gods. Mercury, messenger of the gods, marries Philology, the female embodiment of language. At the wedding, she is presented with seven handmaidens, the Seven Liberal Arts: Grammar, Rhetoric, Dialectic (logic), Geometry, Arithmetic, Astronomy, and Music. The book adopts the Greco-Roman tradition of using female figures to represent human intellectual and creative pursuits, a tradition that would persist in European culture through the early 20th century. Each of the Liberal Arts gives an address at the wedding, summarizing her particular area of knowledge, and these make up most of the text. The speeches of Geometry and Arithmetic may be used to illustrate the depth of knowledge that Martianus Capella presents in his work.

"Geometry" literally means "earth measuring," and most of Geometry's speech describes the regions of the earth—it is really a discourse in late antique geography. However, her discourse ends with a presentation of some of the propositions and definitions of Euclid, without any real sense of proof or mathematical demonstration. Similarly, Arithmetic's account addresses the nature of numbers—odd, even, prime—without the complexities of Greek number theory. She also explains fundamental arithmetical operations. In short, the book provides the reader with an elementary account of the mathematical achievements of the ancient Mediterranean, a set of labels and terms to be mastered and repeated. It does not contain any real sense of Greek mathematical reasoning. The same may be said of the other subjects discussed.

At a time when much of the literary heritage of antiquity was lost in the West, Martianus's book persevered—perhaps because it appeared to provide a convenient summary of a great deal of subject matter. One can imagine Church scribes of the 6th and 7th centuries copying it as a way to cover a lot of intellectual ground. Cassiodorus refers to it in his *Institutiones*, an account of the necessary contents of a library. The monastic schools of the Early Middle Ages used it as the basis of their curricula and divided the seven liberal arts into the *Trivium* (grammar, rhetoric, dialectic) and the *Quadrivium* (geometry, arithmetic, astronomy, music). The studies of grammar and rhetoric were based on the surviving examples of Latin literature and poetry, and that of dialectic focused on the few works of

Aristotelian logic that were translated by Boethius. The resources available for the exploration of the *Quadrivium* were far more sparse. With respect to geometry and arithmetic, monastic scholars could use the actual summaries available in Martianus's book, as well as Boethius's text on arithmetic and the relevant sections of Isidore's *Etymologies*. All of these works shared a common characteristic: they were not written by mathematicians. They merely provide a set of terms and descriptions for the lay person, shorn of any sense of mathematics as a rigorous process of intellectual investigation. Boethius did translate part of Euclid's *Elements*, but apparently only the initial propositions survived.

The concept of the seven liberal arts provided Early Medieval European thinkers with an outline of nonreligious knowledge, a plan for a scholarly structure—but they lacked the intellectual bricks and mortar to actually build that structure. As cathedral schools spread across Western Europe in the 11th century and interest in the *Quadrivium* increased, this inadequacy became clear. The translations of the 12th and 13th centuries finally provided Europeans with the capacity to study the liberal arts in depth. Aristotle's complete works formed the basis for an enhanced understanding of logic (dialectic), and his many treatises on physics, metaphysics, astronomy, and natural philosophy applied to the subjects of the *Quadrivium*. So did the *Almagest* of Ptolemy. Arab commentaries provided useful analysis of these works, rather like textbooks. Finally, Euclidean geometry, Arab numerals, and Arab algebra greatly enhanced the investigation of geometry and arithmetic and inspired the growth of mathematical thought in Western Europe. The seven liberal arts, bolstered by the knowledge of the ancients and the Arabs, lay at the intellectual center of the new universities in Paris, in Oxford, and in Montpellier. A student who completed the arts curriculum successfully at one of these institutions would receive a degree, a license to teach. To this day, university undergraduates around the world work for four years to earn that same degree, a BA—Bachelor of Arts.

LEONARDO OF PISA (FIBONACCI)

Perhaps the most accomplished mathematician of the European Middle Ages, Leonardo of Pisa, was born in the late 12th century, perhaps around 1170. Modern mathematicians and historians often refer to him as "Fibonacci" (short for *filius Bonacci*—son of Bonacci), but he did not use that name while he lived. His father, Guglielmo, was a successful merchant in the Republic of Pisa, and Leonardo traveled with him throughout the Mediterranean. He was educated in Arab North Africa and acquired a knowledge of Indo-Arabic numbers and methods of figuring. He also had access to many of the Greek and Arab mathematical texts that had recently been translated from the Arabic (it is not clear whether he himself could read Arabic). Leonardo was not a member of the clergy, and therefore had no connection with the cathedral schools and universities that dominated European scholarship during his lifetime. Rather, he came to study mathematics for practical purposes, for the furtherance of trade and commerce, and this practicality characterized much of his work. However, at a time when European university faculty were slow to absorb the complexities of Euclidean geometry and Arab algebra, he also came to master those disciplines and to do important original work in pure mathematics. He helped to introduce Arabic numbers and algorism (approaches to computation) to Italy, and he wrote a series of treatises that broke new ground in the studies of algebra and number theory. Yet he never departed from his commercial roots: he used problems in business and currency exchange to illustrate algebraic operations, and he applied Euclidean geometry to worldly problems of land measurement and engineering.

In the *Liber Abaci* (1202), Leonardo adopted Arabic numbers and computational practices; he also used the horizontal bar notation when dealing with fractions (Merzbach and Boyer 2011, 229). He applied these to a series of problems drawn from his mercantile experiences, dealing with monetary transactions and currency conversions. Over the course of the work, he employed the algebraic techniques of al-Khwārizmī and al-Karaji, authors he undoubtedly encountered in his education. It is important to point out that his mathematics did not have a modern appearance: he used al-Khwārizmī's verbal approach to the description of problem-solving, and he often made use of cumbersome unit fractions. Yet the book represents a far more sophisticated understanding of algebra than any other European work of its time. Leonardo also explored the "Fibonacci Sequence" in the *Liber Abaci*, a series of numbers in which each figure is the sum of the two figures preceding it:

$$1, 1, 2, 3, 5, 8, 13, 21. \ldots$$

The sequence had already been described by several Indian authors in the 5th and 6th centuries with reference to the rhythm and meter of Sanskrit verse. Leonardo used it to treat a more earthly problem:

> How many pairs of rabbits will be produced in a year, beginning with a single pair, if in every month each pair bears a new pair which becomes productive from the second month on?
> (Merzbach and Boyer 2011, 230)

The sequence addresses the issue of continuous generation or increase and can be used to describe a variety of phenomena in human society and in the natural world (it has many applications to modern biology, for example). Leonardo introduced it to Western mathematics, and it is named in his honor.

The practical side of Leonardo's intellect is also revealed by his *Practica Geometriae* (*Practical geometry*—1220). The book contains substantial material from Euclid's *Division of Figures*, a work that Leonardo apparently had access to, but was subsequently lost. Here, he examined different methods for the precise partition of geometrical forms:

> And if you wish to divide any triangle by a line parallel to one of the sides of the triangle, the manner of doing so will be demonstrated in the following triangle ABG...
>
> (Leonardo of Pisa 1974, 159)

It is likely that Leonardo was interested in procedures for the exact division of land. He also examined methods for finding the height of structures or for doubling the volume of a cube (Grant 1974, 159; Mahoney 1978, 159–160). Adopting earlier Arab work, he used some algebraic techniques in his geometrical proofs.

Leonardo's later works demonstrated great mathematical sophistication and original thinking. In the *Flos* and the *Liber Quadratorum* (*Book of Squares*), both dated 1225, he pursued complex algebraic problems and indeterminate equations, that is, equations that have more than one solution. Various forms of quadratic and cubic equations interested him, as well as number theory. For example, the second proposition of the *Book of Squares* explores the following question:

> I want to demonstrate why a regular series of squares arises from a regular summation of odd numbers beginning with 1 and going to infinity.
>
> (Leonardo of Pisa 1974, 117)

His techniques reflected the influence of varied Arab mathematicians and the Hellenistic author Diophantus. Interestingly, he continued to use sexagesimal fractions in his more complex mathematics, a practice dating back to ancient Mesopotamia. Clearly, his unique education gave him a degree of mathematical agility that set him apart from other European scholars of his day.

Leonardo of Pisa's work demonstrates the existence of a strain of European mathematical thought outside the cathedral schools and universities of the Church, a strain inspired by commerce and by substantial contact with the intellectual resources of the Arab world. He was appreciated by some in his own time; his patrons included the Holy Roman Emperor Frederick II, another figure with a cosmopolitan background and an unusual education. However, much of his work was beyond the abilities of the other European scholars of his day. He influenced the growing spread of Arabic numbers in Europe in the High Middle Ages, but his more complex ideas would not be widely understood until Europe developed a tradition of advanced algebraic study in the 15th century.

NICOLE ORESME AND THE MOTION OF BODIES

By the 14th century, the European university community had absorbed much of the Greek and Arab mathematical knowledge transmitted in the preceding 200 years. Yet the intellectual priorities of medieval universities affected the way that their resident faculty viewed this knowledge. Late medieval university mathematics was certainly based on Greek and Arab work, but it developed in ways unique to the culture of Western Europe.

The ideas of Aristotle held a central place in university arts curricula and influenced all of the other intellectual enterprises of the schools. For example, the study of medicine assumed an Aristotelian view of matter, while the study of theology adopted Aristotelian logical methods of analysis. The exploration of mathematics would also be affected by this Aristotelian atmosphere.

The work of Nicole Oresme illustrates the intellectual energy of Late Medieval university scholarship in mathematics, as well as its particular character. Oresme was born in Normandy in northern France and educated at the University of Paris. He completed the curriculum and earned a Master of Arts degree; later he received a doctorate in theology. He taught at the university for two decades and then moved on to serve the French royal family and to hold a series of Church appointments. Despite his various professional responsibilities, he had a wide range of intellectual interests and did important work in mathematics, cosmology, physics, and philosophy (in medieval terms, all these fields would be considered philosophy). Oresme was a product of university culture, and his work reflected the Aristotelian emphasis of that culture. That is not to say that he merely mastered Aristotle's thought, for he often challenged Aristotelian arguments and methods of proof. He worked at a time when university scholars felt free to consider new views in a hypothetical sense. Nevertheless, Aristotelian ideas still framed these discussions: Oresme and his colleagues responded to Aristotle's assertions and defined their ideas in relation to his.

The systematic description of natural phenomena had been a priority for Aristotle; much of his work addressed topics in physics, astronomy, cosmology, and the living world. Aristotle used observation and logical analysis to examine these topics; he did not seek to explain nature in mathematical terms. For Aristotle, objects possessed real qualities like color, shape, or texture. Qualities also included the potential for particular action; for example, a pine tree had the capacity to grow. He argued that objects with similar qualities could be placed in groups and described in terms of their common characteristics. Hence, one may say that "All lemons are sour" or that "All falcons fly." Such statements were based on observation and served as the basis for more complex logical arguments about the natural world. Aristotle also recognized that the qualities of things sometimes change over time, and he referred to such change as "motion." In his view, the term referred to more than just change of place; hence, the growth of a tree was a form of motion.

In the 14th century, a number of university scholars in England and France used Aristotle's tools of observation and logical analysis to explore the physical motion of bodies. Aristotle had discussed actual physical motion as a quality that could be quantified, and he understood it as the balance between the movement of a body and the resistance of the medium through which it moves. A group of theorists at Merton College at Oxford studied motion as a quality that changed over time, and they considered the motion of a uniformly accelerating body; that is, an object that is increasing its speed at a uniform rate. They concluded that over a specific period of time, a uniformly accelerating body would cover a given distance, while another body traveling at a constant speed equivalent to the average speed of the first body would cover the same distance in the same period of time. For example, if a car accelerating uniformly from zero to 100 miles an hour covers a given distance in two minutes, another car traveling at a constant speed of 50 miles an hour (the average of zero and 100) would cover that same distance in the same time. The formulation of this "mean speed theorem," was not experimental; rather, it represented an exercise in Aristotelian logic applied to a quantifiable phenomenon. The scholars of Merton College also made extensive use of the mathematics of proportions.

Nicole Oresme built upon the work of the Merton College "Calculators," as they are sometimes called. He adopted a novel approach to the issue of uniformly accelerated motion by representing the phenomenon visually, or as one would say in modern terms, graphically. He drew a horizontal line to represent time, divided into regular increments. He then drew vertical lines of increasing length to indicate the increases in velocity of a regularly accelerating body. Because acceleration was constant, these lines formed a linear slope, a graph of the uniformly increasing speed of the body (Figure 9.3).

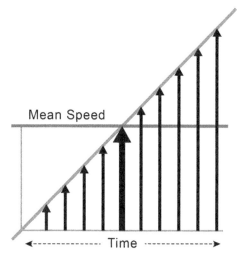

Figure 9.3 Nicole Oresme produced a graphic representation of the mean speed theorem.

The height of each vertical line corresponded to the velocity of the body at that point in time, while the vertical line on the right represented the final velocity of the body. Oresme connected the vertical lines to form a triangle, and he recognized that the area of that triangle was equal to the total distance traveled by the body in motion. He then drew a horizontal line to represent the motion of a second body moving at a constant speed equivalent to the average speed of the first body and showed how that line bisected the hypotenuse of the triangle. It was relatively simple to show that the area of the rectangle formed by the second line was equal to the area of the triangle formed by the slope. In other words, the accelerating body and the body moving at a constant speed covered the same distance in the same period of time. Oresme had proven the mean speed theorem by visualizing physical phenomena in geometrical terms.

Oresme's work, and that of his predecessors, had many important implications. His diagram was the first graphic representation of a mathematical function, and his calculation of velocity had the characteristics of a simple integral. He anticipated both the analytical geometry of Rene Descartes and the kinematics of Galileo Galilei. In fact, Galileo would draw on the mean speed theorem and the ideas of Oresme in the formulation of his laws of naturally accelerated motion in the late 16th century.

Oresme drew on a variety of different elements in the construction of his conceptions: Aristotelian physics, Euclidean geometry, the work of his predecessors at Merton College, and the medieval mathematics of proportions (in turn, based on Euclid). His insights reveal much about the nature of university mathematics in the 14th century. The scholars of England and France exhibited originality and insight as they pursued an understanding of the world based on Aristotelian thought, on the one hand, and Greek or Arab mathematics, on the other. Yet they would continue to refine their command of those Greek and Arab texts and would produce mature traditions of algebra and geometry by the 16th century.

MEDIEVAL ITALIAN COMMERCE

During the High Middle Ages, Italy as a region acquired a vibrant commercial life and a unique set of political institutions. Beginning in the 11th century, Italian merchants established themselves in the Mediterranean, and this increase in overseas trade fueled a corresponding increase in the economic activity of Italy itself. At the same time, self-ruling city-states began to crystallize in the northern part of the Italian peninsula: Genoa, Florence, Lucca, Pisa, Siena, and many more. These two developments supported each other: a growth in the power of the merchant classes fueled the evolution of the city-republics, and the governments of those republics adopted policies that fostered trade and business. This fertile mixture of commercial and political energy had profound cultural implications, for it laid the social foundations for the Italian Renaissance of the 14th and 15th centuries. It also created a unique atmosphere for the advancement of mathematics, one rooted in the practical requirements of commerce and banking.

Venice, Genoa, and Pisa had substantial trading connections in the Mediterranean by the end of the 11th century, and these continued to grow throughout the rest of the Middle Ages. Indeed, it is likely that the Bubonic Plague entered Europe through these trade routes, perhaps on a Genoese ship returning from the Black Sea in 1347. Such commercial links with regions controlled by the Arabs or the Byzantines provided ample opportunity for cultural exchange. Leonardo of Pisa (Fibonacci) acquired his mathematical education in Arab North Africa, and he initially applied that education to the management of commercial affairs. Indeed, it is likely that the adoption of Arabic numeration and algorism (calculating techniques) in Italy over the course of the 13th and 14th centuries was driven by the desire of the commercial community for more efficient means of keeping records and calculating prices and exchange rates.

As a rule, economic activity begets more economic activity, and the commerce of the maritime cities acted as an engine for other Italian economies. Trade within Italy, and with the rest of Europe, increased steadily until the disruption of the plague in the mid-14th century; it recovered again in the 15th. Building and manufacturing thrived as growing prosperity led to increased demand. Florence, for example, made its fortune in the production of woolen cloth. Italy had become the wealthiest region of Europe by the middle of the 13th century, and the scale of commercial activity is illustrated by the emergence of gold coinage in some cities. Florence first minted its florin in 1252; Venice created its ducat in 1284. Italian gold coins were used as a medium of exchange throughout the continent of Europe. As financial transactions grew more elaborate, some of the great merchant families began to act as bankers and provided financing and credit for enterprises across Italy. When the kingdoms of England and France went to war in 1337, King Edward III needed additional funds to pay for his armies, and he eventually borrowed them from two Florentine banking families, the Bardi and the Peruzzi. He defaulted on his loans in 1345, and both banks collapsed. These events indicate just how large the difference in wealth was between Italy and Northern Europe; two Florentine families could underwrite the wars of an English king.

The complexities of commerce, and especially of continent-wide banking, required sophisticated records, and the double entry system of accounting emerged in 14th-century Italy. Its earliest appearance was in Genoa in the 1340s, but it spread throughout Italy by the end of the 15th century.

Double entry accounting requires all transactions to be entered twice in a set of financial books, as a debit in one account and as a credit in another. For example, if a company purchased 500 florins worth of raw wool, the purchase would be recorded as a 500-florin credit in an assets account (the firm now has that much more wool as an asset) and as a 500-florin debit in a capital account (the firm has 500 florins less in cash). In any examination of the company's records, debit and credit accounts should balance. Such techniques were facilitated greatly by the use of Arabic numerals and computational methods, and they allowed Italian banks to handle massive amounts of wealth effectively. In 1494 Luca Pacioli included a description of the double entry system in his summary of mathematics, the *Summa de Arithmetica, Geometria, Proportioni et Proportionalita*. His work spread knowledge of Italian accounting throughout Europe and was thus one of the most influential books of mathematics written in the Middle Ages.

RAPHAEL'S *SCHOOL OF ATHENS* . . . IN PERSPECTIVE

Raphael's *School of Athens* confronts the viewer with a powerful image of Renaissance intellectual and artistic priorities (Figure 9.4). Painted between 1509 and 1511, the fresco is one of three in the Stanza della

Figure 9.4 Raphael painted the *School of Athens* in the early 16th century. The fresco illustrates Humanist reverence for classical learning and shows the techniques of perspective in painting. (Pytyczech/Dreamstime.com)

Numbers in Medieval Europe 215

Segnatura in the Vatican, each depicting a different intellectual discipline. The *School of Athens* was the artist's tribute to philosophy, and Raphael filled his painting with the great thinkers of Greek antiquity and at least one medieval figure. The work bursts with energy and embodies the Italian Humanist enthusiasm for classical culture. It also serves as an excellent example of the applied geometry of single-point perspective in painting, a 15th-century Italian technique for depicting three-dimensional space on a two-dimensional surface.

No definitive guide exists for the identity of the figures in the work, and scholars disagree to some extent. However, it is clear that Raphael included the philosophers and mathematicians whose works had transformed European thought in the Middle Ages and Renaissance. Plato and Aristotle dominate the center of the fresco, presented with equal emphasis, for the recovery of Plato's work during the 15th century challenged the supremacy of medieval Aristotelian thinking. Mathematical authors frame the foreground: Pythagoras on the left, writing in a book, and Euclid on the right, working with a compass. Behind Euclid stands Ptolemy, author of the *Almagest* and the *Geography*, crowned and holding a globe. The woman in white standing by Pythagoras is believed by many to be Hypatia of Alexandria, mathematician and philosopher of the 5th century. Taken as a whole, the fresco captures the intellectual optimism of the Late Italian Renaissance, the optimism of a Europe reunited with the classical past.

The visual structure of *The School of Athens* reflects the techniques of linear perspective developed by Italian painters during the 15th century, techniques that involved the application of optical theory and Euclidean geometry to the representation of physical space. The 14th-century painter Giotto had begun to explore the depiction of three-dimensional space; he departed from medieval painting styles and sought to portray his characters in a more visually authentic manner. In the next century, a series of painters and scholars approached the problem in mathematical terms. In 1407 Ptolemy's *Geography* was translated into Latin in Italy. Ptolemy had wrestled with the problem of projecting the curved surface of the globe onto the flat surface of a map; this resulted in the distortion of some features. This knowledge came to influence Florentine painters, probably through the agency of Paolo Toscanelli, a mathematician interested in geography and mapmaking. Toscanelli was a close acquaintance of Filippo Brunelleschi, a goldsmith, painter, and architect who would later make his reputation as the designer of the great dome of Florence's cathedral. According to sources, Brunelleschi used a visual grid to make a series of drawings of the Florentine Baptistery—a religious building in front of the cathedral—in which the parallel lines of the scene were drawn instead as diagonals, converging on a single point in the center. As in Ptolemaic geography, the

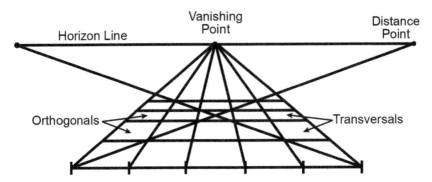

Figure 9.5 In the techniques of Renaissance perspective, orthogonal lines converge on a vanishing point, creating the impression of three-dimensional space.

act of projecting a three-dimensional view onto a two-dimensional surface distorted features of that view in a geometrically predictable way. These lines—orthogonals—defined the space in the drawing and allowed the artist to place objects in that space in a visually accurate way, decreasing in size as the space receded from the viewer (Figure 9.5).

Brunelleschi's drawings have not survived, but they influenced other Italian painters of his time, who copied and refined his techniques. In 1435 the Humanist author and architect Leon Battista Alberti, another acquaintance of Toscanelli, wrote *De Pictura* (*On Painting*), in which he discussed Brunelleschi's work and established formal mathematical principles for linear perspective. Alberti drew on Euclid's geometry in his book, as well as the optical ideas of Ibn al-Haytham (Alhazen). He also was influenced by the practical experiences of artists themselves. Later in the 15th century, the artist and mathematician Piero della Francesca wrote *De Prospectiva Pingendi* (*On Perspective in Painting*), in which he explored the depiction of solid geometrical figures in two-dimensional space. This work in turn inspired Luca Pacioli's *Divina Proportione* (*Divine Proportion*), a study of mathematical proportion in the arts, possibly illustrated by Leonardo da Vinci. The *Divina Proportione* was published in 1509, the same year Raphael began work on *The School of Athens*.

The treatment of physical space in Raphael's fresco reflects the developments in perspective over the previous eighty years of Italian Renaissance art. The enormous scale of the work enhances the effect; Raphael has turned an entire wall into a statement on the history of classical philosophy. A series of Roman arches recedes into the distance, creating an illusion of depth. The lines in the tile floor below and the straight borders of the arches above form a series of converging orthogonal lines that define the space for the painter and the viewer; these lines meet at a vanishing point between the figures of Plato and Aristotle. Other figures are distributed throughout the space, and their size decreases proportionally according to their distance from the viewer. Raphael has created a three-dimensional stage on which his characters may engage in the ongoing disputes of philosophy and mathematics. In a way, Rafael's *School of Athens* serves as an emblem of 15th-century European mathematics, both as an applied discipline and as a subject of academic study. The structure of Raphael's vision of ancient philosophy, and its content, reveal the place of mathematics and geometry in the mind of the Italian Renaissance.

CONCLUSION

Over the course of 1,000 years, Western Europe lost much of its classical intellectual heritage and recovered it again—through Spain and southern Italy during the 12th and 13th centuries and through the efforts of the Humanist scholars of the Italian Renaissance. The order in which books and treatises were recovered had a profound impact on European thought; Aristotelian ideas defined medieval European university culture because Aristotle's body of works, along with many related Arab texts, were translated earliest. Europeans encountered Greek geometry and Arab arithmetic and algebra in the 12th century as well, but these disciplines spread across the continent in a complex manner. The adoption of Indo-Arab numerals and Arab algorism took place relatively quickly, and these ideas had an important effect on European commercial life as well as academic culture. The more complicated aspects of geometry and algebra were disseminated more slowly, partly because Europeans had to develop a community of mathematical thinkers to digest them, and partly because of the dominance of Aristotelian thinking in the schools. By the end of the 15th century, Western Europe had evolved its own mathematical culture, one founded on Greek and Arab work, but with its own unique character. Energized by the confidence of the Italian Renaissance, European mathematicians were poised to break new ground.

PRIMARY TEXT: ROGER BACON, *ON THE IMPORTANCE OF STUDYING MATHEMATICS*

Roger Bacon worked in England in the 13th century. He was not a mathematician, but he wrote a work on the significance of mathematical study. His treatment of the subject reveals the great impact of the transmission of Greek and Arabic texts to Western Europe in the High Middle Ages. The passages included here illustrate the importance of Euclid and Aristotle to medieval European thought. Bacon refers to Aristotle's use of mathematical examples and points out that they cannot be properly understood without the study of Euclid's Elements. Mathematics is thus required for a proper grasp of Aristotle.

He also refers to a passage in the writings of Cicero (Tullius); Cicero is in fact repeating the ideas on geometry expressed by Plato in the Meno (see Chapter 3).

Chapter 3: In Which It Is Proved by Reason That Every Science Requires Mathematics

What has been shown as regards mathematics as a whole through authority, can now be shown likewise by reason. And I make this statement in the first place, because other sciences use mathematical examples, but examples are given to make clear the subjects treated by the sciences; wherefore ignorance of the examples involves an ignorance of the subjects for the understanding of which the examples are adduced. For since change in natural objects is not found without some augmentation and diminution nor do these latter take place without change; Aristotle was not able to make clear without complications the difference between augmentation and change by any natural example, because augmentation and diminution go together always with change in some way; wherefore he gave the mathematical example of the rectangle, which augmented by a gnomon increases in magnitude and is not altered in shape. This example cannot be understood without the twenty-second proposition of the sixth book of the Elements. For in that proposition of the sixth book it is proved that a smaller rectangle is similar in every particular to a larger one and therefore a smaller one is not altered in shape, although it becomes larger by the addition of the gnomon.

Secondly, because comprehension of mathematical truths is innate, as it were, in us. For a small boy, as Tullius in the first book of the Tusculan Disputations, when questioned by Socrates on geometrical truths, replied as though he had learned geometry ...

... Likewise, eighthly, because every doubt gives place to certainty and every error is cleared away by unshaken truth. But in mathematics we are able to arrive at the full truth without error, and at a certainty of all points involved without doubt; since in this subject demonstration by means of a proper and necessary cause can be given ...

Burke, Robert Belle, trans. The Opus Majus of Roger Bacon. Philadelphia: University of Pennsylvania Press, 1928, 117–127.

FURTHER READING

Bacon, Roger. 1974. *On the Importance of Studying Mathematics*. In *A Source Book in Medieval Science,* edited by Edward Grant. Cambridge: Harvard University Press.

Colish, Marcia. 1997. *Medieval Foundations of the Western Intellectual Tradition.* New Haven: Yale University Press, 1997.
Goldstein, Thomas. 1980. *Dawn of Modern Science.* New York: Houghton Mifflin.
Grant, Edward (ed). 1974. *A Source Book in Medieval Science.* Cambridge: Harvard University Press.
Grant, Edward. 1977. *Physical Science in the Middle Ages.* Cambridge: Cambridge University Press.
Isidore of Seville. 1974. *Etymologies.* In *A Source Book in Medieval Science,* edited by Edward Grant. Cambridge: Harvard University Press.
Leonardo of Pisa (Fibonacci). 1974. *Liber Quadratorum.* In *A Source Book in Medieval Science,* edited by Edward Grant. Cambridge: Harvard University Press.
Leonardo of Pisa (Fibonacci). 1974. *Practica Geometriae.* In *A Source Book in Medieval Science,* edited by Edward Grant. Cambridge: Harvard University Press.
Lindberg, David. 1978. *The Transmission of Greek and Arabic Learning to the West.* In *Science in the Middle Ages,* edited by David Lindberg. Chicago: University of Chicago Press.
Mahoney, Michael. 1978. *Mathematics.* In *Science in the Middle Ages,* edited by David Lindberg. Chicago: University of Chicago Press.
Martianus Capella. 1977. *The Marriage of Philology and Mercury.* In *Martianus Capella and the Seven Liberal Arts,* translated by William Harris Stahl and Richard Johnson. New York: Columbia University Press.
Merzbach, Uta and Boyer, Carl. 2011. *A History of Mathematics.* Hoboken: John Wiley and Sons.
Wagner, David (ed). 1983. *The Seven Liberal Arts in the Middle Ages.* Bloomington: Indiana University Press.

10

Numbers in Early Modern Europe

Chronology of Major Figures

Nicolaus Copernicus	1473–1543
Galileo Galilei	1564–1642
Johannes Kepler	1571–1630
Rene Descartes	1596–1650
Pierre Fermat	1607–1665
Blaise Pascal	1623–1662
Baruch Spinoza	1632–1677
Isaac Newton	1642–1727
Gottfried Wilhelm Leibniz	1646–1716

CULTURAL ICON: JOHANNES KEPLER'S LAWS OF PLANETARY MOTION

Historians speak of the "Scientific Revolution" of the 16th and 17th centuries, the time when modern scientific thinking emerged in Western Europe. The disciplines of astronomy and physics played a central role in the thought of the period, particularly in the work of Nicolas Copernicus, Galileo Galilei, and Isaac Newton. It is common to treat the major figures of the Scientific Revolution as if they were modern scientists—after all, their ideas formed the basis of the methodology that characterizes the modern understanding of nature. However, intellectuals in Early Modern Europe lived during a skeptical time when traditional views of learning were called into question, a time when there was not a single, dominant definition of knowledge. Different theorists had different approaches to nature and to truth itself. A single scholar might combine ideas from many different sources and adopt methods or assumptions that a modern mind would consider incompatible, even contradictory. The work of Johannes Kepler—mathematician, astronomer, and astrologer—serves to illustrate the complexities of exploring the natural world in the 16th and 17th centuries, an age of great intellectual energy and change.

Kepler's laws of planetary motion show up in the curriculum of any modern course in astronomy. As a young man, Kepler worked for the Danish astronomer Tycho Brahe, a brilliant observer who collected extensive data on planetary positions. Upon Tycho's death in 1601, Kepler acquired those observations. He also accepted the ideas of Copernicus, who had claimed in 1543 that the universe was heliocentric, that is, that the earth and the other planets traveled around the sun. Kepler spent years trying to derive sun-centered planetary paths from Tycho's data, with particular attention to the motion of the planet Mars. Ultimately, he abandoned the long-accepted notion that heavenly bodies traveled in perfect circles and proposed elliptical orbits instead. In doing so, he rejected the centuries-old tradition of circular astronomy. Kepler's first law, published in the *Astronomia Nova* of 1609, stated that the orbit of each planet is an ellipse with the sun at one of its two foci—the points within an ellipse akin to the center of a circle (Figure 10.1).

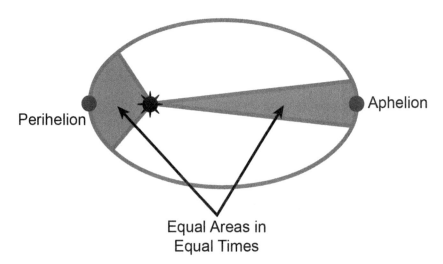

Figure 10.1 Kepler rejected centuries of circular geometry and based his laws of planetary motion on elliptical orbits.

His second law, discussed in the same work, illustrated the uniform character of planetary motion. It was clear to Kepler that planets did not move at a constant rate in their elliptical courses; they accelerated when close to the sun and slowed down when farther away. However, he showed that an imaginary line drawn from a planet to the sun would sweep out equal areas in equal times as the planet moved. Thus, the two shaded areas in Figure 10.1 are equal.

While Kepler embraced the Copernican notion of a central sun, he discarded the Polish astronomer's system of compound circular planetary paths, based on the ideas of the ancient astronomer Ptolemy. Kepler's planets move in simple ellipses. Even more impressively, Kepler's orbital predictions corresponded to observed planetary positions with very little error. Both Copernicus and Ptolemy predicted the motions of the planets with an average error of about 3 degrees of arc. In Kepler's system, that error dropped to almost nothing. Finally, Kepler's third law, published in the *Harmonia Mundi* of 1619, states that the ratio of the square of a planet's period of revolution around the sun (T^2) to the cube of its average distance from the sun (R^3) equals a number (K) that is the *same for every planet*:

$$\frac{T^2}{R^3} = K$$

Again, Kepler had discovered mathematical consistency in the universe; in this case, a constant number K that seemed to tie the motions of the heavens together.

To a modern reader, Kepler's discoveries represent precise natural laws, understood through the mathematical analysis of empirical data. They contribute to a familiar view of nature as mathematically predictable and transparent to the power of human reason. However, Kepler had a different view. He was a devout Lutheran Christian, and he also was influenced by the ideas of the ancient Pythagoreans and the Neoplatonists. For Kepler, the universe revealed the intelligent design of its creator, a design based on mathematical harmony in the Pythagorean tradition. Such harmony bridged the gap between the divine mind and the physical cosmos. The planets swept out equal areas in equal times, and mathematical analysis of their orbits yielded a common constant. These were expressions of this harmony, examples of the divinely conceived order of the universe—an order based on number.

Kepler's mathematical mysticism becomes clear when one looks at other aspects of his work. One of his earliest publications emphasized the supposed relationship between the arrangement of the planets and the five perfect solids. The perfect solids are three-dimensional figures bounded by identical regular polygons. For example, a cube has six identical faces, all

Figure 10.2 The perfect, or "Platonic," solids.

perfect squares, while a tetrahedron has four faces, all equilateral triangles. The ancient Greek mathematician Theaetetus had proven that only five such figures could exist: the tetrahedron, the cube, the octahedron (eight sides), the dodecahedron (twelve sides), and the icosahedron (twenty sides) (Figure 10.2).

In Kepler's sun-centered universe, there were six planets: Mercury, Venus, Earth (now a planet), Mars, Jupiter, and Saturn. Five gaps separated the six planets, and there were five perfect solids. For Kepler, this could not be a coincidence; it reflected divine mathematical harmony. He was convinced that the distances between the planets were based on the perfect solids, that they were spaced out as if the perfect solids filled the gaps. For example, the orbit of Jupiter might be inscribed in a theoretical cube and might circumscribe a tetrahedron that in turn would have the orbit of Mars inscribed in it (Figure 10.3).

Kepler was not saying that actual physical solids filled the universe; rather, he believed that the spatial arrangement of the planets proceeded from the mathematics of those perfect figures. While he did not succeed in linking this vision to observed data, he never abandoned the idea—even as his other, more familiar discoveries were published.

Johannes Kepler embodied the intellectual spirit of the 17th century. His skepticism led him to dismiss the circular astronomy of the Greeks and the Arabs, something Copernicus had not done. On the other hand, he accepted a fascinating variety of ideas; he combined Copernicus' heliocentrism, Pythagorean number mysticism, Tycho's data, Christianity, and his own mathematical insights to transform the astronomy of his day. Some of his ideas are familiar to modern readers: the planets moving around the sun in predictable elliptical orbits. Others seem alien and superstitious: the universe as an expression of the geometry of the perfect solids. As the court astronomer of the Holy Roman Emperor, he conducted research into planetary motion and cast astrological horoscopes for his royal patron—and sincerely believed in the value of both enterprises. Yet all of his intellectual endeavors share a common thread—the idea that nature was bound together by number and that humanity could grasp its secrets through mathematics.

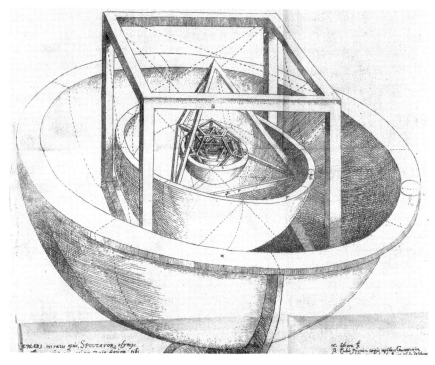

Figure 10.3 Kepler is known for his laws of planetary motion. He also suggested that the spacing of the planets was based on the geometry of the five perfect solids, a view grounded in a form of mathematical mysticism. (New York Public Library)

CULTURAL COMMENTARY: METHOD, SCIENCE, AND NATURAL LAW

In the film *Apollo 13*, the actor Tom Hanks plays James Lovell, the commander of the third U.S. mission to the surface of the moon in 1970. Two days after liftoff, an explosion disables the spacecraft when it is over 200,000 miles from the earth. In order to save power, the astronauts need to turn off most of the equipment in their ship; they coast around the moon and use its gravity to place them on a trajectory back to earth. As they shut down the last system, Hanks turns to his comrades and says, "We just put Sir Isaac Newton in the driver's seat." They are in serious danger; they need to travel a quarter of a million miles back to earth in a damaged craft the size of a large minivan; they need to address multiple technical crises and material shortages. But with all of their concerns, they do not doubt the predictive power of Newton's laws of motion and universal gravitation, and ultimately, they ride these equations home.

In a modern chemistry lab exercise, a student can accurately predict the amount of base necessary to neutralize a given quantity of acid. A flight engineer can predict the path of a missile or a satellite based on its mass and the forces that act upon it. A climate scientist can model the future behavior of the earth's atmosphere. Twenty-first-century human beings live in a world of precision: we measure, we analyze, we predict. We have described nature in terms of mathematical laws, and we have great confidence in the efficacy of these laws. This quantitative, predictive view of the natural world is central to modern scientific thinking. While it has complex origins, some dating back to antiquity, it crystallized in Western Europe during the 16th and 17th centuries, a period often referred to as the Scientific Revolution.

In the early decades of the 1500s, Europe stood at an intellectual crossroads. In the universities, the medieval combination of Aristotelian philosophy and Christianity still dominated, and Aristotle's logic and cosmology defined the natural world. However, outside the universities, other systems of thought were challenging this supremacy. During the late Renaissance, the works of many ancient authors had been reintroduced to Western Europe, and these works provided alternatives to Aristotelian concepts. The re-emergence of books by Plato, the Pythagoreans, Archimedes, Lucretius, and many other writers provided readers with a broad menu of "new" ideas. European thinkers responded by questioning the thought of Aristotle and replacing his methods and views of nature with those of other ancient figures. Such skepticism required real intellectual confidence, as European thinkers analyzed ancient authors critically and embraced or discarded ideas based on their own judgement. Renaissance Humanism served as a source of this confidence, for Humanists stressed their own human potential and sought to stride with the ancients, and perhaps, to surpass them.

While these newly acquired books of the late Renaissance inspired a growing intellectual skepticism, other factors acted to enhance that skepticism in the 16th century. The spread of printing technology in Europe vastly increased the availability of books in a way that encouraged criticism and questioning. As books became more common and less expensive, they lost some of their mystique and ceased to be viewed as objects of overwhelming authority. The voyages of European mariners, from Amerigo Vespucci to Ferdinand Magellan, changed the perception of the globe and invalidated the ideas of ancient and medieval geographers. The work of Protestant reformers, most notably Martin Luther and John Calvin, rejected the authority of the Roman Catholic Church, raised literacy rates, and placed even more emphasis on the judgement of the individual.

It is important to point out that many intellectuals of the 16th century continued to embrace the logical thought of Aristotle, especially in the universities. Yet skepticism continued to grow. As they confronted the crucial questions of the day, scholars mixed ideas and methods from a wide range of sources, ancient and contemporary. Because printing made so many new books available, theorists were able to choose from a considerable range of ideas and to combine them in new ways. The process was highly individual: one writer might adopt Plato's perspectives, while another might choose the views of Archimedes. Mystical and magical traditions played an important part as well. New concepts and methods could mix with those from antiquity. The thought of the Greeks and Romans was still revered by Europeans, but no longer in a manner that demanded complete acceptance of any one author.

In short, during the 1500s and early 1600s, the European medieval mindset was progressively called into question and largely dismissed. As the 17th century proceeded, scholars challenged the authority of the ancients even more substantially; they recognized that if the philosophers of antiquity had disagreed so widely with each other, perhaps they were all flawed to some extent. The notion of starting over, of breaking new intellectual ground, began to take root among European intellectuals. This was especially true with respect to the explanation of natural phenomena. Yet, while Aristotelian views of nature had declined in importance, there was no dominant set of ideas to replace them. Rather, different authors proposed different strategies for explaining the physical world, combining ancient ideas with new ones. What was the best method for gathering knowledge of nature? How did the universe work? What was the character of matter itself? These questions fascinated the authors of the day. Some stressed observation and experiment as a new basis of understanding; others emphasized the purity of mathematical thought; still others searched for the hidden secrets of a natural world that was essentially magical. This time of great conceptual energy—or conceptual chaos—would last for more than a century. While historians refer to it as the Scientific Revolution, it really involved the interaction of a wide variety of ideas—some important to later scientific thinking, some not. Ultimately, the experimental methods and ideas of Newtonian science would grow out of this intellectual dialogue and become the basis of a new European intellectual consensus.

Mathematics played a big part in this period of conceptual transformation; many thinkers incorporated elements of mathematical thinking in their approaches to nature. The notion of a mathematical natural world held great appeal. They began with the work of ancient mathematicians and built upon that work in new ways. European mathematical expertise grew tremendously from 1500 to 1700, and many modern disciplines—analytical geometry, probability theory, infinitesimal analysis, calculus—would emerge during this period. These new techniques would serve as powerful tools in the exploration of natural phenomena.

Nicolaus Copernicus (1473–1543) adopted many of the ideas of the Hellenistic astronomer Ptolemy; he sought to explain the observed positions and motions of the sun, moon, and planets in terms of complex circular geometry. Yet Copernicus shunned the idea of an immobile earth and combined Ptolemy's mathematical techniques with the notion of the sun as the center of the cosmos. Johannes Kepler (1570–1630) embraced Copernicus's heliocentrism but rejected his mathematics. He took advantage of new data and new analytical techniques to produce elliptical paths for the planets, paths that yielded a much higher degree of predictive accuracy than those of Ptolemy or Copernicus. Kepler was also inspired by the ideas of the Pythagoreans, and he saw his elliptical astronomy as a form of the mathematical harmony inherent in the universe.

Copernicus and Kepler both applied mathematics to the motions of the heavens, an intellectual enterprise with a very long history. Galileo Galilei (1564–1642) sought to describe *earthly* motion in a precise mathematical way. Influenced by the work of Archimedes and by the medieval university tradition, he measured the speed of falling bodies and discovered that they all accelerated at a constant rate. He went on to find other mathematical patterns in the behavior of moving bodies. In the process, he contributed to the development of experimental method itself, especially with regard to the collection and analysis of numerical data. He also reflected on the idea of mathematical natural law.

The gathering of data requires mathematically precise observation, that is, measurement. Whether one measures the movement of the planets or the motion of a ball down a ramp, the act depends on the accuracy of one's senses. Rene Descartes (1596–1650) emphasized that human senses could be deceived, and he searched for a different intellectual path towards knowledge of the physical world that would yield complete certainty. Enthralled by the fact that mathematical thought seemed to be beyond doubt, he sought to find an equally certain method for describing nature. Where Kepler and Galileo used math as an analytical tool to discover patterns in the observed behavior of the universe, Descartes used math as a role model, an example of the human ability to know the truth.

Descartes's grand intellectual ambition led to many spirited responses from other mathematicians and philosophers—one may argue that those 17th-century thinkers who followed Descartes had no choice but to react to Descartes. The Jewish philosopher Baruch Spinoza felt that Descartes had not gone far enough in the pursuit of certainty. He constructed a philosophical system, based on Euclid's *Elements*, where every aspect of existence, physical and mental, was proven in geometrically rigorous terms. At the same time, Blaise Pascal took the opposite approach: he stressed the limits of human intellectual capacity and turned away from Descartes's quest for certainty. A gifted mathematician and experimentalist, Pascal was struck by the infinity of the universe and of the divine mind. For him, human knowledge was real but bounded; human powers of observation and reason could lead to real discovery but not absolute understanding. Even the supposed certainties of mathematics rested upon assumptions that seemed to be self-evident but really were not. His intellectual humility was a fascinating product of a skeptical age.

Kepler, Galileo, Descartes, Spinoza, Pascal—it is crucial to point out that as these figures confronted each other over questions of knowledge and nature, they all incorporated elements of mathematics into their work. While some emphasized observation, experiment, and the analysis of measurements, others sought to apply the thought patterns of mathematics itself to the investigation of the physical world—thought patterns that would lead to certainty. While they disagreed on the way in which mathematics should be employed in the exploration of nature, they all recognized the great power of mathematical thinking.

The emergence of a mathematical view of the natural world was a crucial cultural development in early modern Europe. However, the growing intellectual power of European mathematics affected other aspects of culture as well. For example, in the 16th century, Gerardus Mercator drew on the mathematics of perspective and developed a new technique for projecting the surface of a globe onto a flat surface. Mercator's maps accurately represented true bearing or direction and continue to be used in navigation to this day. Applied mathematics influenced the field of architecture, as demonstrated by the 17th-century designs of Christopher Wren—himself a mathematics professor at Oxford. Perhaps most significantly, new methods of computation allowed for the growth of new commercial institutions and business practices. Modern capitalist enterprises appeared in the Dutch Netherlands in the 17th century and spread throughout Europe, vastly expanding the scale of trade and increasing the wealth and power of Western society.

Dozens—even hundreds—of thinkers confronted the issues of knowledge, nature, and method in the 17th century, but one figure in particular stands out. Isaac Newton rejected Descartes's method and his physics and replaced them with a model of nature founded upon experiment and mathematical natural law. Newton explicitly stated that experimentally derived conclusions must be changed in the face of new evidence. His laws of motion and of universal gravitation were unprecedented in their ability to explain and predict the behavior of the physical universe. The great success of Newton's thought resulted in a new intellectual orthodoxy; Newtonian methods and ideas became the model for Western scientific thinking. The age of skepticism had resulted in a view of nature that was perpetually fluid: improved models of the universe could always arise, based on new discoveries. Yet it was also mathematically precise, based on predictive laws that accounted for available evidence—laws that one day would bring the astronauts of *Apollo 13* home again.

PURE MATHEMATICS IN EARLY MODERN EUROPE

The skeptical character of early modern thought actually had little impact on the study of pure mathematics in Europe. Over the period from 1500 to 1700 the capabilities of European mathematicians grew steadily. The knowledge of the Late Middle Ages, based largely upon Greek geometry and Arab algebra, served as the starting point for this process. In the late 15th century, the German mathematician Regiomontanus labored to produce printed translations of mathematical classics; he also authored new treatises on trigonometry and algebra. After 1500 a series of German and Italian algebraists worked to master Arab knowledge and to move in new directions; they experimented with different kinds of notation and problem-solving techniques and wrestled with the complex issues of negative numbers, exponents, and irrational numbers. The Italian mathematicians Geronimo Cardano, Niccolo Tartaglia, and Luigi Ferrari discovered new approaches to cubic and quartic equations, that is, equations with variables raised to the third or fourth power. Printed editions of Hellenistic Greek geometers, particularly Pappus and Apollonius of Perga, became available at this time as well. In France, Francois Viete (1540–1603) continued the development of algebra and introduced elements of trigonometry still in use today. He and his contemporary Simon Stevin also adopted the use of decimal fractions, and the use of decimal point notation became common in the 17th century. John Napier pioneered the use of logarithms as a tool in calculation, and Henry Briggs of Oxford University added to his work by publishing tables of logarithms based on powers of 10 in 1617 and 1624. Logarithms were the basis of the slide rules used as calculating devices until the advent of modern computers in the second half of the 20th century.

In the mid-17th century, a series of French mathematicians laid the groundwork for the modern understanding of analytic geometry, probability theory, and calculus. While the community of thinkers involved was quite large, some figures were particularly notable. In *La geometrie* (*The Geometry*—published 1637), Rene Descartes combined algebra and geometry in exciting new ways, expressing algebraic statements as geometric figures and using geometric techniques to solve algebraic problems. His ideas led directly to the practice of representing an equation as a set (or locus) of points plotted on a graph, the foundation of analytic geometry. Because of the great significance of his work, the familiar rectangular coordinate system based on perpendicular X and Y axes is referred to as the Cartesian coordinate plane (Figure 10.4).

Descartes's contemporary Pierre Fermat developed similar techniques; in many ways he could be credited with the codiscovery of analytic geometry. In addition, his studies of the equations of curves brought him very close to a formulation of differential calculus, and he did important work in number theory. Blaise Pascal wrote a brief treatment of conics as a very young man, influenced by ancient works on the subject readily available in printed editions. He went on to pursue the topic in much greater detail as an adult (Merzbach and Boyer 2011, 333–334). Pascal also did pioneering work in probability theory. In 1655 he published his *Treatise on the Arithmetical Triangle*, an analysis of the triangular array of numbers studied earlier by Omar Khayyam in Persia, as well as by various Indian and Chinese scholars. Pascal pointed out new ways to apply the triangle to problems of probability; the "arithmetical triangle" is often called "Pascal's triangle" in his honor.

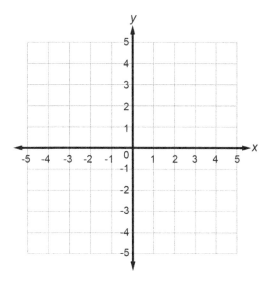

Figure 10.4 Descartes's combination of algebra and geometry lies at the basis of the Cartesian coordinate plane.

Infinitesimal analysis may be defined as the attempt to discover the size of an unknown quantity or area by dividing it into ever-smaller known quantities or areas and adding those together. For example, one might approximate the area between a given curve and a straight line (perhaps the X axis) by dividing that area into a series of rectangles: the more rectangles used, the more accurate the approximation (Figure 10.5).

An infinite number of rectangles would produce the precise area; the question is, how does one produce a mathematical statement that would treat an infinite number of rectangles? The roots of such thinking may be found in the "method of exhaustion" first employed by Eudoxus in the 5th century BCE and later by Archimedes. A number of European mathematicians explored infinitesimal analysis in the 16th and 17th centuries. Johannes Kepler used a version of it in his work on elliptical orbits. Pierre Fermat devised a method to find the areas under certain kinds of curves, and Bonaventura Cavalieri devoted a book to the subject in 1635, his *Geometria Indivisibilibus Continuorum*. The study of infinitesimal analysis was crucial, for it led European mathematicians towards the discovery of calculus.

While some precursors of calculus had been explored by Indian and Middle Eastern mathematicians in the Middle Ages, a truly unified view of

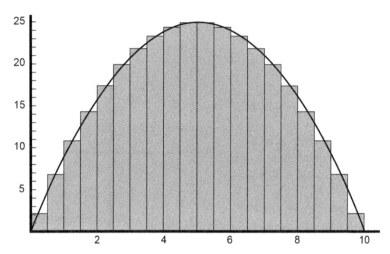

Figure 10.5 The division of the area under a curve into a series of ever-smaller rectangles is an example of infinitesimal analysis.

calculus was discovered independently by two European authors in the second half of the 17th century: Isaac Newton and Gottfried Wilhelm Leibniz. Calculus is the study of change in mathematical statements or functions. Differential calculus treats the rate at which a function changes, while integral calculus addresses the results of change. Newton discovered a version of differential calculus in 1666 when he was only twenty-three years old as an extension of his work on infinite series. He continued to develop what he referred to as his "method of fluxions" and added the techniques of integral calculus but he did not publish any discussion of the subject until 1693. He did employ some of his methods in his treatment of physics, the *Principia*, published in 1687.

Leibniz apparently developed the techniques of integral calculus in 1675, influenced by the work of Pascal on conics (Merzbach and Boyer 2011, 384). He moved on to differential calculus and published his studies in 1684. The terminology and notation used in modern calculus are those of Leibniz. Newton and Leibniz apparently knew something of each other's work, but a bitter dispute broke out between the two men in 1711 over credit for the discovery. The Royal Society of London, a body dedicated to the study of natural philosophy, ruled in Newton's favor in 1713, but Newton's position as president of the Royal Society at the time certainly influenced the decision. Modern historians accept the likelihood that both men made their discoveries independently. Despite the drama surrounding its origins, one might argue that the discovery of calculus was the inevitable result of the developments in analytical geometry and infinitesimal analysis over the previous half-century. The procedures of calculus would inspire great mathematical creativity in succeeding centuries; they would also provide a powerful tool for the mathematical description of natural processes.

NICOLAUS COPERNICUS

Nicolaus Copernicus was born in 1473; he was nineteen when Columbus first sailed across the Atlantic and forty-four when Martin Luther began the Protestant Reformation. Born in an age of skepticism, he would embrace it and challenge the 2,000-year-old Greek tradition of earth-centered astronomy. His heliocentric (sun-centered) view of the universe, published in 1543, influenced many of the later thinkers of the Scientific Revolution. Yet while Copernicus rejected important elements of ancient astronomy, his work incorporated much of the complex circular geometry employed by the late Hellenistic astronomer Ptolemy.

Medieval Europe had access to two dominant models of the universe: Aristotelian and Ptolemaic. While both featured a central earth, they differed in other respects. Aristotle's system was based on a series of concentric material spheres that carried the sun, moon, and planets in their various motions, while Ptolemy argued that the planets traveled in compound circular paths around the earth. Medieval scholars, for the most part, accepted Aristotle's view as representative of nature; they saw Ptolemy's system as a useful mathematical approach to predicting the positions of the planets. Some historians argue that Ptolemaic astronomy was incompletely understood for much of the European Middle Ages. The text was first translated by Gerard of Cremona in the 12th century; a subsequent version was produced by William of Moerbeke. However, at that time, Europeans had only recently recovered the text of Euclid's *Elements* in Latin translation, and it took them several generations to digest that material and develop a sophisticated mathematical culture. Real command of Ptolemy's astronomy and the associated geometry was probably not possible until the Late Middle Ages.

Ptolemy had accepted the ancient Greek assertion that all heavenly motion was circular in nature; he also assumed a central earth. As a mathematical astronomer, he sought to explain and predict the observed positions of the sun, moon, and planets in precise terms, using complex circular geometry (see the entry on Ptolemy in Chapter 4). Briefly, he employed three main techniques. An *epicycle* was a compound circular path followed by a planet in its course around the earth; the planet moved in a circle around a point that in turn moved in a circle (or deferent) around the earth (Figure 10.6).

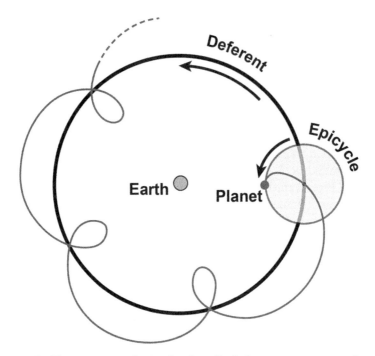

Figure 10.6 The geometry of epicycles described planetary movements in terms of two different circular motions.

An *eccentric* was a circular path centered on a point other than the earth; the earth would lie within the circle, but not at its exact center. Finally, an equant was a point in space that served as a center of constant motion. Ptolemy's system was based on his observations of planetary positions, and his elaborate mathematics could account for those positions with an average error of about 3 degrees of arc. His work, the *Almagest*, reflected the sophistication of Hellenistic Greek geometry, and any substantial critique of its contents would require similar sophistication.

Such a critique took place in 13th-century Persia, as astronomers of the Marāgha observatory pointed out difficulties in the system and proposed geometrical solutions (see the entry on the Marāgha observatory in Chapter 8). As European mathematics continued to develop in the 15th century, the German mathematician Regiomontanus produced a critical commentary on Ptolemy's *Almagest* as well. Copernicus drew on these predecessors in his analysis of planetary motion, but his work also proceeded from his own substantial mathematical expertise. Essentially, he produced a systematic review of Ptolemy's applied geometry from a heliocentric point of view. His books, the *Commentariolis* and the *De Revolutionibus*

(*On the Revolutions*) also demonstrated the skeptical nature of European thought in the 16th century: by placing the sun in the center of the universe, Copernicus was discarding the geocentric character of nearly all ancient Greek astronomical systems.

While Copernicus's heliocentrism was revolutionary, he retained many of the other assumptions essential to ancient astronomy. He accepted the notion that celestial motions were circular, and he continued to use epicycles and eccentrics to account for observed planetary positions (he did largely discard the concept of the equant). Modern readers are often surprised to find that Copernicus's system did not feature planets moving in simple circular paths around the sun. His detailed reworking of Ptolemaic geometry to accommodate a central sun revealed his mathematical expertise and his deep understanding of Ptolemy's text. Like the ancients, he also believed that the universe was finite, although he expanded its dimensions considerably for mathematical reasons. In a heliocentric system, the earth moves with respect to the stars as it revolves around the sun, and this change in position should produce some change in the apparent orientation of the stars as viewed from earth. No such change, or parallax, is detectable with the naked eye. Copernicus argued that the stars were so far away that the motion of the earth around the sun was negligible in comparison—a view that is still accepted today.

Copernicus's reasons for adopting a heliocentric approach to the universe are unclear. He did not have any new evidence; indeed, much of his work was based on the planetary observations found in Ptolemy's book. His sun-centered astronomy did not yield a significant increase in accuracy either, for like Ptolemy, his predictions of planetary positions had an average margin of error of about 3 degrees. Copernicus may have been attracted to the greater simplicity of celestial motion inherent in heliocentrism. In a geocentric system, the earth is stationary in the center of the cosmos, and all heavenly bodies must revolve around it once per day (in addition to their other motions). Copernicus's moving earth rotated on its axis every twenty-four hours, explaining the rising and setting of the sun, moon, and stars and vastly reducing the amount of motion in the universe. His choice to move the sun to a central position may also have been inspired by Renaissance natural magic—ancient mystical texts recovered by Renaissance Humanists refer to the sun and its light as sources of divine influence in the cosmos. The Italian philosopher Marsilio Ficino made similar claims in his treatise *On the Sun*, written in 1492.

Whatever his motivations, Copernicus contributed substantially to the transformation of European astronomy in the Early Modern period. While his book did not initially cause a stir when it was published in 1543, it had a major effect on later authors, particularly Johannes Kepler and Galileo Galilei. Copernicus's *De Revolutionibus* stands as testimony to the mathematical abilities of its author and to the skeptical intellectual atmosphere of Europe in the 16th century.

GALILEO GALILEI

"And yet it moves." The words of Galileo ring out over the centuries, a rallying cry for the freedom of thought and the intellectual independence of scientific inquiry. The story is well-known. Condemned by the Inquisition of the Roman Catholic Church for asserting the motion of the earth around the sun, Galileo muttered his defiance even as he was forced to formally renounce his written ideas.

Like many great stories, the tale combines truth and falsehood. Galileo was as much a provocateur as a martyr; he could not, in fact, prove the earth's motion around the sun; and he had many intellectual allies among the Church hierarchy. In many ways, the events of Galileo's trial obscure the more important details of his life and work. While his stature as a martyr for modern scientific thinking is indeed debatable, his role in establishing the conceptual basis for such thinking is not. Galileo played a crucial role in the development of experimental methodology, and he broke new ground by describing the behavior of moving bodies in exact mathematical terms. His analysis of terrestrial motion, that is, the motion of objects on earth, contributed greatly to the culture of precision that continues to characterize our understanding of the natural world. Further, both his heliocentric astronomy and his mathematical approach to motion reflect his skepticism toward traditional intellectual authority. Galileo confidently advanced his own views of physical nature and of the pursuit of knowledge itself.

Galileo was born in Florence in the year 1564. A bright young man, his family sent him to the University of Pisa with the hope that he would study medicine. However, young Galileo did not care for the standard Aristotelian university curriculum of his day, and he left the school after two years. He pursued the study of mathematics and returned to the University of Pisa as a mathematics professor in 1589. After three years, he moved to the University of Padua in the Republic of Venice, and there he began the work on terrestrial motion that would interest him for the rest of his life.

Aristotle had discussed two kinds of motion: natural motion that occurred on its own and violent motion that was imparted to an object. The best example of natural motion was the falling or rising of bodies—a stone would fall to the ground when dropped, and an air bubble would rise to the surface of water when released. Aristotle used the idea of "natural place" to explain these phenomena. All earthly objects were composed of a mixture of four elements—earth, water, air, and fire. One of these elements predominated in any given object, and the object would move on its own to the natural place of that element in the cosmos. For example, a stone was composed mainly of the element earth, and it would drop to the ground when released. Similarly, a bubble of air released in a tank of water would move upwards to the air. Aristotle based his arguments on observation, logic, and the assumption of a consistent universe; he watched solid bodies fall and maintained that all such "earthy" bodies would do so. His view of natural motion was predictive, but not quantitative—he did not describe the motion in terms of number or quantity. Rather, he confidently predicted that certain kinds of natural motion would always take place.

Galileo challenged both the views and the methods of Aristotle. As a skeptic in a skeptical age, he rejected the opinions of written authorities and chose to confront nature directly through experiment. Years later, he described his view in *The Assayer* (1623):

> ...I do not wish to be counted as an ignoramus and an ingrate toward Nature and toward God; for if they have given me my senses and my reason, why should I defer such great gifts to the errors of some man?
>
> (Galileo 1954, 272)

Galileo went beyond Aristotelian observation and logic; he designed structured experiments to examine specific phenomena. As an early advocate of such methods, he used them in a variety of ways. Sometimes he employed experiment to confirm his existing ideas, while on other occasions he began with experimental evidence and moved towards a more general understanding. Galileo often used thought experiments, theoretical illustrations of his principles. He also recognized the limits of his methods and wrote that human experiments and observations could never yield conclusions that were absolutely true, for nature was vast, and new information might always alter existing theory. Thus, in a skeptical age, he endorsed a fluid intellectual approach that could produce ever-changing views of the world. Crucially, much of his work had a quantitative character; his observations often took the form of measurements. He assumed that physical phenomena would exhibit some mathematical order, and he directed his studies to find that order. Galileo's view on mathematics in nature may have been influenced by the engineering work of Archimedes, but he directed his attention towards the analysis of motion.

To investigate the question of naturally accelerated motion, Galileo constructed a ramp marked with regular linear distances. He recognized that any mathematical study of motion required a method to measure elapsed time and designed a means to do so using a constantly flowing stream of water into a collecting vessel. He then rolled balls down the ramp, measuring their velocity as they moved from point to point. He discovered that the balls accelerated at a uniform rate (9.8 meters per second, per second, in modern units) and argued that the same rate would apply to all falling bodies. Galileo certainly benefitted from the work of Nicole Oresme and other late medieval natural philosophers on motion. However, his data and his analysis allowed him to predict the position and speed of a ball at each moment of its descent, a degree of precision unprecedented in the study of moving bodies.

During his eighteen years at the University of Padua, Galileo explored a number of other problems in the physics of motion. Most significantly, he argued that the path of any moving projectile (a thrown ball, an arrow, or a bullet) took the form of a parabolic curve and that the horizontal and vertical components of that curve were independent of each other. In other words, the fact that a body was moving horizontally did not affect the fact that it was also rising or falling vertically. His mathematical studies of moving bodies proved to be the most significant aspect of his thought, but not the most famous (Figure 10.7).

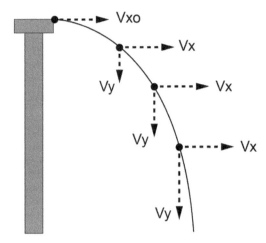

Figure 10.7 Galileo demonstrated that the horizontal and vertical aspects of projectile motion produced a parabolic curve.

In 1609, Galileo constructed a telescope of his own design, and his observations of the heavens made him the most celebrated mathematician and astronomer of his day. For the next twenty-four years, he advocated for the Copernican system. He argued with Aristotelian university professors, lobbied Church officials, and drew on his patrons in the Medici family for their support. His book on the subject ultimately drew the attention of the Roman Inquisition, and his famous trial took place in 1633. Yet throughout these decades, he never forgot his mathematical investigations of physical motion and referred to them frequently in the process of advancing his heliocentric astronomy. After his trial, he was imprisoned in his own house for the last eight years of his life, and he spent his time organizing his research on moving bodies into his last great book, *The Discourses and Mathematical Demonstrations Relating to Two New Sciences*. Published in 1638, the *Two New Sciences* made his conclusions and his methods available to the rest of Europe, and they would have a profound impact on the thought of Isaac Newton later in the 17th century.

Galileo's *The Assayer*, written in 1623, was a response to a rival astronomer's views on the nature of comets. Yet he used the treatise to express his ideas on other topics as well. At the end of the essay, he discussed the nature of physical matter and argued that the only real characteristics of matter were size, shape, position, and motion. In other words, he believed that matter was made up of physical particles and that the sensations of color, odor, and taste were merely human interpretations of the size, shape, and motion of such particles. It is important to note that size, shape, position, and motion can all be measured; they are quantifiable. For Galileo then, matter itself could only really be understood through mathematics. As he wrote in the Assayer:

> Philosophy is written in this grand book, the universe, which stands continually open to our gaze. But the book cannot be understood unless one first learns to comprehend the language and read the letters in which it is composed. It is written in the language of mathematics, and its characters are triangles, circles, and other geometric figures without which it is humanly impossible to understand a single word of it . . .
> (Galileo 1954, 237–238)

RENE DESCARTES

The thought of Rene Descartes (1596–1650) had a profound effect on European mathematics, philosophy, and views of nature. Descartes represented the height of early modern European skepticism. He dismissed medieval approaches to the definition of truth and the nature of matter, he challenged the intellectual authority of classical authors, and he even doubted the reliability of sense information itself. Yet he also had great intellectual confidence. To replace the ideas he discarded, he sought a new philosophy of nature, a philosophy as certain as the proofs of Euclidean geometry. Indeed, geometry and mathematics were his points of departure. Descartes's mathematical abilities were impressive; he published the first real treatment of the discipline of analytical geometry in 1637. This mathematical prowess shaped the rest of his thought, for he believed that because his mathematical knowledge was certain and self-evident, he could formulate a method to produce knowledge of the natural world that was equally certain and self-evident. Mathematical thought, then, was his intellectual role model. Descartes's method of inquiry, and the views of nature that he derived from it, would have a dramatic effect on European philosophy and science in the second half of the 17th century.

At a time when intellectual authorities of all kinds were coming under skeptical attack, Descartes wished to start over, to seek out a strategy for absolute certainty. Interestingly, he employed skepticism and doubt as tools in his quest for certain knowledge. In 1628 he wrote the *Regulae*, a set of rules for intellectual inquiry, and he refined his ideas in the *Discourse on Method*, published in 1637. Here, Descartes presented a method based on rigorous doubt as a path to certainty. Confident in the ability of the human mind to recognize pure truth, he argued that in pursuing a particular question, one should dismiss any idea that was uncertain. Once all uncertainties were discarded, one would be left with basic intuitive truths—what Descartes called "clear and distinct ideas." One could then build on these truths to produce more complex understanding. His method was based on his view of Euclidean geometry, and his "clear and distinct ideas" were akin to the axioms used in a geometrical proof. In such a proof, one starts with simple, self-evident ideas, axioms, or definitions, like "Parallel lines never meet." These are used to prove more complex concepts in a step-by-step manner. Descartes believed that a similar intellectual process could yield certain conclusions about subjects other than geometry.

Descartes's understanding of physical matter provides a good example of the application of his method to questions about the natural world. He rejected ancient discussions of matter as uncertain—especially the ideas of Aristotle. He also dismissed human senses as a source of knowledge about matter, for he argued that sense impressions could be uncertain as well. For him, the only thing about matter that could not be doubted was the fact that it occupied physical space, that it was extended in three dimensions. For him, this was a "clear and distinct idea." It could not be doubted that matter was solid and took up space, just as in geometry it could not be doubted that parallel lines never met. He thus sought to explain natural phenomena only in terms of matter taking up space. Building on this simple first principle, he maintained that physical nature must be based on the motion and collisions of material particles—which took up space. These particles of matter also obeyed laws that he felt were equally certain and could be derived from his first principle:

1. Unless collision with another part occurs, each part of matter retains the size, shape, motion, or rest it originally has.
2. One part of matter can only gain as much motion through collision as is lost by the part colliding with it.
3. The motion of any moving body tends to be rectilinear, even if in fact it is circular or curved through collision.

It is crucial to point out that Descartes's model of particulate matter was not based on observed physical evidence. Rather, he used his method of doubt to arrive at these principles—an exercise in abstract thinking akin to mathematical proof. Beginning with the ordinary experience of material objects, he progressively eliminated all the factors he considered uncertain—color, taste, and smell, for example—and arrived at the notion that the essential nature of matter consisted of its extension in space. Once Descartes established this "certain" starting point, he concluded that natural phenomena could only result from the effect of one extended particle upon another, so he postulated a universe full of colliding particles of matter. Finally, he went on to deduce the rules of such motion.

Descartes's view of matter clearly related to his understanding of geometry; he conceived of matter as filling geometrical space. However, while he was a mathematician, his physics did not include the kind of mathematical natural laws emphasized by Galileo (and, ultimately, by Newton). The formulation of such laws required data based on observation, and Descartes considered sense observations to be uncertain. His work illustrates the broad variety of 17th-century approaches to knowledge and truth. Other theorists, including Galileo, also suggested a particulate view of matter as a way to explain observed evidence. Descartes's notion of matter and motion, on the other hand, was grounded firmly in a rigorous process of pure thought, modeled on mathematical reasoning. For him, the particulate character of matter was as self-evident as the definition of a circle.

Published with the *Discourse on Method*, Descartes's *Geometry* represented the author's main contribution to the study of pure mathematics. Here, he combined algebra with geometry to an unprecedented degree and showed how algebraic statements could be expressed as geometrical figures. The work revealed Descartes's mathematical approach towards space itself, a view also reflected in his definition of matter as extension in three dimensions. In responding to the ideas of the Hellenistic mathematician Pappus, Descartes laid the groundwork for describing mathematical functions in terms of spatial coordinates—points on an X axis and a Y axis, for example. His concepts served as the foundation for modern analytic geometry, and the "Cartesian coordinate plane" is named for him.

Descartes's method and his conception of nature attracted considerable criticism, and he responded in the *Meditations* (1641). Here, he defended his view that the human mind had the capacity for certain knowledge. He began by proving the existence of his own mind ("I think therefore I am"). His most famous philosophical arguments aimed to justify the existence of "clear and distinct ideas" in the mind, the kind of ideas that formed the basis of his physics and his mathematics. In the end, Rene Descartes wanted to show why mathematical thinking was fundamental to human thought.

The cultural significance of Descartes lay in his use of mathematics as the model for a view of truth based on pure thought, inherent in the mind itself. At a time when European thinkers were exploring different paths to understanding, he took an extreme position. Ultimately, many of his ideas would be dismissed in favor of Newtonian physics and experimental science. However, Descartes played a crucial role in framing 17th-century debates, for later thinkers, including Newton, had to address the questions he posed. Descartes's approach to nature proved unconvincing, and it contributed to the acceptance of experimental methods that treated certain knowledge as something beyond reach.

BLAISE PASCAL

In mid-17th-century Europe, mathematics filled a number of different intellectual roles. The growth of pure mathematics proceeded steadily, as scholars discovered new techniques and new fields of study. The work of Galileo and Kepler inspired the application of mathematics to the study of physical nature through the analysis of numerical data. On the other hand, Descartes had treated mathematics as a model of pure deductive thinking, an example of the human mind's capacity for certainty. He argued that a similar degree of certainty could be achieved through a deductive approach to the operations of the physical universe. All of these varied kinds of mathematical understanding existed at the same cultural moment in the West, and all of them influenced the thought of Blaise Pascal.

Pascal was born in 1623 to a minor noble family in central France. He was educated by his father, Etienne, perhaps because his poor health forced him to remain at home. As a young man, he exhibited an extraordinary talent for mathematics, and his father encouraged his studies. At the age of sixteen, Pascal authored a short treatise on conics, and later in his life he wrote on the subject at greater length. His longer work on conics has been lost, but it may have had an important influence on Leibniz's development of calculus (Merzbach and Boyer 2011, 435). Pascal constructed a mechanical calculating machine when he was nineteen, and is thus considered a contributor to the development of artificial computing—the programming language *Pascal* is named for him.

Arguably, Pascal's most significant mathematical innovations came in the field of probability; indeed, he is often considered the founder of probability theory. In an exchange of letters with the French mathematician Pierre Fermat in 1654, Pascal examined the "Problem of Points," a scenario in which two players had begun a game of coin-tossing. Each had advanced a wager, and the game would end when one of the players achieved a certain number of wins, or points. Fermat and Pascal discussed how to determine a fair distribution of the money if the game were interrupted when the players had amassed different numbers of points—that is, if one player was ahead. Pascal demonstrated that the problem could be solved in exact mathematical terms; he took into account the number of possible combinations, as well as the value of those combinations with respect to the money involved.

The "Arithmetical Triangle," is a triangular array of numbers in which each figure is the sum of the two figures above it. It appears in a number of mathematical traditions: the Chinese, Indians, and Arabs were aware of many of its properties, and the Persian mathematician and poet Omar Khayyam described it as well. In his *Treatise on the Arithmetical Triangle*, written in 1654, Pascal addressed the uses of the array in detail, repeating earlier work and building upon it. He also discovered important new applications, employing the triangle to solve problems of combination in probability. He discussed his approach to the problem of points in the same treatise, published posthumously in 1665 (Edwards 2003, 43–48). Pascal's insights inspired subsequent work in probability theory, particularly the ideas of Jacob Bernoulli in the 18th century.

While Pascal excelled in pure mathematics, he did not neglect the experimental methods that were gaining adherents during his lifetime. A student of Galileo, Evangelista Torricelli, had performed a series of experiments on the phenomena of air pressure, and Pascal resolved to study the issue. Torricelli had filled a narrow glass tube with mercury, closed one end, and inverted the other end in a vessel filled with mercury. Only some of the mercury in the tube flowed out; the rest remained suspended in the tube—apparently held there by the pressure of the air (Figure 10.8).

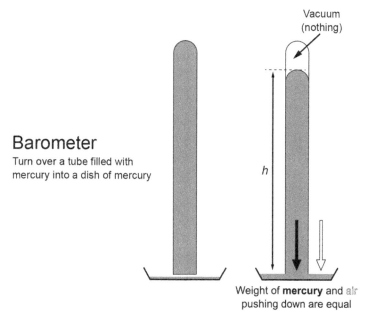

Figure 10.8 Both Torricelli and Pascal demonstrated that the weight of the air could suspend a vertical column of mercury.

Pascal designed a number of different glass instruments to examine the question further, some filled with mercury, some with wine. His experiments confirmed the role of air pressure in supporting a column of liquid in an enclosed tube. He also argued that the empty space left at the top of such a tube represented a true vacuum, contrary to the ideas of Aristotle and Descartes—both of whom had suggested, for quite different reasons, that vacua were impossible in nature. In addition, he measured the weight of the air pressing down on the mercury and showed that such weight was reduced at higher altitudes. Throughout all this work, he adhered to the idea that quantified experimental evidence could lead to a mathematically precise understanding of nature.

The wide scope of Pascal's interests and methods was notable; he adopted a number of different approaches to knowledge: pure mathematics (with respect to his conic studies), applied mathematics (with respect to probability), and experiment. His range of thought led him to reject Cartesian views of human certainty, modeled on deductive mathematical thinking. In the *Pensees*, published after his death, Pascal criticized Descartes's attempts to explain natural phenomena in terms of detailed particulate mechanisms:

> We must say summarily: "This is made by figure and motion," for it is true. But to say what these are, and to compose the machine, is ridiculous.
>
> (Pascal 1941, 30)

In 1654, while he was immersed in the study of probability, he had a mystical experience that he recorded on paper. After that experience, he became convinced of the limitations of human understanding. For Pascal, true certainty could only be achieved through religious faith; it was a matter of the will, rather than the mind. He immersed himself in religious contemplation for much of the rest of his life. He continued to respect other intellectual pursuits, but reflected on their boundaries. Like Galileo, Pascal wrote that experimentation was limited by the vast scope of nature, for one could not make definite conclusions based on observations that were, by their very nature, incomplete. New evidence might always be found. He also maintained that mathematical judgements were constrained by the character of such initial concepts as number and geometrical space—these seemed to be self-evident to human beings, but could not be considered absolutely true. Thus, for Pascal, Descartes was wrong: mathematics was not certain, it merely represented the extent of human intellectual powers. The mind of god, for example, might hold a superior understanding of mathematics. Human certainty, then, was impossible:

> This is our true state; this is what makes us incapable of certain knowledge and absolute ignorance. We sail within a vast sphere, ever drifting in uncertainty, driven from end to end.
>
> (Pascal 1941, 25)

While he denied the power of both mathematics and experiment to yield absolute truths, Pascal defended the existence of human understanding itself. The human mind had great capacity to know, but that knowledge had to be accompanied by humility, by an awareness of the limits of thought. In many ways, Pascal was the most modern of 17th-century mathematicians, for he recognized the revisionary aspects of scientific methodology and anticipated the questioning tone of 20th-century philosophy.

ISAAC NEWTON

In many ways, the intellectual controversies of the 17th century reached their climax in the work of Isaac Newton. His methods set a standard for later experimental practice, and his physics revolutionized the European understanding of nature. Newton was a mathematician, first and foremost, and his work illustrated the varied roles played by the study of mathematics in a time of skepticism and great conceptual creativity. On the one hand, he contributed immensely to the steady evolution of pure mathematics, building on the ideas of Descartes, Fermat, Pascal, and others and discovering the techniques of calculus. At the same time, he broke new ground in applying mathematics to the study of nature. Like Galileo, Newton advocated the mathematical analysis of data acquired through experiment and observation; like many of his English contemporaries, he rejected Descartes's physics and the Cartesian method of pure intellectual deduction. In the end, the success of his work heralded the rise of modern scientific thinking in Western Europe.

In 1665 Isaac Newton was a student at Trinity College in Cambridge University. The school was forced to close because of an outbreak of Bubonic Plague, and Newton returned home. Over the next two years, he discovered differential calculus and did preliminary work on optics and planetary astronomy. Once he returned to Cambridge, he revealed his mathematical work to one of his professors, Isaac Barrow. Barrow supported Newton's candidacy for the Lucasian Professorship in mathematics, and Newton joined the faculty at Cambridge before he was thirty. Newton's mathematical discoveries were of great importance, but he did not publish his version of the calculus until 1693. He did employ some of his new mathematical techniques in his great study of the physics, the *Principia*.

The growing acceptance of heliocentrism in the second half of the 17th century had created something of a crisis in European astronomy and cosmology. Descartes had claimed that the planets were kept in their orbits by a vortex of particles that moved around the sun like a continuous whirlpool. This idea conformed to his view that all the phenomena of nature needed to be explained by particles and impact. Other theorists doubted the concept and sought a different explanation for Kepler's elliptical sun-centered orbits. In England, Robert Hooke maintained that an attractive force emanated from the sun and kept the planets in their paths, but he did not develop the idea mathematically. In 1685 Edmund Halley asked Isaac Newton for his opinion of the problem, and that conversation led Newton to write his most important book, the *Mathematical Principles of Natural Philosophy*. Published in 1687, the *Principia* inspired the growth of experimental methodology and the acceptance of a mathematical view of natural law in Western thought.

Newton opened the *Principia* with a discussion of his methodological choices, his "Rules of Reasoning in Natural Philosophy." He made it clear that he based his judgements on experiments and that he accepted his experimental results as typical of the behavior of nature. In other words, he stated that nature acted consistently. He also described the qualities of matter: "extension, hardness, impenetrability, mobility, and force of inertia (mass)." Newton embraced the idea of particulate matter, but he added inertia, or mass, to the list of its qualities. It is important to point out that extension in three dimensions (length, width, and height) and mass are measurable qualities; Newton thus advanced the view that matter may be described in mathematical terms. Finally, he made a statement about the nature of his conclusions:

> In experimental philosophy we are to look upon propositions collected by general induction from phenomena as accurately or very nearly true, notwithstanding any contrary hypotheses that may be imagined, till such time as other phenomena occur, by which they may either be made more accurate, or liable to exceptions.
>
> (Newton 1953, 5)

Here, Newton states that his ideas would proceed from general induction, that is, the construction of general principles based on the observation of natural phenomena. In his case, those observations would often take the form of quantified measurements, and his conclusions would be mathematical statements. He goes on to say that such conclusions should be accepted until "such time as other phenomena occur, by which they may either be made more accurate, or liable to exceptions." Newton believed in a mathematical approach to nature, but like Galileo and Pascal, he recognized the limitations on human observations. New evidence or data could always arise and force the revision—or dismissal—of established models or theories. Newton's understanding of the universe would be experimental, mathematical, and subject to revision in the face of new evidence. Unlike Descartes, he did not seek certain truths; he pursued valid generalizations consistent with observed evidence. Newton's "Rules of Reasoning," in brief, was a statement of modern scientific thinking.

In the first part of the *Principia*, Newton discussed his three laws of motion. Here, he expanded upon the work of Galileo, who had conducted experiments on the behavior of moving bodies earlier in the century. Galileo's studies lay in the field of kinematics—the description of motion itself in mathematical terms. Newton moved on to the study of dynamics, the examination of the forces acting on a moving body. Newton's first law stated that the forces acting on a body at rest, or in a constant state of motion, must add up to zero. His second law proceeded from the first: for bodies that are accelerating or decelerating, the sum of the forces acting on the body must equal the mass of that body multiplied by the acceleration of the body:

$$F = m \times a$$

Finally, Newton's third law maintained that every action has an equal, and opposite, reaction; that is, if a body applies a force to another body, the same force is applied to the first body in the opposite direction. For example, if a person pushes against a car with a given force, that same force is applied to the person's hands.

All three of Newton's laws of motion could be expressed as equations, and he confirmed his conclusions through the mathematical examination of experimental data. The laws of motion transformed the way that European thinkers conceived of physical phenomena. Galileo had discovered mathematical laws that described certain kinds of motion observed on the earth. Newton's laws had the power to describe all motion in the universe with complete precision.

The laws of motion alone would have made Newton the most significant mathematician and natural philosopher of his age. In the second part of the *Principia*, he went on to treat the subject of universal gravitation. Combining the work of Kepler and Galileo, he demonstrated the existence of an attractive force between all material bodies. The magnitude of the attraction between two objects could be described by the following equation:

$$f(g) = \frac{Gm_1 m_2}{r^2}$$

where $f(g)$ was the attractive force, m_1 and m_2 the mass of the two objects, r^2 the square of the distance between them, and G the gravitational constant (which he calculated). In many ways, Newton's law of gravitation served as the culmination of 17th-century European natural philosophy. It explained Kepler's laws of planetary motion and Galileo's work on falling bodies and on the parabolic paths of projectiles. It settled the issue of heliocentrism and created a new standard of mathematical precision in physics and astronomy. Finally, it demonstrated the power of the Newtonian method to derive mathematically predictive natural laws from observed evidence.

It is tempting to define Newton as a modern scientist because of the *Principia*, but to do so would be simplistic and anachronistic. Newton was very much a product of the eclectic age in which he lived, he embraced a number of different intellectual traditions, and his particular understanding of the Christian religion played a central role in his thought. The *Principia* certainly highlighted his commitment to experimental methodology and the mathematical analysis of nature. Yet in the same book, he claimed that the law of universal gravitation was evidence of the divine design of the universe, and he implied that the force of gravity itself was a form of divine action. He believed in prophecy and natural magic and enthusiastically devoted his time and expertise to the pursuit of alchemy. However, over time, the methods and conclusions in his published works—the *Principia* and the *Opticks* (1705)—came to dominate European ideas of nature. The great intellectual variety of the 17th century gave way to a new intellectual orthodoxy in the 18th, an orthodoxy grounded in reason, natural law, mathematical analysis, and experimental science. The thought of Newton, shorn of its mystical and magical elements, became the great symbol of that orthodoxy.

DUTCH CAPITALISM

By the middle of the 17th century, one half of all the ships in Europe were Dutch. This was remarkable in itself, for the Dutch Republic was a relatively small state in northwestern Europe. But the Dutch dominance of European shipping represented much more than just a commitment to maritime enterprise. Rather, during the early modern period the Dutch took existing commercial institutions and methods of accounting and combined them with new financial practices to carry out business transactions on an unprecedented scale. Dutch public life was dominated by the practical mathematics of investing money, and many of the institutions of modern capitalism developed during this period.

One key to Dutch financial success during this period was their ability to concentrate wealth, or investment capital, for the purpose of supporting business ventures. They recognized that they could combine the resources of many small investors by selling shares of stock in a given enterprise; this would raise the funds necessary to carry out that enterprise. For example, a trading voyage to southeast Asia might cost 50,000 guilders, but fifty investors contributing 1,000 guilders each could finance the project. Each investor would then receive 1/50th of the resulting profits. Earlier commercial societies in Europe—Renaissance Italy and late medieval Flanders—had pursued a similar strategy, but on a smaller scale. Stock sales provided the Dutch with an effective way to direct existing wealth towards business transactions that would produce more wealth. These techniques allowed the Dutch to engage in substantial seaborne trade, for the cost of shipbuilding was distributed among many investors. The profits from these voyages could be used to fund future ventures. Over time, Dutch merchants amassed a huge merchant fleet, largely funded by stock sales.

As Dutch merchants and investors grew wealthier, such concentration of capital allowed them to engage in real economies of scale. Instead of sending two or three ships across the Baltic Sea to purchase Scandinavian timber, Dutch merchants would buy the rights to a forest and send a small fleet to pick it up. By doing so, they were able to dominate markets in a variety of commodities: salt, fish, timber. The sheer number of ships owned by the Dutch also allowed them to control the carrying trade—foreign merchants often had to hire them to ship their products. The Dutch were heavily immersed in commerce with India and southeast Asia, and they came to control the East Indies—modern Indonesia. The conduct of trading voyages to Asia was very expensive, but the Dutch financed them easily and made tremendous profits from the spice trade.

Profits from commerce fueled consumption in Dutch cities. Dutch merchants bought houses, consumer products, fine art—all the symbols of financial success. In the process, they contributed to the wealth of other members of society, who in turn could invest their profits in commercial ventures. Rapid financial growth would characterize the Dutch Republic for much of the 17th century.

New Dutch institutions grew in response to the growing volume of commercial activity. The *Wisselbank* was established in 1609 as a means to ensure secure transfer of funds. This bank had an elite group of depositors drawn from the most prosperous business leaders of Dutch society. Merchants could pay each other or transfer funds for investment through the bank; it would move funds from one account to another and charge fees to account holders for such services. This created great confidence among the Dutch business community; "bank money"—the transfer notes of the *Wisselbank*—was literally as good as gold.

The *Bourse*, the Dutch stock market, served as a center of investment and speculation. Merchants and companies could raise funds there by selling stock to investors, and those shares could be traded in turn. A merchant who invested 1,000 guilders in a trading voyage might not be willing to wait for the end of that venture. He could sell his interest in the voyage—his stock—to another investor for a small profit. That second investor would now have the rights to the profits of the voyage—unless he sold the stock to someone else. Hundreds of such transactions made up the business of the stock exchange. Fortunes could be made, and lost, as investors and speculators sought financial advantage.

In addition to the concentration of capital and the trading of stock, the Dutch actively sought to minimize risk in their business activities. To a certain extent, the pooling of capital does that—each investor only risks a small amount on a given enterprise. Economies of scale serve to reduce the risk of competition. To further reduce risk, the Dutch expanded the use of maritime insurance. Insurers accept risk for a price; they agree to pay for mishaps in return for payment of a premium.

Investment, stock trading, banking, insurance—all of these complex transactions required a significant level of confidence in the Dutch financial system. Investors needed to believe that they would be repaid; bank depositors needed to be sure their deposits would be secure and available for them to draw upon. The republican character of Dutch government served as one source of this confidence; Dutch investors recognized that a government made up of people engaged in business would work to ensure the security of business. For business pervaded Dutch society. Dutch men and women were acutely aware of the financial aspects of their lives. Their sense of identity was bound up with the intricate role played by wealth in their society, bound to bank accounts and interest rates and stock transactions. Like their counterparts in modern societies around the world, their lives were defined in part by the practical mathematics of money.

CONCLUSION

The Scientific Revolution was really a period of tremendous conceptual exploration, as European intellectuals wrestled with different methods of investigation, different definitions of truth and knowledge, different notions of nature and matter. Number and mathematics lay at the center of many of these discussions. Magical views of nature stressed mathematical harmony; Cartesian views of nature stressed mathematical certainty; Newtonian views of nature stressed mathematical data and precision. The story of mathematics in early modern Europe is very much the story of modern Western thought itself. In the end, the ideological disputes of the 17th century resulted in the rise of scientific methodology, perhaps the most powerful cultural force in human history.

PRIMARY TEXT: ISAAC NEWTON, *MATHEMATICAL PRINCIPLES OF NATURAL PHILOSOPHY*

In the third book of his Mathematical Principles of Natural Philosophy *(1687), Isaac Newton discussed his "Rules of Reasoning in Philosophy," his basic assumptions with respect to experimental reasoning and the characteristics of matter. His third rule addressed the qualities of matter. Note that he argued that matter has inherent qualities of size, shape, position, and inertia (mass), all of which may be measured and expressed in numerical terms. He also discussed the force of gravity as a phenomenon related to quantity. Newton thus established the mathematical basis of the material universe.*

RULE III

The qualities of bodies, which admit neither intension nor remission of degrees, and which are found to belong to all bodies within reach of our experiments, are to be esteemed the universal qualities of all bodies whatsoever.

For since the qualities of bodies are only known to us by experiments, we are to hold for universal, all such as universally agree with experiments; and such as are not liable to diminution, can never be quite taken away. We are certainly not to relinquish the evidence of experiments for the sake of dreams and vain fictions of our own devising; nor are we to recede from the analogy of Nature, which is wont to be simple, and always consonant to itself. We no other way know the extension of bodies, than by our senses, nor do these reach it in all bodies; but because we perceive extension in all that are sensible, therefore we ascribe it universally to all others, also. That abundance of bodies are hard we learn by experience. And because the hardness of the whole arises from the hardness of the parts, we therefore justly infer the hardness of the undivided particles not only of the bodies we feel but of all others. That all bodies are impenetrable we gather not from reason, but from sensation. The bodies which we handle we find impenetrables and thence conclude impenetrability to be a universal property of all bodies whatsoever. That all bodies are moveable, and endowed with certain powers (which we call the forces of inertia) or persevering in their motion or in their rest, we only infer from the like properties observed in the bodies which we have seen. The extension, hardness, impenetrability, mobility, and force of inertia of the whole result from the extension, hardness, impenetrability, mobility, and forces of inertia of the parts: and thence we conclude that the least particles of all bodies to be also all extended, and hard, and impenetrable, and moveable, and endowed with their proper forces of inertia. And this is the foundation of all philosophy. Moreover, that the divided but contiguous particles of bodies may be separated from one another, is a matter of observation; and, in the particles that remain undivided, our minds are able to distinguish yet lesser parts, as is mathematically demonstrated. But whether the parts so distinguished, and not yet divided, may, by the powers of nature, be actually divided and separated from one another, we cannot certainly determine. Yet had we the proof of but one experiment, that any undivided particle, in breaking a hard and solid body, suffered a division, we might by virtue of this rule, conclude, that the undivided as well as the divided particles, may be divided and actually separated into infinity.

Lastly, if it universally appears, by experiments and astronomical observations, that all bodies about the earth, gravitate toward the earth; and that in proportion to the quantity of matter which they severally contain;

that the moon likewise, according to the quantity of its matter, gravitates toward the earth; that on the other hand our sea gravitates toward the moon; and all the planets mutually one toward another; and the comets in like manner towards the sun; we must, in consequence of this rule, universally allow, that all bodies whatsoever are endowed with a principle of mutual gravitation. For the argument from the appearances concludes with more force for the universal gravitation of all bodies, than for their impenetrability, of which among those in the celestial regions, we have no experiments, nor any manner of observation. Not that I affirm gravity to be essential to all bodies. By their inherent force I mean nothing but their force of inertia. This is immutable. Their gravity is diminished as they recede from the earth. (Newton 1953, 3–5)

> Motte, Andrew, trans. *The Mathematical Principles of Natural Philosophy*. London, 1729.

FURTHER READING

Cohen, I. Bernard. 1985. *The Birth of a New Physics*. New York: W.W. Norton.
Edwards, Anthony and Fairbank, William. 2003. *Pascal's Work on Probability*. In *The Cambridge Companion to Pascal*, edited by Nicholas Hammond. Cambridge: Cambridge University Press.
Galilei, Galileo. 1957. *The Assayer*. In *Discoveries and Opinions of Galileo*, edited by Stillman Drake. New York: Anchor Books.
Gingerich, Owen. 2005. *The Book Nobody Read*. New York: Penguin Books.
Knight, David. 2014. *Voyaging in Strange Seas*. New Haven: Yale University Press.
Merzbach, Uta and Boyer, Carl. 2011. *A History of Mathematics*. Hoboken: John Wiley and Sons.
Newton, Isaac. 1953. *Mathematical Principles of Natural Philosophy*. In *Newton's Philosophy of Nature*, edited by H.S. Thayer. New York: Hafner Publishing.
Pascal, Blaise. 1941. *Pensees*. New York: Random House.
Westfall, Richard. 1978. *The Construction of Modern Science*. Cambridge: Cambridge University Press.

11

Numbers in 18th- and 19th-Century Europe

CULTURAL ICON: CHARLES BABBAGE'S DIFFERENCE ENGINE #1

The Difference Engine was the first mechanical calculating device capable of solving complex problems (Figure 11.1). Designed by the British mathematician Charles Babbage in 1821, it was never completed. However, about 15 percent of the device was constructed, and that portion functioned successfully. Babbage wanted to build a mechanical calculating machine capable of producing mathematical tables, for example, tables of logarithms. His design was based on the method of divided differences, an approach to solving polynomial equations first described by Isaac Newton. The method allowed polynomials to be solved through processes of tabulation and addition, operations well-suited to mechanical computation. While the mathematics behind Babbage's machine were well-known, his innovative design and his vision of building such an intricate device using the growing technical capacity of industrial Britain represented real insight. His work on the Difference Engine led Babbage to conceive of a more capable machine, the Analytical Engine, which had the capacity to perform many of the functions of a true computer.

Charles Babbage was born in 1791 to a wealthy London family. He was well-educated and demonstrated an interest in mathematics as a young

Figure 11.1 Charles Babbage's Difference Engine was a mechanical device capable of solving complex polynomial equations. Only a portion of the machine was completed, but it functioned effectively. (World History Archive/Alamy Stock Photo)

man. At Cambridge University, he cofounded the Analytical Society with other students interested in rigorous mathematical study. Following his graduation, he pursued a variety of mathematical projects and helped to found the Royal Astronomical Society. His work there led him to conceive of the Difference Engine as a way to perform astronomical calculations quickly and with consistency. In 1828 he rose to the Lucasian Professorship of Mathematics at Cambridge (Isaac Newton's position) and held the post for more than ten years.

The British government funded Babbage's design for the Difference Engine in 1823, and he began work on its construction. The completed device would have weighed four tons and included a mechanical printer. However, the technical demands of the design prevented its successful completion. He did build a working portion of the machine, but the project grew increasingly expensive and was eventually abandoned.

Babbage moved on to larger projects; he designed an Analytical Engine in 1834 that was far more elaborate. It would be programmable, using a punch card system to encode mathematical operations. The Jacquard loom, invented in France in the early 19th century, was the model for this programming technology. The Analytical Engine would include a mechanical computing module (the "mill"), a form of memory (the "store"), and a printer. Babbage wrote some programs for the machine, and his friend and associate Ada Lovelace, the daughter of the Romantic poet Lord Byron, developed a process for finding Bernoulli numbers with the device. Some historians consider Lovelace to be the author of the first real computer program, while others question that judgement. She did recognize the potential of the programmable Analytical Engine to solve problems far more profound than mathematical calculations.

None of Babbage's machines were ever completed while he was alive. However, a working version of his second Difference Engine was built by the London Science Museum in 2002, and it functions as its designer intended. Babbage's work is considered by many to be the starting point for modern computing. His ability to conceive of working designs, using the technology of his day, made him one of the crucial 19th-century figures in the history of information technology. (The other, George Boole, the innovator who conceived of Boolean algebra, will be discussed in Chapter 12.)

CULTURAL COMMENTARY: NATURAL LAW, PRECISION, AND POWER

It is common to think of the European Enlightenment of the 18th century in terms of political thought. After all, Enlightenment ideas of equality and government by consent formed the intellectual basis of the American and French revolutions and continue to play a central role in contemporary political values. However, in a broader sense the Enlightenment was a response to the extraordinary success of Newtonian physics and mathematics. Isaac Newton's *Principia*, published in 1687, demonstrated the ability of the human mind to discover the fundamental laws of nature, mathematical laws that predicted the behavior of physical bodies with great accuracy. Other thinkers adopted Newton's methods and built upon his concepts as they confidently searched for additional insights into the workings of the universe. This sense of confidence, this notion that natural law existed and that human reason was capable of understanding it—this was the real essence of the Enlightenment.

Of course, the entire Enlightenment was not just about Newton. For example, the philosophy of John Locke (1632–1704) shaped a great deal of 18th-century thought. Locke maintained that human ideas were based mainly on observation; human beings could find patterns in those observations through the use of reason. This view of knowledge fit in perfectly with the experimental methods of Newton. Locke also wrote about politics; he argued that all men were equal (it is possible that he felt that women were included), that they all had basic rights, and that governments drew their power from the consent of the people they governed. Thus, Isaac Newton and John Locke served as twin pillars of Enlightenment thinking. Voltaire, an important figure in the early French Enlightenment, worked to spread Locke's ideas of government in France; he also published a popular treatment of Newton's ideas, *The Elements of the Philosophy of Newton*, in 1738.

The search for rational natural law lay at the center of the Enlightenment. Clearly, that search would be directed at the physical processes of nature itself. Newton used the term "experimental philosophy" to describe his own efforts to understand the natural world, but it is quite appropriate to refer to the efforts of his Enlightenment successors as "science." Scientific thinking is based on observation and experiment, and it employs mathematical methods when possible. Further, it avoids statements of absolute truth—scientific ideas are all subject to revision and rethinking in the face of new evidence. Newton stated these precepts clearly in his *Principia*, and Enlightenment scientists embraced them. During the 18th century, they would apply the principles of Newtonian physics to a wide variety of phenomena. For example, Daniel Bernoulli published *Hydrodynamica*, a mathematical study of fluid mechanics. Pierre-Simon Laplace explored the intricate gravitational relationships of the bodies in the solar system in his massive work *Celestial Mechanics*. Joseph-Louis Lagrange also worked on issues of gravity and established the rules for the interaction of three gravitating bodies. Antoine Lavoisier adopted a quantitative approach to chemistry, established a system for naming chemical compounds, and discovered the nature of oxidation. All of these figures, and many more, sought an understanding of natural law through the mathematical analysis of observed evidence.

Yet the Enlightenment search for rational natural law was not limited to the physical universe. Seventeenth-century political theorists, including Locke, had applied reason to the origins of political authority and government. Enlightenment thinkers would expand on this trend; they would seek the fundamental natural laws that regulated human society. The roots of the modern social sciences may be found in their efforts. In 1748 Charles-Louis de Secondat, the Baron de Montesquieu, published the *Spirit of the Laws*, an attempt to explain different systems of human law in rational terms. He began with a general definition of law, including natural law:

> Laws in their most general signification, are the necessary relations derived from the nature of things. In this sense, all beings have their laws, the Deity has his laws, the material world it laws, the intelligences superior to man have their laws, the beasts their laws, man his laws.
>
> (Montesquieu 1977, 98)

Other authors also tried to use reason to explain or critique human institutions. In his *Discourse on the Origins of Inequality* (1755), Jean Jacques Rousseau gave a rational account of the origins of human social structures, with an emphasis on the disparities of wealth and power that characterized the society of his day. Cesare Beccaria argued against the use of torture and the death penalty in *On Crimes and Punishments* (1764) and suggested that criminal punishments should be based on reason. Most significantly, in 1776, Adam Smith published *The Wealth of Nations*, the founding text of modern economics. Smith dismissed the attempts of nations to artificially control their economies as counterproductive. He advocated free trade and argued that individual people, acting in their own self-interest, would be the most efficient producers of wealth. Smith's arguments suggested a view that natural principles governed the origins and movement of wealth, and he tried to explore those principles rationally. *The Wealth of Nations* inspired a host of later works that adopted a mathematical approach to economic laws; for example, David Ricardo would explore the concept of free trade systematically in *Principles of Political Economy and Taxation* (1817).

Voltaire wrote about Newton because he believed that an understanding of the English physicist's work was necessary for other intellectual endeavors, if only because mathematical natural law served as a powerful model of intellectual achievement. For him, Newton symbolized the vast potential of human reason. Other Enlightenment intellectuals pursued a wide variety of different subject matter, but they all drew inspiration from Newtonian physics and mathematics. For example, the laws of physics had an effect on the concept of human equality, so important to Enlightenment political theorists. If natural law applied equally to all particles of matter, should not human law apply equally to all people? Traditional hierarchies seemed foolish in the face of Enlightenment science. Similarly, the notion of virtue itself acquired a mathematical character during the Enlightenment. Jeremy Bentham, a philosopher of the late Enlightenment, suggested that actions and policies could be determined as right or wrong based on the concept of the greatest good for the greatest number. Essentially, he tried to create an equation for morality and attempted to numerically rank the pleasures and displeasures that could arise from personal actions or choices (he referred to this as his "hedonistic calculus"). Bentham's ideas became the foundation of Utilitarianism, one of the major theories of ethics in the modern world. Indeed, many of the ideas of Enlightenment thinkers—in physics, in chemistry, in political theory, in economics—would lay the groundwork for the modern idea of a rational universe, transparent to human understanding.

While the concept of mathematical natural law did much to inspire the ideas of the Enlightenment, a more practical application of mathematical knowledge helped to bring about the First Industrial Revolution of the late 18th and early 19th centuries. The Industrial Revolution transformed European methods of manufacturing; as a result, the production of goods increased dramatically. In addition, industrial technology enhanced the military and naval power of European nations and allowed them to build global empires during the second half of the 19th century. The rise of industry also had profound effects on relations between workers and their employers; industrial workers labored in harsh conditions and were treated in an impersonal manner.

The Industrial Revolution grew from major technical innovations in the late 18th century, particularly the development of new machines for spinning thread and weaving cloth and the invention of an efficient steam engine. Spinning frames and power looms increased the rate of cloth production tremendously, but they required an external source of mechanical power. This could be provided by water wheels, but it was the steam engine that truly revolutionized the cloth industry. The earliest steam engines emerged in Britain at the close of the 17th century; Thomas Newcomen built a successful one in 1712. In 1776 the Scottish inventor James Watt introduced a far more efficient machine, in partnership with Matthew Boulton. Their engines were often used to pump water out of mines. As more powerful steam engines emerged in the early 19th century, they were used to power cloth-making machinery in early British factories, and this marked the real start of industrial production.

The central feature of a steam engine is a hollow cylinder with a moving piston inside. In early steam engines, including Watt's, steam condensed to water inside the cylinder and formed a vacuum that moved the piston. In later, more powerful engines, high-pressure steam moved the piston directly. In both cases, boiling water produced the steam; steam engines thus converted heat energy into mechanical motion (Figure 11.2).

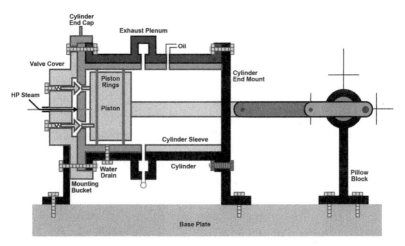

Figure 11.2 The piston lies at the heart of steam engine technology. It is moved by the energy of compressed steam.

However, the production of the cylinder and piston were crucial, for the piston had to fit snugly into the cylinder. Any significant gaps would allow the steam to escape the cylinder and render the machine useless. Exact measurements were required, and precise techniques for boring the cylinders and producing the pistons had to be developed. Watt and Boulton drew on the expertise of John Wilkinson, who had designed effective processes for the boring of cannon barrels. In 1774 Wilkinson built a machine for boring steam engine cylinders to acceptable tolerances; his expertise made Watt's engine a success.

Wilkinson had produced a machine tool, a powered device that performed a specific manufacturing function to a high degree of mathematical precision. As the Industrial Revolution proceeded, it resulted in the development of more steam-powered machine tools, capable of making mechanical components according to increasingly demanding mathematical specifications. This technology could produce the parts for intricate machinery: from railroad locomotives, to artillery pieces, to the next generation of machine tools. A culture of industrial precision emerged, with huge implications for the growth of technology. Such technology lay at the heart of the dramatic growth of European and American economic and military power during the 19th century. The modern discipline of mechanical engineering evolved during this period; the first Institution for Mechanical Engineers was founded in Britain in 1847.

The increasing levels of production made possible by European industry in the early 19th century came at a substantial human cost. Factories were new economic institutions, largely unregulated by government. As a result, working conditions in factories were extremely harsh by modern standards. Further, as agricultural methods were modernized, farm workers were forced off the land and flocked to the growing industrial cities. This surplus of available labor had the effect of keeping wages low. In essence, human labor was treated as a commodity, for its price fluctuated with demand. The modern notion of a minimum wage would ultimately address this issue by creating a level below which wages could not fall. However, such ideas did not arise until the late 19th century. The "Iron Law of Wages" discussed by such economists as David Ricardo and Ferdinand Lasalle maintained that wage levels would always seek the lowest possible level. A variety of workers' movements would grow in response to the stark mathematics of wages during the Industrial Revolution, including the Chartist campaign in Britain and the socialist ideas of Pierre Joseph Proudhon, Karl Marx, and others.

One could argue that over the course of the 19th century, industry, commerce, and technology transformed Western Europe into a society of numbers. Numbers punctuated the lives of human beings in unprecedented ways: the hourly wages of factory workers, the timetables of railroads and steamships, the prices in department stores. Watches, timetables, and bank accounts were not new, but as Europe grew richer, they affected more people than ever before. A burgeoning middle class became increasingly dominant, a middle class that defined status by numerical wealth more than by aristocratic titles. Consider an image of modernity: an attorney checking his watch as he waits for his train, while reading about the performance of the stock market or the sports scores in a newspaper—a common sight in 1890, or in 1990. Many familiar aspects of our world find their roots in the growing significance of numbers in 19th-century European life.

The ideals of the Enlightenment and the realities of the Industrial Revolution both sprung, in part, from a mathematical view of the universe. Over the course of the 19th century, scientific developments would continue to demonstrate the intellectual potential of quantitative methods. In chemistry, the organization of the periodic table of the elements and the discovery of the laws of thermodynamics contributed to a growing atmosphere of prediction and precision. In natural history, the adoption of a geological time scale created a sense of the distant past. In physics, the exploration of electromagnetism revealed a new realm of mathematical natural law, culminating in the work of James Clerk Maxwell. One event in particular embodied this culture of exactitude: the discovery of the planet Neptune. William Herschel had discovered the planet Uranus in 1781, and astronomers had observed it as it moved in its eighty-four-year orbit around the sun. The planet deviated from its predicted path, and it was suggested that the gravity of another, unknown planet might be causing this phenomenon. Urbain Le Verrier in France and John Couch Adams in Britain both used the irregularities of Uranus's orbit to calculate the position of the new planet, and in September 1846, Neptune was discovered based on Le Verrier's predictions, a fitting exhibition of the power of Newtonian science.

EUROPEAN PURE MATHEMATICS IN THE 18TH CENTURY (AND BEYOND)

The discovery of differential and integral calculus in the late 1600s shaped the character of mathematics in Europe during the next century, as European mathematicians adopted the new techniques and expanded upon them. While Isaac Newton and Gottfried Wilhelm Leibniz had both developed calculus independently, the fluid character of Leibniz's notation ensured that his version of the new mathematics would serve as the standard in continental Europe. In fact, one could argue that because British mathematicians continued to use the more awkward Newtonian notation and procedures, most new developments in calculus took place on the European continent.

The emergence of calculus and analytic geometry allowed 18th-century scholars to focus on processes of mathematical analysis. The word "analysis" may be employed as a general term to refer to the rigorous study or the rational investigation of something; it has been used this way in this book. However, "analysis" has a more technical meaning in mathematics and describes the study of problems that may involve an infinite range of values, a set of parameters tending towards infinity or towards zero (these might be termed infinitesimal values). For example, to find the area under a curve, one could divide the space into a series of rectangles. The more rectangles one creates, the closer the estimate of the area. A mathematical statement that would describe an infinite number of rectangles, each with a width approaching zero, would yield the precise area under the curve. Thus, mathematical analysis includes the study of limits, infinite series, derivatives, integrals—operations that allow the description of continuously changing phenomena over an infinite range of values. Much of the material in a modern calculus course would fall into the category of analysis. The slope of a line tangent to a curve or the velocity of an accelerating body serve as good examples. A given curve contains a potentially infinite number of points, and a mathematical expression for the slope of a line tangent to teach of those points must then address an infinite set of values.

Some of these problems had been treated geometrically in the past; the Greeks had used the "method of exhaustion" to find the area of curved figures, for example. In the late 17th century, Newton still used geometrical methods to address issues of planetary motion in his *Principia*. However, the calculus of Newton and Leibniz and the resulting rise of analysis allowed mathematicians to solve more varied and more complex problems, and to express those problems as equations and functions rather than as geometrical proofs.

Among the numerous mathematicians of 18th-century Europe, some figures had particular prominence. The various members of the Bernoulli family of Switzerland contributed a great deal to the spread of Leibniz's approach to calculus. Jacques Bernoulli worked in the late 17th century and conducted a long-term correspondence with Leibniz; he suggested the term "integral calculus" to the German mathematician. He applied Leibniz's calculus to a host of new problems, including the study of infinite series and the mathematical analysis of various curves. His brother Jean wrote on related subjects. Jean's three sons all became mathematics professors, part of a growing community of thought on the continent of Europe in the early decades of the 18th century. Nicolaus and Daniel both held posts at St. Petersburg Academy, a sign of Russia's new presence in the European intellectual sphere. In addition to his work in mathematics, Daniel Bernoulli published *Hydrodynamica* in 1738, a study of the physics of fluids. He described the inverse relationship between the speed of a moving fluid and its potential energy, a relationship known to modern science as Bernoulli's principle.

Daniel Bernoulli left the St. Petersburg Academy and returned to Switzerland in 1733. His position as chief mathematician at the academy went to Leonhard Euler, who was only twenty-six at the time. Euler would spend eight years in St. Petersburg and another twenty-five at the Berlin Academy. Despite difficulty with his sight (he went blind from cataracts at the age of fifty-nine), he would write more than 800 books and papers, many published after his death in 1783. He is generally considered to be the most accomplished mathematician of the 18th century and one of the most important figures in the history of ideas.

Euler's work touched many aspects of modern mathematics. He worked tirelessly expanding the techniques of mathematical analysis and applying them in new ways. He did substantial work on mathematical series and defined trigonometric functions in analytical rather than geometrical terms. Euler also developed the basic methods for the treatment of differential equations, that is, equations that link multiple mathematical functions. The scope of his work was truly impressive, and his impact was magnified by the textbooks he published in his lifetime. As a result, he also had a significant effect on mathematical notation. The use of "e" for the base of the natural logarithm, of "π" for the ratio of the circumference of a circle to its diameter, of "i" for $\sqrt{-1}$, of "Σ" to refer to summation, and of "$f(x)$" to refer to a function of variable "x" all reflect his influence (Merzbach and Boyer 2011, 408–409).

France was an important center of mathematical study in the 18th century, and French mathematicians contributed a great deal to the Enlightenment and the French Revolution. Two of the most prominent mathematicians in France during the period were Joseph-Louis Lagrange (1736–1813) and Pierre Simon Laplace (1749–1827). Lagrange was actually Italian, born and educated in the city of Turin. He worked for years in Prussia, but moved to France in 1787. Lagrange devoted himself to the discipline of analytical mathematics and sought out new creative techniques. He also worked extensively on the application of calculus to physics. In 1788 he published *Méchanique analytique*, a survey of Newtonian mechanics using purely analytical methods: a calculus-based mechanics (as opposed to Newton's frequent use of geometry). He also did important work on the gravitational interaction of three different bodies. Laplace was born in the French province of Normandy and educated in Caen; as a young scholar, he caught the attention of d'Alembert, who assisted him in his career. Laplace pursued an extremely broad range of interests: he made crucial contributions to the field of probability and worked on quantitative approaches to heat with the chemist Lavoisier. However, he is best known for his massive study of celestial mechanics, the *Mécanique celeste*. Here, he examined the gravitational interactions of the solar system and mathematically demonstrated its stability. In a famous exchange with Napoleon Bonaparte, Laplace was supposedly asked how God fit into his system and responded that he "had no need of that hypothesis." The conversation may never have taken place, but the story does illustrate the confidence of Enlightenment thought and the shrinking role played by religion in late 18th-century views of nature. It is notable that both Lagrange and Laplace used 18th-century mathematical analysis to build upon Newton's work, employing the techniques of calculus to express complex Newtonian principles.

Carl Friedrich Gauss (1777–1855) opened up new avenues of mathematical thought. While still a young man, he published the *Disquisitiones Arithmeticae*, a highly innovative study of number theory. Gauss pioneered the field of differential geometry—the mathematical exploration of surfaces—an area that would play a central role in the study of electromagnetism in the 19th century. He also anticipated the emergence of non-Euclidean geometry, but did not publish on the subject. His exceptional powers of calculation allowed him to excel in the computation of the orbits of celestial bodies, and he served as the director of the observatory at Göttingen for more than forty years (Merzbach and Boyer 2011, 470). While he was educated in the 18th century, Gauss's insight and creativity moved mathematics in new directions, and his work inspired many of the groundbreaking ideas of the 19th and 20th centuries.

ÉMILIE DE BRETUIL, MARQUISE DU CHÂTELET

Modern Western thought emerged in a recognizable form during the Enlightenment of the 18th century. While Enlightenment ideas grew in many regions of western and central Europe, France played a particularly important part in the spread of new worldviews. The Enlightenment in France may be conceived of as an intellectual response to 17th-century intellectual innovations, particularly the physics and mathematics of Isaac Newton and the philosophy and political thought of John Locke. It may also be viewed as a project of reform adopted by a community of scholars with close personal and intellectual ties to each other. The life and work of Émilie de Bretuil, the Marquise du Châtelet, illustrates both of these aspects of the French Enlightenment.

Émilie de Bretuil (or du Châtelet) was born in 1706 to a noble family in the court of King Louis XIV of France. While she was educated in languages and the classics, she demonstrated a great talent for mathematics at a young age, and she was tutored in the subject. Marriage was considered a priority for a young noblewoman in the early 18th century, and she was wed at the age of nineteen to the Marquise du Châtelet. She had three children in the early years of her marriage, but she returned to the study of mathematics in her mid-twenties. Because of her social standing, she had contact with many of the leading scholars of the early French Enlightenment. Her instructors in mathematics included Pierre Louis Maupertuis and Alexis Claude Clairaut, a devotee of Newtonian physics. Her most important intellectual and personal relationship was with Voltaire, one of the most prominent authors of the day and a leader of the Enlightenment movement.

The relationship between du Châtelet and Voltaire lasted from 1733 to 1749; he was with her when she died at the age of forty-three. In addition to their romantic attachment, they shared strong bonds of mutual intellectual interest and scholarly collaboration. When Voltaire's work aroused the resentment of the authorities in Paris, the pair moved to the chateau of Cirey-sur Blaise, a possession of her husband's family. They lived there for fifteen years and maintained close ties with other Enlightenment thinkers.

During these years, du Châtelet continued her mathematical work and studied Newtonian physics as well, probably because of the influence of Clairaut. Newton's thought occupied a central place in the intellectual agenda of the Enlightenment, and in 1738, Voltaire published a summary of it, *Elements of the Philosophy of Newton*, with significant contributions from du Châtelet (Schiebinger 1989, 60–61). She began the challenging task of translating Newton's signature work of physics, the *Principia*, into French. Influenced by some of the German mathematician Leibniz's criticisms of Newton, she added an important commentary on the text in which she discussed the principle of the conservation of energy. Her translation of the *Principia* was published after her death and had a lasting effect on the acceptance of Newtonian views of method and mathematical natural law in France. It still serves as the standard French-language version of Newton's book (Schiebinger 1989, 63).

Du Châtelet produced a number of shorter scientific treatises during her time at Cirey, as well as a longer work, the *Institutions de Physique*. The *Institutions* proved to be a source of some conflict, for one of du Châtelet's tutors tried to claim credit for its authorship. Published in 1740, the book gave an account of the major developments in physics in the 17th century, with emphasis on the ideas of Descartes, Leibniz, and Newton. Du Châtelet placed this material in the context of earlier thought and considered different strategies for the treatment of matter, space, and time (Osen 1974, 62–63). In many ways, the *Institutions* may be considered an early work in the history of science and mathematics, for it illustrated du Châtelet's understanding of Enlightenment physics as a product of the revolutionary ideas of the previous century.

Émilie du Châtelet died in 1749 as a result of complications from childbirth. She spent much of the second half of her life pursuing the intellectual goals of the Enlightenment as a scientist, a translator, a commentator, and a historian. She had interactions with many of the influential thinkers of her day, and she and Voltaire had real influence on each other's work. Her focus on Newton's mathematical approach to physics may be understood as part of a larger enterprise: the pursuit of a rational understanding of the laws of nature. Émilie du Châtelet's dedication to this enterprise made her a vibrant member of the Enlightenment community of minds in the mid-18th century.

GEOLOGICAL TIME

Modern scientists estimate that the earth is 4.5 billion years old. Life has existed on the planet for 3.5 billion years; complex organisms evolved around 600 million years ago; human ancestors appeared 6 million years ago; the human species itself is about 200,000 years old. These are familiar concepts to an educated person living in the 21st century. Modern people see the earth's past as a vast span of geological and evolutionary time; the physical features of the planet and the species that inhabit it are understood to be the results of processes that unfolded over millions of years. Human beings cannot easily relate to such immense periods of time, but they recognize the immense difference in scale between natural history and human history.

The modern view of geological time originated during the European Enlightenment of the 18th century and continued to develop over 200 years of scientific thought on the subject. Until the Enlightenment, European scholars accepted a time scale based on the book of Genesis. The story of the seven days of creation maintained that the earth, and the universe, were created only days before humanity; natural history and human history thus shared the same time scale. James Ussher, an Anglican bishop and biblical scholar in the first half of the 17th century, studied the books of the Bible and concluded that the creation took place in the year 4004 BCE. While his figure was not universally accepted, it was representative of the pre-Enlightenment mindset in Europe: the world was about 5,500 years old.

In a private letter written to Thomas Burnet near the end of the 17th century, Isaac Newton pointed out that the language of Genesis was imprecise and that the lengths of the "days" of creation were unclear. However, these were private musings, and Newton remained deeply devoted to the truths of Judeo-Christian scripture. The first widely published study of the age of the earth appeared in the *Natural History* (*L'Histoire Naturelle*) of

Georges-Louis Leclerc, the Comte de Buffon (1707–1778). Buffon was a French nobleman who sought to publish a systematic description of all animals, plants, and minerals known to European science—an immense project. Over the course of his life, he published thirty-six volumes; another seven were added by his associates after his death. In 1778 he wrote a section of the greater work entitled *The Epochs of Nature*, in which he presented an account of the emergence of the natural world that departed from the text of Genesis.

Buffon approached the question of the age of the earth through a combination of experiment and mathematical analysis. He had an iron foundry on his family estate, and he had a series of iron spheres created of different sizes. He placed these in the furnace until they were red hot and then recorded the amount of time it took for them to cool. The cooling times increased rapidly as the spheres increased in size, for the volume of the spheres increased in proportion to the cubes of their radii. Once he established the pattern of increase in the time of cooling, he estimated the amount of time it would take for a sphere the size of the earth to cool from a near-molten state.

This study included some interesting assumptions. Buffon assumed that the earth was largely composed of iron, presumably because it had a magnetic field. He assumed that it had formed in a molten state and had cooled to the point where the surface was inhabitable by plants and animals. Finally, he assumed that he could estimate the earth's age by measuring the cooling time of smaller spheres. He arrived at a figure of 75,000 years and published that in the *Epochs of Nature*. He divided that period into seven epochs and discussed the developments during each of them. Buffon's figure seems quite small by modern standards, but it had great intellectual significance. He had dismissed the biblical time scale and greatly expanded the age of the earth. He had arrived at his conclusions through a rational process of observation and mathematical analysis, setting a precedent for future study. In his private notes, he wrote that millions of years were probably necessary for the formation of the planet.

The Enlightenment was an age of rational thought, and other authors followed Buffon in applying observation and reason to the problem of the earth's formation. In 1788 James Hutton published a treatise called *Theory of the Earth*, in which he embraced the idea that the processes that had formed the features of the earth operated very much like the processes at work in the present day. He wrote of the immense amounts of time required for the deposition of rock layers, the elevation of mountains, and the erosion of canyons through ordinary physical means. His estimates ran to millions of years, and he viewed the earth's origins as shrouded in the mists of time. Hutton ended his discussion with a famous conclusion, "... we find no vestige of a beginning, no prospect of an end."

Hutton's work supported a philosophical agenda; as a deist, he believed that a rational supreme being was the author of natural law and that the earth's regular processes were evidence of that creator's rational plan. However, his work helped to inspire secular views of geology and natural history. Over the course of the 19th century, the scientific study of rock strata, geological processes, and fossils produced a fascinating and ever-evolving picture of the earth's past, a picture with a steadily increasing time scale. In the 20th century, the technology of radiometric dating yielded precise values for the age of rocks, and the earth acquired its 4.5-billion-year history. Geological time has become a part of modern culture; it inspires awe at the distant past and confronts modern human beings with a sense of their own fragility and transience.

ADAM SMITH AND CLASSICAL ECONOMICS

In 18th-century Europe, intellectuals applied the Enlightenment notion of natural law to the structures and institutions of human society. The modern social sciences evolved significantly during this period. For example, many historians view *An Inquiry into the Nature and Causes of the Wealth of Nations*, published by Adam Smith in 1776, as the first work of analytical economics. Smith sought to describe human economic activity in terms of rational principles, often with a quantitative character. He defined different aspects of production and commerce and analyzed their roles in the creation and distribution of wealth.

Smith identified the division of labor, the partition of a complex task into individual components, as the most important element of production. He gave the example of the various steps in the manufacture of pins and pointed out that the performance of these steps by different workers vastly increased the number of pins produced. Labor played a central role in his thought, and he argued that labor was the source of added value in manufactured commodities: pins were worth more than the steel wire they were made of because the labor of workers enhanced their worth. This "labor theory of value" represented an attempt to analyze the origins of wealth and value. Smith went on to define the relationship between workers who provided labor and the owner of an enterprise who employed them:

> In all arts and manufactures the greater part of the workmen stand in need of a master to advance them the materials of their work, and their wages and maintenance till it be completed. He shares in the produce of their labour, or in the value which it adds to the materials upon which it is bestowed, and in this share consists his profit.
>
> (Smith 1973, 586)

Here, Smith emphasizes the need for a source of invested wealth—capital—to make a productive enterprise possible and notes that labor "adds value to the materials upon which it is bestowed." He also asserts that the owner's role entitles him to a share of this added value, which he defines as "profit." Smith's analytical approach to the subject provided the reader with a rational description of economic phenomena.

The notion of natural law had a profound influence on Smith's ideas; he believed in a natural economic order that arose when people pursued their own self-interest in an honest manner. For example, when a person bought bread from a baker, it was in the buyer's interest to exchange money for the bread, while it was in the baker's interest to exchange the bread for money. If the buyer considered the price too high, then it was no longer in her self-interest to purchase the bread. For Smith, the cumulative effect of all of these transactions was an orderly economy that matched supply with demand. He referred to such order as the "invisible hand," the natural economic equilibrium that led to the efficient production of wealth. Smith argued that governments should interfere as little as possible with natural economic processes, for such interference reduced the production of wealth. However, he was aware that unfettered economic forces could create suffering for ordinary workers, and he recognized that government had an obligation to protect them.

The Wealth of Nations was one of the most influential books of the 18th century, and it inspired a great deal of further economic analysis. The period that followed its publication is often referred as the time of "classical economics." In 1798 Thomas Malthus, an English scholar and clergyman, published *An Essay on the Principle of Population*, in which he explored human population growth in a rational way. He pointed out that increases in food supply would lead to population increases and that such increases would rapidly outpace the human ability to produce more food. According to Malthus, food supplies increased arithmetically; that is, they would grow by addition as new land was cultivated or as new forms of agriculture were employed. Population, on the other hand, increased geometrically; unless it was restrained, it would double and double over time (he claimed that it could double in twenty-five years). Malthus predicted that as the population expanded, starvation, disease, and struggle for available resources would result, to the great detriment of society. He recommended different measures to control population growth, including the delay of marriage. Malthus worked on other issues in economics as well, but his essay on population was his most important work and continues to have relevance in the modern world.

David Ricardo was born in Great Britain in 1772. He grew wealthy through banking and investment and became an early contributor to economic theory. In 1817 he published his *Principles of Political Economy and*

Taxation, in which he sought to analyze issues of trade, wages, and profits in quantitative terms. Like Smith, he adopted a labor theory of value, arguing that a commodity derived its value from the amount of labor required for its production. His views on labor and value contributed to his arguments on the economic advantages of free trade. He pointed out that nations should import products from other nations if those nations could fashion those products with less labor; each nation should focus on producing those commodities that it could make most efficiently. Trade conducted in this manner could be mutually beneficial; the wealth of each nation would increase as a result. Ricardo also analyzed the mathematical relationship between profits and wages, and his ideas, combined with those of Malthus, contributed to the notion of the "Iron Law of Wages" that arose later in the 19th century. This concept maintained that wages would always seek their lowest possible level based on the availability of labor; if a surplus of labor existed, wages would necessarily decrease. Such ideas had an important influence on the ideas of Karl Marx and other reformers.

Adam Smith and the other economists of the classical period explained the production and movement of wealth in a systematic way and defined the quantitative relationships between labor, value, wages, and population. They also explored the character of commerce and considered how governments could act (or refrain from acting) in order to maximize prosperity. In 1890 Alfred Marshall published the *Principles of Economics*, in which he introduced the notion of equilibrium between the forces of supply and demand and represented these ideas graphically as intersecting curves. His views, so important to contemporary economic theory, proceeded from those of the authors of classical economics, the heralds of Enlightenment social science.

MATHEMATICS AND THE FRENCH REVOLUTION

The French Revolution shook the entire European continent. It transformed the politics of the most populous nation of Western Europe, challenged the institutions of monarchy and aristocracy, and plunged the continent into twenty-five years of warfare. The ideals of the Enlightenment—human reason, natural law, and political equality—gave intellectual direction to a revolutionary movement largely fueled by economic crisis and age-old class resentments. The Revolution began in June 1789, and by 1790 a short-lived constitutional monarchy had been established. In the summer of 1792, that regime was replaced by a National Convention dedicated to the writing of a republican constitution. However, the nation suffered from a year of constant political infighting, and by late 1793 the revolutionary government had descended into a period of bloody extremism. Eventually, stability was restored, and the constitution was completed in 1795. However, despite the fractious character of revolutionary politics, the varied governments all worked to remove aspects of French law and society based on kingship, Christianity, or custom and to replace them with new institutions grounded in reason.

Perhaps because mathematics was viewed as a pure form of reason, a number of mathematicians achieved positions of real leadership during the French Revolution. Marie Jean Antoine Nicolas de Caritat, the Marquis of Condorcet, wrote several works on calculus as a young man, but he grew increasingly interested in social and political issues. He applied his mathematical knowledge to different questions of social policy and once made a quantitative argument in favor of smallpox inoculation (Merzbach and Boyer 2011, 429). In 1785 he wrote a mathematical analysis of the behavior of juries and legislatures. Condorcet believed fervently in equality; he strongly opposed slavery and advocated for the education of women. He actively supported the early stages of the French Revolution and was elected to the Legislative Assembly in 1791. Condorcet saw the French Revolution as a golden opportunity to build a state based on reason and rose to a position of leadership in the legislature. However, with the rise of the National Convention in 1792, his fortunes began to wane. His proposals were viewed with suspicion, and when he opposed the constitution of 1793, he was denounced. After months in hiding, he was eventually arrested and died mysteriously two days later in prison.

Other mathematicians also found important places in the revolutionary regime. Gaspard Monge (1746–1818) worked as Minister of the Marine and played a major role in the founding of the École Polytechnique in 1794. The school would rapidly become one of Europe's leading institutions for the teaching of mathematics and engineering, and Monge served on the faculty. Later, he would become a devotee of Napoleon Bonaparte and would accompany his army on its Egyptian expedition. Lazare Nicolas Marguerite, the Count Carnot, studied mathematics and engineering as part of his military education. During his lifetime, he would publish works on engineering, geometry, and military fortification. Like Condorcet, he was elected to the Legislative Assembly in 1971 and also sat as a member of the National Convention in 1792. Carnot rose to a place on the Committee of Public Safety in 1793–1794 and in this capacity, he reorganized the French army during the early wars of the Revolution. In 1795 he was chosen to be one of the five directors, or chief executives, of the new French revolutionary government, a position of great power. After the rise of Napoleon, Carnot held the post of minister of war.

While some individual mathematicians achieved positions of authority during the revolutionary period, the governments they served sought to create new national institutions based on reason. Perhaps the most enduring product of their efforts was the metric system of weights and measures. Traditional French units of measure varied widely by locale and involved a dizzying array of different mathematical conversions. For example, the *toise*, a unit of length, contained six *pieds* (roughly equivalent to an English foot). In 1790 the new French government established a commission to create a more rational approach to measurement. Inspired by revolutionary fervor and by the Enlightenment emphasis on natural law, the commission wished to establish units based on natural phenomena. The meter was thus defined as one ten-millionth of the distance from the North Pole to the Equator, along a given line, or meridian, of longitude. In addition, the system of measures would be purely decimal in nature; units would be related to each other by multiples of 10.

The commission led to the work of Jean-Baptiste Joseph Delambre and Pierre François André Méchain (see the Introduction for a more complete discussion). The two mathematicians were tasked with measuring the exact distance of a section of the meridian running through Paris and extending through much of France and Spain. This would lay the foundation for the actual length of the meter. Errors in some of Méchain's data resulted in the adoption of a flawed value; indeed, one may argue that variations in the shape of the earth made the enterprise essentially impossible (Alder 2002, 297–98). However, the efforts of the two men led to the establishment of the standard meter in 1799.

The other units of the metric system proceeded from the meter. One liter was equal to a cubic decimeter; that is, the volume of a cube with a side one-tenth of a meter in length. One gram was equal to the weight of one cubic centimeter (or one milliliter) of water. The system reflected the Enlightenment priorities of the revolutionary French government, but it was not used for long. Napoleon Bonaparte dismissed it and returned France to its traditional measures in 1812 (Alder 2002, 316). The metric system would not be established again in France until later in the 19th century. It would become the global standard for weights and measures in the 20th century, and with some changes, it continues to serve as the foundation of the SI system, the International System of Units.

NON-EUCLIDEAN GEOMETRY

In the 1660s and 1670s the philosopher Baruch Spinoza wrote the *Ethics*, a systematic treatment of existence, knowledge, and human thought. Published in 1677, the book consists of 253 propositions, treated sequentially and based on a series of definitions, axioms, and postulates. Spinoza wished to construct a philosophy of certainty based on pure reason, and the model he used in shaping his work was considered a paragon of reasoning in the 17th century: Euclid's *Elements*.

The *Elements* was one of the most influential books in world intellectual history. It represented the foundations of the geometrical tradition of Hellenistic Greek culture, it played an important role in medieval Arabic mathematics, and it formed an essential part of the educational framework of medieval and early modern Europe. Many of the European philosophers and scientists of the 17th and 18th centuries used Euclid as an example of the power of deductive thought. In addition to Spinoza, David Hume employed a Euclidean example to illustrate the operations of pure mind, and Immanuel Kant argued that the Euclidean concept of space represented a form of reasoning inherent to all human beings. Kant went on to base much of his philosophy of knowledge upon such reasoning. Thus, in addition to its mathematical importance, Euclid's work had great significance in Western culture as an educational standard and as a hallmark of rational intuitive thinking.

Euclid's *Elements* begins with the definitions and axioms necessary for the rest of the work. The author also includes five postulates, statements that serve as the basis for further reasoning. Euclid's fifth postulate states:

> That, if a straight line falling on two straight lines makes the interior angles on the same side less than two right angles, the two straight lines, if produced indefinitely, meet on that side on which are the angles less than the two right angles.

In other words, if a line crosses two other lines in such a way that it forms acute angles with each on the same side, the two lines will ultimately cross. The statement has important implications. For example, if the two lines form right angles with the first line, then they will be parallel and never meet. As a result, the fifth postulate is also referred to as the parallel postulate (Figure 11.3).

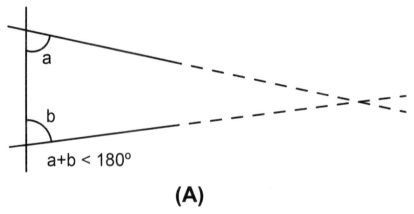

Figure 11.3 The proof of Euclid's parallel postulate challenged mathematicians for centuries.

The parallel postulate received considerable attention from notable Middle Eastern mathematicians during the Middle Ages. Ibn al-Haytham (Alhazen) attempted to prove the postulate in the early 12th century, and Nasir al-Din al-Tusi tried again in the 13th century. In the modern period, Girolamo Saccheri published a work on Euclid in 1733 that included a detailed examination of the parallel postulate. Saccheri constructed a quadrilateral with equal sides perpendicular to a common base and explored the possibilities that resulted (Figure 11.4).

Figure 11.4 Saccheri's attempts to prove the parallel postulate were ultimately unsuccessful.

In Euclidean geometry, if the two equal sides cross a second line, they will form right angles to that line, and the line will be parallel to the base of the quadrilateral. Saccheri recognized two other possibilities: that the sides would intercept a second line at acute angles (less than 90 degrees) or obtuse angles (greater than 90 degrees). He attempted to show that both of these cases were logically contradictory and that the Euclidean view was the only one that was mathematically possible. He believed that he had succeeded, but in fact, his analysis was flawed, and he had unknowingly opened the door to non-Euclidean approaches to geometry (Merzbach and Boyer 2011, 403–404). Later in the 18th century, a German mathematician, Johann Heinrich Lambert, also tried to prove the parallel postulate but did not succeed. In the process, he recognized that attempts to prove the postulate seemed to assume its validity. That is, mathematicians, including himself, were biased in favor of the Euclidean position; they could not conceive that alternative geometries were possible (Merzbach and Boyer 2011, 421).

In the first half of the 19th century, Nikolai Ivanovich Lobachevsky took the step of constructing a non-Euclidean approach to geometry, using the parallel postulate as a starting point. He began with a derived version of the postulate: given a line "L" and a point "p" that was not part of line "L," there is only one line on a plane that could be drawn through "p" that would not intersect line "L" (clearly, this second line would be parallel to "L") (Figure 11.5).

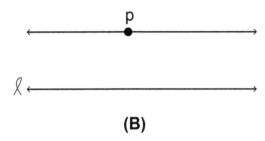

(B)

Figure 11.5 Lobachevsky explored a correlate of the parallel postulate. His work revealed that non-Euclidean views of space were logically possible.

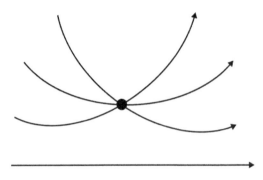

Figure 11.6 In curved space, multiple lines through a given point may be conceived that do not cross a second line.

As Lobachevsky tried to prove this statement, he realized that an alternative geometry was conceivable in which multiple lines could be drawn through point "p" that did not intersect line "L" (Figure 11.6).

In Lobachevsky's new geometry, space itself was conceived as curved, and in curved space, parallel lines had different characteristics. (When one represents such lines in Euclidean flat space, as in Figure 11.6, they appear as curves). This geometry proceeded from different initial assumptions about the nature of space, but it had no internal logical contradictions; it was not mathematically impossible as Saccheri had argued. Euclidean geometry had defined space in a specific way, namely, in a manner in which two lines perpendicular to the same line would always be parallel (the parallel postulate). This definition of space appeared to conform to human intuition and sensory experience. Lobachevsky's work demonstrated that other notions of space and geometry were internally consistent, even if they seemed to depart from traditional thought. He first published his views in 1829, and in 1835 he wrote his *New Foundations of Geometry*.

Other 19th-century thinkers pursued similar ideas. In Hungary, Janos Bolyai wrote a non-Euclidean response to the parallel postulate, but published it after Lobachevsky. Most importantly, G.F.B. Riemann established a broad view of non-Euclidean geometry and linked it to other forms of complex mathematical analysis. Riemann introduced his concept of geometry in a famous lecture at the University of Göttingen in Germany in 1854. He argued that multiple geometries were possible, based on rules applied to sets of parameters and dimensions. For Riemann, Euclidean geometry is a particular case in which the parameters that define space form a specific pattern. His ideas form the basis of the geometrical and topographical studies that still interest modern mathematicians.

The discovery of non-Euclidean geometry had significant cultural implications. Euclid had served as a towering example of the power of human reason, and the idea of multiple geometries challenged this picture. As discussed earlier, the philosopher Immanuel Kant had used Euclidean geometry as an example of the universal character of human rational thought, an idea crucial to the rest of his philosophy. The rise of non-Euclidean geometry complicated such a view. One could argue that the new geometries might apply to the natural world, that actual physical reality might not be Euclidean. Indeed, Riemann's work provided scientists with powerful mathematical tools for the study of nature. In the early 20th century, such analysis would form the basis of Albert Einstein's general theory of relativity, a theory that would revolutionize the human understanding of the universe itself.

KARL MARX AND MARXISM

The ideas of Karl Marx (1818–1883) proceeded from both the Enlightenment and the Industrial Revolution. Enlightenment ideology stressed the importance of rational, predictive natural law, as modeled by the physics of Isaac Newton. Newton's achievement demonstrated the ability of the human mind to discover such mathematical natural laws. The Industrial Revolution transformed European methods of production and vastly increased the output of material goods, but it in the process it treated human labor as a mere commodity. Workers were regarded as a production expense by management; their wages were subject to the laws of supply and demand and their identities reduced to a money value paid per hour. Marx, in the spirit of the Enlightenment, wished to establish natural laws for human history. He also drew upon the developing social sciences; he described the relationship between industrial workers and their employers using the ideas of Adam Smith and other early economists. At the same time, he rejected the harsh treatment of industrial workers and proposed a theory of industrial production and wealth distribution that he believed would significantly enhance their standard of living. His ideas embodied the Enlightenment concept of predictive natural law and the quantitative analysis of economic structures pioneered by the classical economists that preceded him.

Karl Marx was German, but because of his ideas he lived in exile in London for much of his life. Modern readers justifiably associate him with the origins of socialist thought, for his writings inspired many of the socialist and communist regimes of the 20th century. Yet in many ways, Marx was a historian first. Affected by the Enlightenment and by the German philosophies that shaped his education, he wished to explain historical events as expressions of natural law. Marx emphasized the notion of historical materialism, the concept that events are caused by material factors. In other words, history could be explained in terms of people seeking wealth and the necessities of life.

In simple terms, Marx's view of history would explain the past as the product of material causes. For example, he might claim that the Crusades of the Middle Ages, the wars between Christians and Muslims over possession of Jerusalem, were inspired more by the desire for land among European nobles than they were by religion. However, Marx's materialism went further than that. He defined social classes in terms of their relationship to property. The upper classes owned the "means of production"—the farmland, mines, and factories that served as the basis of agriculture and industry. The lower classes—factory workers and agricultural peasants—did not. Rather, they worked for members of the upper classes in return for an hourly wage. Marx claimed that as members of the lower classes came to recognize the injustice of their situation, they would develop "class consciousness." This would lead them to band together and challenge the dominant classes. Such class struggles could only end with the victory of the lower classes and the establishment of a society in which all shared wealth equally. Marx believed that class struggle was inevitable, and he described the historical stages through which a society must pass in its march towards the ultimate triumph of the industrial working class (the proletariat, in his terminology).

Marx's approach to history was essentially predictive; he provided rational explanations for its progress in a particular direction. He did not predict individual events; rather, he laid out the general path that events must follow as society moved towards the socialist world he envisioned, a world where all enjoyed equal access to material wealth. To reach this stage, the private property of the upper classes would be confiscated as part of the greater class struggle. In order to justify this new order of society, Marx proposed a labor theory of value.

In *Das Kapital* (1861), Marx linked his notion of historical materialism to an analysis of capitalist economics. He adopted many of the ideas of Adam Smith and other classical economists and directed them towards his notion of class struggle. Central to his argument was the concept of surplus value—the difference between the sale price of an item and the cost of producing it (including wages, the cost of labor). For example, in a tire factory, the owner pays for rubber, factory expenses, and the wages of workers and produces tires. Those tires are sold for more than it took to produce them. The owner (or owners—the stockholders of the company) keep this profit, or surplus value, a reward for their enterprise.

For Marx, this capitalist reality represented a systematic oppression of industrial workers. Marx adapted a version of the labor theory of value present in the work of Smith and Ricardo; he argued that surplus value originated with labor. In other words, workers' labor was the source of the extra value present in the tires, the value that became profit. Without labor, that extra value would never be produced. When owners kept the lion's share of profits and paid labor only a small hourly wage, they were essentially stealing surplus value from their workers.

Das Kapital was an immensely complex work, but Marx intended it to provide economic justification for his revolutionary understanding of history. For him, the sharing of material wealth among workers in an ideal society was a just response to the wrongs of capitalism. Because of this, he worked hard to bring about the revolution that he predicted. While he saw a socialist society as inevitable, he wished to bring it about as soon as possible. Other socialist theories existed in the 19th century, but Marx's unique combination of historical natural law, economic analysis, and personal analysis made his ideas particularly appealing to many. His concepts would form the basis of most socialist and communist politics in the 20th century.

JOHN SNOW AND STATISTICAL EPIDEMIOLOGY

Cholera is primarily a water-borne disease; it is often spread through the contamination of water by sewage. The disease is caused by *Vibrio cholerae*, a bacterium often found in shellfish. Symptoms include vomiting, diarrhea, and muscle contractions, and the disease can lead to fatal dehydration. The first known large-scale outbreak of cholera began in South Asia in 1817; it spread throughout India, China, and southeast Asia before it subsided in 1824. Mortality from the epidemic was horrific, and hundreds of thousands died.

A second pandemic started in the late 1820s, again in South Asia. It moved through the Middle East and Russia, reached Great Britain and France in 1832, and crossed the Atlantic to North America. Cholera outbreaks continued around the world throughout the 19th century. European physicians believed that the disease was spread through poisons in the air, and they had no effective treatments for it. However, during the pandemic of the 1850s, John Snow, an English physician, used statistical mapping to correctly identify cholera's mode of transmission and to recommend effective public health measures to control its spread. While his work was considered controversial at the time and was not immediately accepted, he is considered to be one of the founders of the science of epidemiology.

Snow was born in 1813 to a working-class family in the city of York. As a young man, he worked as a medical apprentice and eventually went to medical school at the University of London. He earned his medical degree at the age of thirty-one. Snow was an innovative young physician and embraced the growing use of anesthesia in medicine to reduce or eliminate the pain of surgery. He became an expert in the administration of ether and chloroform, and he provided anesthesia to Queen Victoria during the births of her last two children in the 1850s.

Snow's willingness to break new ground in medicine was also evident in his response to cholera. He rejected the prevailing opinion that cholera was an airborne disease and argued that the disease was spread by contact with human waste and tainted water. When the disease broke out in London in 1854, he put his new ideas to the test. A large cluster of cholera deaths had affected the Soho region of London. Snow gathered data on cholera incidence in the area, noting the number of cases and their locations. His analysis, combined with his ideas on the waterborne transmission of the disease, allowed him to identify its likely source. Snow argued that a water pump on Broad Street appeared to be associated with the majority of cases and prevailed upon the authorities to deactivate the pump by removing the handle. His arguments proved correct, and the number of cases in the area dropped significantly. Snow also prepared a statistical map of cholera occurrence in London, which included graphic data on the number of cases in different neighborhoods (Figure 11.7).

Figure 11.7 John Snow mapped the location and frequency of cholera cases in 1854. His work represents the first systematic use of statistics in public health. (Wellcome Collection, London)

London had grown very quickly in the late 18th and early 19th centuries, and the city had not constructed sanitary systems to respond to increased population pressures. Often, raw sewage flowed into the Thames River. Snow noticed from his statistical maps that neighborhoods close to the river had greater incidence of cholera than those located at higher elevations. He also designed an experiment to determine the effect of Thames River water on cholera incidence. He compared the rates of disease among clients of two different waterworks companies. The Lambeth Company obtained its water from the Thames upstream from the densely populated part of the city, while the Vauxhall Company drew water from the river in London itself. The latter company had a substantially higher rate of cholera among its customers:

> I accordingly asked permission at the General Register Office to be supplied with the addresses of persons dying of cholera, in those districts where the supply of the two Companies is intermingled in the manner I have stated above. Some of these addresses were published in the "Weekly Returns," and I was kindly permitted to take a copy of others. I commenced my inquiry about the middle of August with two sub-districts of Lambeth, called Kennington, first part, and Kennington, second part. There were forty-four deaths in these sub-districts down to 12th August, and I found that thirty-eight of the houses in which these deaths occurred were supplied with water by the Southwark and Vauxhall Company, four houses were supplied by the Lambeth Company, and two had pump-wells on the premises and no supply from either of the Companies.
>
> (Snow 1855)

Snow's statistical analysis made a convincing case for the waterborne transmission of the disease. While his ideas initially met resistance, continued outbreaks of cholera, coupled with increasing pollution of the Thames, led to the construction of a new system of sewers in the city in the 1860s and 1870s under the guidance of Joseph Bazalgette, the chief engineer of London's Metropolitan Board of Works.

John Snow pioneered the practice of examining the spread of a disease in its geographical and social context and gathering numerical data in order to design preventative measures. At a time when physicians were largely powerless to cure cholera patients, he strove to understand the transmission of the disease and to reduce its incidence. His work did much to establish the data-driven disciplines of epidemiology and public health, and 21st-century people living in a time of pandemic can especially appreciate the relevance of his efforts.

PERIODIC TABLE OF THE ELEMENTS

Modern chemistry is a quantitative science; it describes the behavior of matter in terms of measurable parameters and predictive natural laws. The periodic table of the elements serves as a potent symbol of this mathematical character. In the contemporary version of the table, the elements are arranged in order of increasing atomic number, that is, the number of protons in the nucleus of an atom. In an electrically neutral atom, the number of electrons surrounding that nucleus corresponds to the number of protons within, and these electrons determine the chemical reactivity of the element. The rows and columns of the table reflect the properties of the elements they contain, based on the distribution of electrons within the atom. The periodic table also tells the atomic weight of each element, the sum of the protons and neutrons in the nucleus; atomic weight increases steadily with atomic number, with some minor variations. Finally, the periodic table is often grouped with additional tables that contain more quantitative information about the elements and their combinations: electronegativity, solubility curves, densities, boiling points, etc.

The application of quantitative measurements to chemical processes has grown over hundreds of years. Throughout antiquity and the Middle Ages, practical chemistry around the world was based on the knowledge that materials had to be mixed in specific proportions to produce desired products: in metallurgy, in cooking and baking, in the production of gunpowder, in the making of soap. On a less mundane level, the alchemists and chemists of China, the Middle East, and Renaissance Europe also described their procedures in quantitatively specific terms. However, the notion that chemical phenomena could be described in terms of precise mathematical natural law arose in the 17th century, when Robert Boyle, with the assistance of Robert Hooke, described the inverse relationship between the volume of a gas and its pressure. Boyle's law, like Newton's laws of motion, had a predictive character. In the late 18th century, the French chemist Antoine Lavoisier discovered the process of oxidation and linked it to the combustion of fuel. He attempted to determine the amount of heat that would be produced by the generation of a given amount of carbon dioxide (or "fixed air," in 18th-century usage). However, the idea that the structure of matter itself could be described mathematically really originated in the work of John Dalton (1766–1844).

The idea of atoms as the basis of matter originated in ancient Greece. The Greek atomists of the 5th century BCE believed that the physical universe was composed of indivisible particles, or *atomos*, and the Epicurean thinkers of the late 4th century adopted this idea. Epicurean ideas were transmitted to Western Europe during the Renaissance and inspired a new understanding of material nature. Over the course of the 17th century, a number of European natural philosophers proposed particulate views of matter. For example, Robert Boyle wrote that an atomist approach explained his experimental results well and suggested that matter might be composed of indivisible particles characterized by shape. In other words, an "alphabet" of different particle shapes existed, and these particles could combine to produce more complex bodies, just as the letters in the alphabet combined to produce words. During the Enlightenment of the 18th century, the notion of shaped atoms became a dominant theme in European chemistry.

John Dalton broke new ground by discarding the emphasis on the shape of atoms and focusing on weight as the essential factor in their identity. During the 18th century, European chemists had isolated a number of new elements, that is, substances that could not be broken down into simpler components. Dalton argued that each element was composed of atoms with a characteristic weight, a weight that could be measured. Atoms, the basic building blocks of matter, could thus be described in mathematical terms, and the identity of elements was grounded in a measurable quantity—atomic weight. When atoms combined to produce compounds, the basic unit of the compound could be described in terms of the combined weights of its component atoms. Dalton's approach would result in a quantitative conception of matter, a conception that would transform chemistry in the 19th century.

Other chemists built on Dalton's foundations. Figures such as Jons Jacob Berzelius and Amadeo Avogadro sought to establish the proportions of atomic combination in compounds and to refine the understanding of atomic weight. Berzelius also introduced the modern system of symbols for the elements and produced accurate tables of atomic weights, using the weight of hydrogen as a base unit (Leicester 1971, 156–160). As more new elements were discovered in the 19th century, chemists began to explore different systems for arranging them. John Alexander Reina Newlands noted that as the elements were placed in order of increasing atomic weight, they could be grouped according to similar properties and that these similarities seemed to repeat themselves every eight elements (Leicester 1971, 193). The periodic law, the notion that the elements exhibited a mathematical pattern of repeating physical and chemical properties, was firmly established by the work of Lothar Meyer and Dmitri Ivanovich Mendeleev.

Working independently, both Meyer and Mendeleev categorized the elements based on common properties, while also arranging them in order of weight. Each recognized the periodic nature of the elements based on extensive laboratory work. Mendeleev published his ideas, an early form of the periodic table of the elements, in 1869. He had such confidence in the patterns he found that he predicted the discovery of new elements based on gaps in the table, and his predictions were ultimately confirmed. The periodic table thus represented a quantified ordering of matter itself, as well as a testament to the power of experimental science and observation.

It is crucial to point out that Mendeleev did not provide an underlying theory for the periodic law; his periodic table proceeded from the analysis of observed data. In 1897 the electron was discovered by J.J. Thompson; protons and neutrons were discovered in the early decades of the 20th century. Knowledge of subatomic particles provided a deeper understanding of the periodic table: the identity of the elements proceeded from the number of protons contained in their nuclei, and the periodicity of chemical and physical properties was largely explained by the distribution of electrons in the individual atom. The modern periodic table of the elements embodies the mathematical character of contemporary chemical—and physical—knowledge. Yet its foundations lay in the Enlightenment culture of 18th- and 19th-century science.

ELECTROMAGNETISM AND JAMES CLERK MAXWELL

Throughout the Enlightenment of the 18th century, Newton's laws of motion and universal gravitation served as the model for scientific inquiry. European scientists worked to extend the scope of these laws, to describe nature in terms of Newtonian mechanics. However, some phenomena did not fit neatly into this picture of the universe. For example, the behavior of light was a matter of scientific dispute. Newton believed that light was made up of particles, but his contemporary, the Dutch theorist Christiaan Huygens, proposed a wave theory of light in 1678. He believed that light consisted of waves moving in all directions from a given source. Huygen's ideas were reinforced by the work of Thomas Young in the early 19th century. Young did experiments on the interference patterns created by light moving through parallel slits; these patterns suggested that light was a wave phenomenon. Between 1815 and 1822, Augustin Fresnel worked to produce a mathematical description of the behavior of light waves. His work resulted in the acceptance of the wave theory of light throughout the 19th century.

Electricity and magnetism also posed difficult questions. In 1800 Alessandro Volta invented a type of chemical battery; his "Voltaic Pile" allowed for the study of electrical current. Hans Christian Oersted discovered the relationship between electricity and magnetism in 1820, when he found that a magnetic compass needle was affected by an electrical current. However, the figure most important to the study of electricity in the 19th century was Michael Faraday.

Michael Faraday designed and conducted dozens of experiments on electricity and magnetism from the 1820s to 1860s. He used iron filings to study the "lines of force" that surrounded magnets and electromagnets, and he measured their intensity. This work led him to suggest the existence of magnetic and electrical fields. In 1831 he discovered that an electrical current moving through a coil of wire would create current in an adjacent coil. He also found that a magnet moving inside or around a coil of wire would create, or induce, an electrical current in that wire. Such results led him to formulate the concept of electromagnetic induction, the notion that a moving magnetic field creates an electrical field. Faraday was a brilliant experimenter, but he lacked the mathematical training to thoroughly analyze his data. In the 1860s, Faraday's work would form the basis of a mathematical description of field theory published by James Clerk Maxwell.

Maxwell was born in Edinburgh, Scotland, in 1831. He was a brilliant young student and wrote his first original paper on geometry when he was fourteen. He studied natural philosophy and mathematics at the University of Edinburgh and at Cambridge University and became a member of the faculty of Marischal College in Aberdeen, Scotland. While he was there, he did a mathematical analysis of the rings of Saturn that established his scholarly reputation. In 1860 he took an academic post at King's College in London, and it was there that he focused his attention on electromagnetism.

In 1861 Maxwell published *On Physical Lines of Force*, a paper that analyzed Faraday's results and expressed them as a series of differential equations. He had begun the work when he was still a student at Cambridge. In essence, Maxwell created a mathematical model describing the relationship between electricity and magnetism. He went on to demonstrate mathematically that electromagnetic fields would produce waves that traveled through space at the speed of light and suggested a link between such waves and light waves. In other words, Maxwell maintained that light was part of a broader set of electromagnetic phenomena. He published his work in 1873 under the title *A Treatise on Electricity and Magnetism.*

The mathematical models of James Clerk Maxwell became one of the foundations of modern physics. His equations were condensed into four partial differential equations by Oliver Heaviside in 1881, and they served as the basis for most later work in electromagnetism. Later in the 1880s, Heinrich Hertz would perform a series of experiments that confirmed Maxwell's prediction of the existence of electromagnetic waves (in this case, radio waves) and their relationship with visible light. In addition, discrepancies between the predictions of Maxwell's mathematics and the work of later experimenters would inspire Albert Einstein to re-examine the nature of electromagnetic waves and their motion. Einstein would set aside some of the fundamental assumptions of a Newtonian view of the universe and propose his special theory of relativity. Maxwell's mathematical model had unified 19th-century research in light and electromagnetism and made further insights possible.

CONCLUSION

In the 18th and 19th centuries, a growing command of mathematics and its applications altered European culture in fundamental ways. The pursuit of scientific knowledge through Newtonian methods resulted in a new understanding of nature, while the growth of the social sciences changed the way European humanity thought about society itself. The development of powered machinery put Western Europe in an unprecedented position of wealth and influence. Yet these changes had a dark side: the exploitation of labor, the invention of military technologies with great destructive power, the expansion of imperialism, the beginnings of industrial pollution, and the emergence of a depersonalized world, where people themselves might be referred to as mere names—or numbers—on a list. These developments, positive and negative, would continue in the 20th century, as Western ideas and cultural trends would acquire a truly global influence.

FURTHER READING

Alder, Ken. 2002. *The Measure of All Things*. New York: The Free Press
Eliot, Simon and Stern, Beverly. 1979. *The Age of Enlightenment*. New York: Harper and Row.
Gay, Peter. 1967. *The Enlightenment*. New York: Alfred Knopf.
Gay, Peter, ed. 1973. *The Enlightenment, A Comprehensive Anthology*. New York: Simon and Schuster.
Heilbroner, Robert. 1999. *The Worldly Philosophers*. New York: Touchstone Books.
Hobsbawm, Eric. 1962. *The Age of Revolutions*. New York: Vintage Books.
Leicester, Henry. 1971. *The Historical Background of Chemistry*. Mineola: Dover Publications.
Merzbach, Uta and Boyer, Carl. 2011. *A History of Mathematics*. Hoboken: John Wiley and Sons.
Montesquieu. 1977. *The Spirit of the Laws*. Berkeley: University of California Press.
Osen, Lynn. 1974. *Women in Mathematics*. Cambridge: MIT Press.
Schiebinger, Londa. 1989. *The Mind Has No Sex*. Cambridge: Harvard University Press.
Smith, Adam. 1973. *The Wealth of Nations*. In *The Enlightenment, A Comprehensive Anthology*, edited by Peter Gay. New York: Simon and Schuster.
Snow, John. 1855. *On the Mode of Communication of Cholera*. London: John Churchill. UCLA Department of Epidemiology, Fielding School of Public Health, https://www.ph.ucla.edu/epi/snow/snowbook3.html

12

Women and Numbers

CULTURAL ICON: MARYAM MIRZAKHANI'S FIELDS MEDAL

The Fields Medal is perhaps the most distinguished award in mathematics (Figure 12.1). The International Mathematical Union (IMU) presents the award every four years; winners must be under the age of forty. In 2014, the IMU honored Professor Maryam Mirzakhani of Stanford University for her work on the mathematics of Riemann surfaces. Mirzakhani was the first woman to win the Fields Medal.

Maryam Mirzakhani was born in 1977 in Tehran, Iran, and grew up during the period of the Iranian Revolution. While she had many intellectual interests as a young woman and attended an elite school, she did not study mathematics seriously until her final year. At that time, she trained for the International Mathematical Olympiad, won a place on the Iranian team, and took a gold medal at the Olympiad competition in Hong Kong. She repeated her performance the next year in Toronto.

Mirzakhani enrolled at Sharif University in Iran and excelled as she pursued her mathematics degree. She graduated in 1999 and went to Harvard University in the United States for her graduate study. While at Harvard, she worked on Riemannian geometry, a form of non-Euclidean geometry, and wrote a ground-breaking dissertation in which she approached the solution of a Riemannian problem in terms of another branch of geometry, moduli spaces. She received her PhD in 2004, and her success led to a faculty position at Princeton University.

Figure 12.1 The Fields Medal is awarded every four years to outstanding mathematicians. Maryam Mirzakhani won the Fields Medal in 2014. (Stefan Zachow/National Science Foundation)

While at Princeton, Professor Mirzakhani continued her research in Riemannian geometry, collaborating with Alex Eskin at the University of Chicago. They published a highly successful paper in 2013. Her work at Princeton, and later at Stanford, earned her great academic acclaim. In addition to her Field Medal in 2014, she was elected to a number of prestigious scientific societies, including the Paris Academy of Science and the National Academy of Sciences. However, she tragically died of cancer in 2017 at the young age of forty. She left a legacy of outstanding achievement and great intellectual creativity, and she serves as a role model for aspiring mathematicians around the world.

CULTURAL COMMENTARY: WALKING ON THE GRASS

In *A Room of One's Own*, published in 1929, Virginia Woolf wrote of her experiences visiting the campus of a major British university. Lost in her thoughts while walking, she happened to stray off the path and onto the grass and was quickly confronted by the university staff:

> It was thus that I found myself walking with extreme rapidity across a grass plot. Instantly a man's figure rose to intercept me. Nor did I at first understand that the gesticulations of a curious-looking object, in a cut-away coat and evening shirt, were aimed at me. His face expressed horror and indignation. Instinct rather than reason came to my help, he was a Beadle; I was a woman. This was the turf; there was the path. Only the Fellows and Scholars are allowed here; the gravel is the place for me. Such thoughts were the work of a moment. As I regained the path the arms of the Beadle sank, his face assumed its usual repose, and though turf is better walking than gravel, no very great harm was done. The only charge I could bring against the Fellows and Scholars of whatever the college might happen to be was that in protection of their turf, which has been rolled for 300 years in succession they had sent my little fish into hiding.
>
> (Woolf 1989, 6)

Woolf's reaction to the encounter was relatively benign; she expressed disappointment at having been distracted from her train of thought (her "little fish"). Later in the same visit, she approached one of the university's libraries but was refused entry:

> ... but here I was actually at the door which leads into the library itself. I must have opened it, for instantly there issued, like a guardian angel barring the way with a flutter of black gown instead of white wings, a deprecating, silvery, kindly gentleman, who regretted in a low voice as he waved me back that ladies are only admitted to the library if accompanied by a Fellow of the College or furnished with a letter of introduction.
>
> That a famous library has been cursed by a woman is a matter of complete. indifference to a famous library. Venerable and calm, with all its treasures safe locked within its breast, it sleeps complacently and will, so far as I am concerned, so sleep for ever. Never will I wake those echoes, never will I ask for that hospitality again, I vowed as I descended the steps in anger.
>
> (Woolf 1989, 8–9)

Clearly, this second episode of rejection inspired a stronger reaction from the author, and she referred to her own anger. Woolf told this story early in her essay because she wished to make a point about the deliberate exclusion of women from academic and literary circles in the early 20th century. She went on to point out that men controlled the sources of wealth in British society and that they used that wealth to support intellectual institutions—particularly universities—that were reserved for their own participation. Women had relatively little wealth, and this allowed them to be barred from such institutions in the name of tradition. As an author of novels, Woolf was primarily concerned with the literary world. She maintained that "a woman must have money and a room of one's own" in order to write fiction. However, the experience she related in *A Room of One's Own* serves to epitomize the general exclusion of women from institutions of higher learning in Europe, and indeed, most of the world, until the late 19th and 20th centuries. This exclusion would have a significant impact on the ability of women to pursue work in natural philosophy, science, and mathematics.

European universities first emerged during the High Middle Ages as establishments of the Roman Catholic Church. Intellectual life in Western Europe before the 11th century was centered in monastic communities that emphasized study and the copying of books. While male monasteries dominated such activities, double monasteries—communities with both male and female components—also played a role. During the Carolingian period of the 9th century, such houses were discouraged, and the intellectual life of female monastics declined as a result (Noble 1992, 100–104). In subsequent centuries, the church hierarchy placed heavy emphasis on clerical celibacy and further reduced the interaction of male clergy with women. The cathedral schools and universities of the high medieval ages (1000–1350) were founded in this atmosphere of suspicion towards women. This ensured that they would be purely male institutions, staffed by male faculty largely isolated from female contact (Noble 1992, 146–148). While they would cease to be church institutions in the modern period, European universities would retain this male-centered ethos until the 20th century, as demonstrated by the experiences of Virginia Woolf.

The intellectual energy of Italian cities during the Renaissance provided an alternative to the world of the universities. Italian Humanists, with their positive view of human potential, would challenge many aspects of earlier medieval intellectual life. For example, Christine de Pisan (1364–1430) used Humanist ideals to question the status of women in the literary culture of her time. As a well-read woman in the Late Middle Ages, Christine rejected the negative image of the feminine presented in many works of classical literature and medieval theology. She mounted a defense of women's virtue and capabilities in *The Book of the City of Ladies*. Speaking through the voice of Lady Reason, a female personification of scholarly virtue, she pointed out the nobility of women and asserted their intellectual and moral equality with men:

> ... she was created in the image of God. How can any mouth dare to slander the vessel which bears such a noble imprint? But some men are foolish enough to think, when they hear that God made man in his image, that this refers to the material body. This was not the case, for God had not yet taken a human body. The soul is meant, the intellectual spirit which lasts eternally just like the Deity. God created the soul and placed wholly similar souls, equally good and noble in the feminine and in the masculine bodies.
>
> (Christine de Pisan 1982, 23)

According to Christine de Pisan, men and women had similar souls, similar "intellectual spirits," created in the image of the Christian god. Yet while Italian Humanist ideas certainly inspired her views, such ideas did not substantially alter the access of women to education and scholarship in Renaissance Italy.

Renaissance scholarship resulted in the reintroduction of many works of Classical Greek literature and philosophy to Western Europe in the 15th and 16th centuries, and these inspired great intellectual changes. During the "Scientific Revolution" of the 17th century, medieval ideas lost their appeal, and European thinkers embraced a variety of new approaches to knowledge and to nature. In this period of innovation and ferment, women participated in the philosophical conversations of the day to a significant extent. Noble female patrons supported scholars and engaged them in their work (Schiebinger 1989, 44–47). A number of female authors—Margaret Cavendish, Anne Conway, Mary Astell—wrote extensively on issues of natural philosophy. In her *A Serious Proposal to the Ladies* (1694) Mary Astell suggested the establishment of separate colleges for women, and in the second part of her treatise (1697) she commented on the extent of human intellectual powers:

> To sum up all: we may know enough for all the purposes of life, enough to busy this active faculty of thinking, to employ and entertain the spare intervals of time and to keep us from rust and idleness, but we must not pretend to fathom all depths with our short line.
> ... Therefore to be thoroughly sensible of the capacity of the mind, to discern precisely its bounds and limits and to direct our studies and inquiries accordingly, to know what is to be known, and to believe what is to be believed is the property of a wise person.
> (Astell 1994, 105)

Astell spoke of the powers and the limitations of *human* reason in a treatise addressed to women; she apparently saw no difference between the capabilities of men and women. While she pointed out differences in human abilities, she did not believe that those differences were correlated with sex.

However, while some elite 17th-century women found a place in European scholarship, the male-oriented character of European intellectual institutions would remain. This was especially true with respect to the growing disciplines of experimental philosophy, disciplines that would plant the seeds of modern scientific thinking. Isaac Newton did his work in mathematics and physics in the all-male confines of Cambridge University. The Royal Society, an association of natural philosophers founded in 1660 and chartered by King Charles II, was an exclusively male institution; its prominent members included the chemist Robert Boyle and the microscopist Robert Hooke. When the female philosopher Margaret Cavendish wished to come to a meeting of the society in 1667, her request caused great controversy (Noble 1992, 231). Other scientific societies, like the Academie Royale des Sciences in Paris, also excluded women.

This situation persisted throughout the 18th century. Italy was something of an exception, for a number of Italian women found a place in the academic community. In 1678 Elena Cornaro Piscopia was granted a doctoral degree in philosophy by the University of Padua; she also studied theology and mathematics. Laura Bassi received a doctorate in natural philosophy from the University of Bologna in 1732. She served on the faculty for over forty years and did much to spread Newtonian ideas in Italy. Maria Angela Ardinghelli and Celia Borromeo were also both known for their mathematical abilities.

In northern Europe, academic and scientific institutions were exclusively male, and women interested in mathematics and science could only pursue their interests through associations with men (Schiebinger 1989, 65).

For example, Émilie de Bretuil, the Marquise du Châtelet, spent a lifetime studying Newtonian physics and mechanics, but her access to the French intellectual community depended significantly on her relationship with Voltaire and other male scholars. Further, because women had no access to university study, they were limited in their ability to acquire high-level training in fields of interest. In the late 18th century, Sophie Germain had to use an assumed male name to submit her mathematical work to the Ecole Polytechnique in Paris; she used the same name in her correspondence with the German mathematician Carl Friedrich Gauss.

Enlightenment thinkers embraced John Locke's views on human equality and fundamental rights, but they did not apply these notions of equality to women. In 1798 Mary Wollstonecraft published *The Vindication of the Rights of Woman* in England. Early in the book, she explicitly rejected the idea that women were subject to male authority:

> I love man as my fellow; but his sceptre, real, or usurped, extends not to me...

Wollstonecraft did not make an open statement of female equality, but she pointed out that women could not participate effectively in society at large because they were denied a proper education. She insisted that the notion of feminine virtue that prevailed at her time robbed women of autonomy and independence:

> My own sex, I hope, will excuse me, if I treat them like rational creatures, instead of flattering their fascinating graces, and viewing them as if they were in a state of perpetual childhood, unable to stand alone.
>
> (Wollstonecraft 2008, xxviii)

Education played a significant role in her discussion, and she deplored the idea that the education of young women should have a domestic focus. Instead, she suggested that the young men and women be schooled together. Such an "experiment" would allow women to demonstrate their capabilities and contribute to society:

> Let an enlightened nation then try what effect reason would have to bring them back to nature, and their duty; and allowing them to share the advantages of nature and government with man, see whether they will become better, as they grow wiser and become free. They cannot be injured by the experiment; for it is not in the power of man to render them more insignificant than they are at present.
>
> (Wollstonecraft 2008, 229)

The late 18th and early 19th centuries were times of great change in Europe and the United States. The ideas of the Enlightenment and the Romantic movement, the events of the French Revolution, and the rise of industrial technology contributed to the erosion of traditional social structures. As the middle classes became an important social and political force, new voices called for the removal of traditional barriers to women's education and political participation. In 1848 the Seneca Falls Convention met in New York State, the first women's rights convention in history. The British philosopher John Stuart Mill published *The Subjection of Women* in 1869 and called for female equality in politics, education, and the family. In the United States, collegiate education became increasingly available to women. Georgia Female College (now Wesleyan College) opened in 1836. In 1837 Mount Holyoke College for women was founded in Massachusetts, and Oberlin College in Ohio began to admit female students as a coeducational institution. A series of women's colleges opened in the middle of the century: Vassar College in 1861, Wellesley in 1870, Smith in 1871. After the Civil War, colleges for African American women were founded; perhaps the most famous of these is Spelman College, founded in 1881 as the Atlanta Baptist Female Seminary. Europe was slower to develop higher education for women; Bedford College was founded in London in 1849.

While women's colleges provided a forum for undergraduate education, advanced training in science and mathematics was still largely closed to women for much of the 19th century. The career of Sonya Corvin-Krukovsky Kovalevsky illustrates the point. Born in Moscow in 1850, she showed great mathematical promise as a young woman and sought a university education. Russian universities did not admit women at the time (the late 1860s), so she entered into a marriage of convenience and traveled to Germany to study. Kovalevsky took classes at the University of Heidelburg, but she needed permission to do so; she then wished to study with Professor Karl Weierstrass at Friedrich Wilhelm University in Berlin, but could not be admitted because of her sex. Weierstrass agreed to teach her privately, and she earned her doctorate in 1874. However, no university would offer a faculty position in mathematics to a woman at that time; she was finally offered a professorship at the University of Stockholm in 1883. In 1888 she won a major prize offered in mathematics by the French Academy of Sciences, but she was not allowed to join the academy itself.

The First and Second World Wars had a profound effect on the status of women in Europe and the United States. A large number of men were required for military service, and women filled a wide variety of industrial and professional jobs. This transformed their role in society. Many countries gave women the right to vote during the First World War or shortly afterwards: Britain in 1918, Germany in 1918, the United States in 1920. During the Second World War, women in the United States with mathematical training were hired by the government as "human calculators" and tasked with generating tables of figures for military purposes, particularly the aiming of artillery. By the end of the war, the U.S government developed ENIAC, the first programmable electronic computer, to do such military calculations. The machine was built at the Moore School of Engineering at the University of Pennsylvania, and six women drawn from the calculating staff completed its initial programming. Women were also hired to do the complex mathematics necessary for wartime aeronautical testing by the National Advisory Committee for Aeronautics (NACA). In 1958 the NACA was merged into the newly formed National Aeronautics and Space Administration (NASA), and its female mathematicians performed computations necessary for the missions of the U.S. space program in the 1960s.

The United States enjoyed a period of tremendous prosperity after the Second World War, and this prosperity helped to fuel movements for racial equality and gender equality in the 1950s, 1960s, and 1970s. Many American universities first opened their doors to women during this period. Advocates for women's rights sought to remove social and political obstacles that stood in the way of female participation in the professional life of the country. During the 1970s and 1980s, women's roles changed considerably, as they entered the workforce in substantial numbers and sought academic and professional training. By 1990 more than half of the college degrees awarded in the United States went to women, and this percentage has continued to increase. American feminism has also had a global impact, and women around the world have been inspired to challenge traditional boundaries and limits on their freedom and professional accomplishments.

Yet despite the achievements of the women's movement, mathematics, science, and engineering continue to be largely male professions in both academic and industrial settings. A number of factors explain the trend, particularly the manner in which young women are educated. Advocates for women in mathematics and science maintain that girls are socialized to avoid these subjects in school, that real bias is directed against women who enter these fields, and that insufficient professional role models exist to guide women through their academic training. As the number of female mathematicians and scientists increases, these problems are being addressed. However, issues of access still exist. Virginia Woolf is allowed on the grass, and in the library, but she is not yet completely welcome.

HYPATIA OF ALEXANDRIA

While it certainly had a turbulent history, Alexandria retained its status as the center of Mediterranean mathematics and astronomy from the time of Euclid to the 5th century CE, spanning the Hellenistic, Roman, and Late Antique periods. Hypatia of Alexandria lived a life of scholarship and study in the tradition of her intellectual forbears. Educated in mathematics and philosophy, she wrote commentaries on the works of a number of Hellenistic mathematicians and astronomers. She also embraced the Neo-Platonic philosophy of Plotinus, and she taught extensively. Caught up in the religious politics of the city in the 5th century, she was murdered by a Christian mob in the year 415 CE. One might argue that the death of Hypatia of Alexandria represents the final decline of the vibrant intellectual atmosphere that had thrived in the city for seven centuries.

Our sources on the life and work of Hypatia are sparse. None of her books survive in their original form, but segments of existing works have been attributed to her by scholars. Hypatia was probably born around the year 370. Her father, Theon of Alexandria, was a mathematician and astronomer in his own right, best known for editing Euclid's *Elements* and part of Ptolemy's *Almagest*, removing centuries of copying mistakes from the texts. The sources also refer to him as a member of the Mouseion, possibly a Late Antique version of the vanished Hellenistic institution. Hypatia was probably educated by her father and shared his interests in philosophy, mathematics, and astronomy, although she clearly surpassed him as a teacher and scholar.

Hypatia achieved great fame as a teacher of mathematics and philosophy, and she clearly played a central role in the intellectual life of the Alexandria of her day. She authored a number of mathematical books, but none have survived. She wrote a commentary on the *Arithmetica* of Diophantus, a collection of algebraic problems written in the 3rd century BCE. Some historians argue that parts of her work were incorporated into Diophantus's text itself and can be identified. She also produced a commentary on the *Conics* of Apollonius of Perga, and she edited the third book of Ptolemy's *Almagest*, and perhaps other parts as well. Philosophically, she embraced the ideas of Plotinus, a Neo-Platonic thinker of the 3rd century CE. Plotinus adopted Plato's notion of pure, abstract truths, but he stressed the existence of the One, the Good, in which all truths reside. The material universe emanates necessarily from the One's existence, but the One does not create the universe through an act of will. Such ideas may have appealed to Hypatia as a mathematician and astronomer; they address the issue, so important to the Greeks, of why the universe exhibits mathematical patterns. At a time when Christianity was growing dominant in Alexandria, she adhered instead to a Neo-Platonic view of pure Divine Mind.

As a prominent figure in Alexandrian society, Hypatia played a significant role in public life. She apparently acted as an adviser to Orestes, the Roman prefect of Alexandria (at this point in time, Alexandria was under the control of the Eastern Roman Empire). In the year 415, Orestes challenged the policies of Cyril, the new Christian bishop of Alexandria. Because of her association with the prefect, Hypatia was attacked in the streets by a group loyal to Cyril and brutally murdered. Her death has often been interpreted by historians as symbolic of the decline of the intellectual spirit of Alexandria in the 5th century in the face of the growing power of Christianity. For example, Edward Gibbon, the 18th-century author of the *Decline and Fall of the Roman Empire*, argued that her murder cast a shadow on the Church in Alexandria. Modern feminist scholars point out that her status as a female intellectual made her unacceptable to the male-dominated hierarchy of the Christian church. To say that Alexandrian philosophy and mathematics died with Hypatia in 415 is an oversimplification; schools of philosophy continued to operate there throughout the 5th century. However, her murder does serve as a potent sign that the intellectual focus of the city, and indeed, of the entire Mediterranean, was turning towards the Christian religion at the close of classical antiquity. The "Age of Faith" was dawning, and the era of classical philosophy and mathematics was drawing to a close.

ÉMILIE DE BRETEUIL, MARQUISE DU CHÂTELET

Modern Western thought emerged in a recognizable form during the Enlightenment of the 18th century. While Enlightenment ideas grew in many regions of western and central Europe, France played a particularly important part in the spread of new worldviews. The Enlightenment in France may be conceived of as an intellectual response to 17th-century intellectual innovations, particularly the physics and mathematics of Isaac Newton and the philosophy and political thought of John Locke. It may also be viewed as a project of reform adopted by a community of scholars with close personal and intellectual ties to each other. The life and work of Émilie de Bretuil, the Marquise du Châtelet, illustrates both of these aspects of the French Enlightenment.

Émilie de Bretuil (or du Châtelet) was born in 1706 to a noble family in the court of King Louis XIV of France. While she was educated in languages and the classics, she demonstrated a great talent for mathematics at a young age, and she was tutored in the subject. Marriage was considered a priority for a young noblewoman in the early 18th century, and she was wed at the age of nineteen to the Marquis du Châtelet. She had three children in the early years of her marriage, but she returned to the study of mathematics in her mid-twenties. Because of her social standing, she had contact with many of the leading scholars of the early French Enlightenment. Her instructors in mathematics included Pierre Louis Maupertuis, the first French author to write extensively on Newton, and Alexis Claude Clairaut, a devotee of Newtonian physics. Her most important intellectual and personal relationship was with Voltaire, one of the most prominent authors of the day and a leader of the Enlightenment movement.

The relationship between du Châtelet and Voltaire lasted from 1733 to 1749; he was with her when she died at the age of forty-three. In addition to their romantic attachment, they shared strong bonds of mutual intellectual interest and scholarly collaboration. When Voltaire's work aroused the resentment of the authorities in Paris, the pair moved to the chateau of Cirey-sur Blaise, a possession of her husband's family. They lived there for fifteen years and maintained close ties with other Enlightenment thinkers.

During these years, du Châtelet continued her mathematical work and studied Newtonian physics as well, probably because of the influence of Clairaut. Newton's thought occupied a central place in the intellectual agenda of the Enlightenment, and in 1738 Voltaire published a summary of it, *Elements of the Philosophy of Newton*, with significant contributions from du Châtelet (Schiebinger 1989, 60–61). She began the challenging task of translating Newton's signature work of physics, the *Principia*, into French. Influenced by some of the German mathematician Leibniz's criticisms of Newton, she added an important commentary on the text in which she discussed the principle of the conservation of energy. Her translation of the *Principia* was published after her death and had a lasting effect on the acceptance of Newtonian views of method and mathematical natural law in France. It still serves as the standard French-language version of Newton's book (Figure 12.2) (Schiebinger 1989, 63).

Émilie du Châtelet produced a number of shorter scientific treatises during her time at Cirey, as well as a longer work, the *Institutions de Physique*. The *Institutions* proved to be a source of some conflict, for one of du Châtelet's tutors tried to claim credit for its authorship. Published in 1740, the book gave an account of the major developments in physics in the 17th century, with emphasis on the ideas of Descartes, Leibniz, and Newton. Du Châtelet

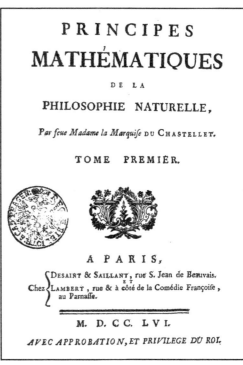

Figure 12.2 Émilie de Breteuil translated Newton's *Principia* into French and added commentary. Her translation is still used today. (Émilie de Breteuil, Marquise du Châtelet. *Principes Mathématiques de la Philosophie Naturelle*, 1756)

placed this material in the context of earlier thought and considered different strategies for the treatment of matter, space, and time (Osen 1974, 62–63). In many ways, the *Institutions* may be considered an early work in the history of science and mathematics, for it illustrated du Châtelet's understanding of Enlightenment physics as a product of the philosophical ideas of the previous century.

Émilie du Châtelet died in 1749 as a result of complications from childbirth. She spent much of the second half of her life pursuing the intellectual goals of the Enlightenment as a scientist, a translator, a commentator, and a historian. She had interactions with many of the influential thinkers of her day, and she and Voltaire had real influence on each other's work. Her focus on Newton's mathematical approach to physics may be understood as part of a larger enterprise: the pursuit of a rational understanding of the laws of nature. Émilie du Châtelet's dedication to this enterprise made her a vibrant member of the Enlightenment community of minds in the mid-18th century.

SOPHIE GERMAIN

The French Revolution was a time of political and social upheaval. Institutions that had lasted for centuries were dissolved, borders were changed, class and family relationships were altered, and new ideas thrived. In the early days of the revolution in Paris, the customary limits on female behavior seemed to change as well, and some women added their voices to the political discourse of the time. However, the revolutionary government ultimately turned against such activity, and later in the 1790s Napoleon Bonaparte actively supported traditional gender roles. The mathematical career of Sophie Germain followed a similar pattern; while she benefitted from supportive intellectual relationships in a time of great change, her opportunities were ultimately limited because of her sex.

Sophie Germain was born in 1776 to a middle-class family in Paris. At a young age, she became fascinated with the books of mathematics in her family's library; she was especially moved by the story of the death of Archimedes. Despite her parents' objections to her mathematical studies, she continued to work on her own. The late 18th century was an exciting time in the history of European mathematics, and a young student would be confronted with many challenging new developments. Yet because she was a woman, Germain did not have access to high-level mathematical training. For example, when the new Ecole Polytechnique opened in the suburbs of Paris in 1794, it did not allow female students. Determined to advance her mathematical investigations, Germain obtained lecture notes from the school. Using the name of a male student, M. le Blanc, she sent some of her work to the noted mathematician Joseph-Louis Lagrange, a member of the faculty. Lagrange responded positively to her submission, and when he discovered that the work came from a woman, he actively encouraged her studies.

Perhaps the most innovative mathematician of the age was Carl Frederich Gauss of Germany. Germain read his *Disquisitiones Arithmeticae* and sent him some of her own work, again under a male name. She and Gauss exchanged letters for some years. In 1807, as the troops of Napoleon occupied northern Germany following their victory at the Battles of Jena and Auerstadt, Germain became concerned that Gauss might be in some kind of danger and used her family connections with the French military to enquire about him. In the process, Gauss learned her true identity. He continued to write to her about complex mathematics, complemented her work, and pointed out her abilities to other mathematicians.

In addition to her interests in pure mathematics and number theory, Germain did work in new fields of physics. In the early 19th century, the French Academy of Sciences sought papers on the mathematical patterns created by flexible surfaces when they vibrate. Germain became interested in the subject and submitted a series of entries, improving her understanding of the complex mathematics of the phenomenon with each attempt. Her lack of formal training in physics and higher mathematics clearly hampered her efforts. Her third paper, *Memoir on the Vibrations of Elastic Plates* (1819), was accepted by the academy, and she was awarded a prize for achievement. Despite this, she was still not allowed to attend formal sessions of the Academy of Sciences. However, her work brought her to the attention of the wider French intellectual community, and she was admitted to the meetings of the prestigious *Institut de France* (Osen 1974, 90).

Working largely on her own, Sophie Germain had entered the circle of accomplished European mathematicians. After 1819, she pursued work in number theory and contributed significantly to the ongoing effort to prove Fermat's last theorem, a puzzle that had challenged the mathematical community since the death of the famous French scholar in 1665. Other scholars would draw on her work as they continued to wrestle with the problem. Germain suffered from breast cancer at the end of her life, but she continued working until she died in 1831. Despite a lifetime of accomplishment in mathematics, she never held a formal academic position. Excluded from the educational institutions of her day, she was forced to develop her skills independently, with the support of some notable mentors. Schools and academic prizes are now named in honor of a woman who advanced knowledge largely through her own dedication and initiative.

SONYA CORVIN-KRUKOVSKY KOVALEVSKY

In the late 19th century, universities and other advanced intellectual institutions remained largely closed to female applicants. However, social and political reform movements were growing across Europe during this period, and intellectuals were questioning traditional values. Rising levels of female education in Europe and America led to a desire among some young women for higher-level academic training. Sonya Corvin-Krukovsky Kovalevsky, the first European woman to receive a PhD in mathematics, confronted the obstacles faced by academically ambitious women during this period, and her success proceeded from the rigor and creativity of her thought.

She was born in 1850 to a military family in Moscow; her father was a general in the army of the Russian tsar. In the family estate at Pablino, young Sonya had a room that had been wallpapered with printed lectures on calculus, and she later wrote that this stimulated her interest in mathematics. Her family arranged her education, and she excelled in mathematics and calculus as a teenager in St. Petersburg. While she wished to proceed with her education, universities in Russia did not admit female students at the time. As a result, she arranged a marriage of convenience with a young student of paleontology, Vladimir Kovalevsky. This allowed her to travel with her new husband to Western Europe, where she had a greater chance of attending a university.

Sonya Kovalevsky and her husband went to Germany. He pursued a degree in paleontology, and she took classes in mathematics and physics for two years at the University of Heidelberg. On the recommendation of her professors, she sought the instruction of Karl Weierstrass, professor of mathematics at Frederick William University in Berlin. Unfortunately, female students were not admitted there, so Kovalevsky requested that Weierstrass teach her privately. When she demonstrated her ability to solve complex problems in an elegant manner, he accepted her as a student and guided her mathematical studies for the next four years. As she developed her abilities, she wrote a series of successful research papers on topics in pure and applied mathematics: partial differential equations, Abelian integrals, and Laplace's description of the rings of Saturn. With the help of Weierstrass, she received a doctoral degree in 1874 from the University of Göttingen based on this original research, the first doctorate in mathematics to be earned by a woman.

Despite her academic success, no European university would give a faculty position to a woman in 1874. Kovalevsky returned to Moscow, where she immersed herself in literary activities and started a family, but she missed her mathematical work. She returned to Western Europe, lived in Berlin and Paris, and did research on optics. In 1883 her husband committed suicide, probably because of business failures. Kovalevsky moved to

Sweden, where the University of Stockholm offered her a position as a lecturer. She would teach there for the rest of her life and was the first woman to achieve the rank of professor in a European university. In 1888 Kovalevsky submitted a paper on the mathematics of rotating irregular bodies to the French Academy of Sciences. She was awarded the *Prix Bordin*, one of the most prestigious academic prizes in Europe, and she was also made a member of the Russian Academy of Sciences.

The mathematical achievements of Sonya Krukovsky Kovalevsky were part of a larger trend in her life, one of dismissing traditional limitations. She lived during an exciting time, when views of authority were being challenged across Europe. She had substantial literary abilities, and these brought her into contact with a circle of intellectuals who supported social transformation. Sonya's sister Anna was an active socialist; she was married to Victor Jaclard, a French socialist leader. Sonya herself embraced the ideas of Russian Nihilism, an intellectual movement that rejected established cultural values. Her mathematical successes were a product of her profound intellectual abilities, but they also proceeded from her determination to confront inequitable institutions and customs. At a time when democracy was growing in Europe, when Marxism was attracting adherents, when women began to demand political power, Sonya Kovalevsky sought membership in the largely male world of European mathematics. Her untimely death from influenza in 1891 interrupted a fruitful life that had been devoted to research, scholarship, and change.

EMMY NOETHER

Emmy Noether was the most accomplished and talented female mathematician of the early 20th century. She lived at a time when women had gained access to some academic institutions and could obtain high-level training. However, they were still excluded from much of academic life, especially in mathematics and the natural sciences. Despite these obstacles, Noether obtained her doctorate in mathematics and taught for years at one of the most prestigious universities in Germany. She made striking contributions to fields of modern physics and abstract algebra, and her work influenced generations of later mathematicians and scientists. She also maintained strong political opinions during a time of great unrest in Europe and ultimately left Germany for the United States in response to the rise of the Nazi Party.

Noether came from a Jewish family immersed in mathematics and science; her father was a professor of mathematics at the University of Erlangen in southern Germany. Emmy and her brother Fritz also found academic careers in mathematics, while another of her brothers became a chemist.

She was born in 1882 and grew up in the university town where her father taught. Noether took classes at the University of Erlangen, despite policies that strongly discouraged women from attending. After she passed her exams in 1903, she continued her studies at the University of Göttingen and then came back to Erlangen to do her graduate study under the guidance of Paul Gordon. In 1907 Noether completed her doctoral dissertation on invariant analysis, a form of abstract algebra. This was Gordon's main area of interest; it is clear that she felt a great deal of loyalty towards him.

University professorships were not available to female mathematicians in the first decade of the 20th century. After she received her doctorate, Noether gave lectures at Erlangen, but did not have an official position at the university. She continued to develop her interests in abstract algebra under the influence of Ernst Fischer. After the start of the First World War, Noether was asked to come to the University of Göttingen by the mathematician David Hilbert, despite the resistance of some members of the faculty. Once more, she was not offered a formal university post, but she taught classes under Hilbert's sponsorship. Hilbert valued her expertise in invariant analysis, and she quickly distinguished herself in 1915 by proving the theorem that any symmetrical system in nature is linked to a law of conservation (such as the conservation of momentum or energy). This theorem, now known as Noether's theorem, played an important role in theoretical physics.

The First World War had overturned many long-standing social conventions concerning gender. Women had worked to maintain the economies of the warring nations, and this increased their economic and political stature. In Germany, for example, women acquired the right to vote in 1918. Some of the restrictions on women in universities were also removed, and in 1919 Emmy Noether was awarded her habilitation, her formal status as a member of the university faculty of Göttingen. She was promoted to the professorship in 1922, but she was not paid for her teaching until another promotion brought her a stipend. Noether was a dedicated teacher with great intellectual energy, and she guided the work of a number of promising graduate students—young men and women who were devoted to her. She also continued to do pioneering work in abstract algebra during an exciting period in the history of mathematics and became one of the most accomplished mathematicians at Göttingen. During her career, she published more than thirty papers and influenced an important textbook, B.L. Van der Waerden's *Modern Algebra*.

The 1920s were a turbulent time in the history of European politics. The Russian Revolution had resulted in the establishment of the Soviet Union, and Noether shared some of the values of the new communist regime. She gave a series of lectures at Moscow State University in 1928, and she maintained contacts with Soviet mathematicians. In 1933 the National Socialist Party came to power in Germany, and Adolf Hitler became chancellor.

The Nazis had little tolerance for Jewish scholars or for those with links to the Soviet Union, and Noether lost her position at Göttingen. Like a host of other European intellectuals, including Albert Einstein and Enrico Fermi, she emigrated to the United States. She accepted a faculty position at Bryn Mawr College outside of Philadelphia for a larger salary than she was ever awarded in Germany. However, Noether taught there for less than two years, for she died of complications from surgery in 1935. During the fifty-three years of her life, she had become one of the most prominent mathematicians in the world, and her creative approach to modern algebra had a profound influence on the development of mathematics and theoretical physics in the 20th century. Following her untimely death, her colleagues paid tribute to her accomplishments and her dedication to her discipline. Albert Einstein wrote of her:

> In the judgement of the most competent living mathematicians, Fraulein Noether was the most significant creative mathematical genius thus far produced since the higher education of women began.
>
> (*New York Times*, May 4, 1935, p. 12, in Osen 1974, 151)

DOROTHY VAUGHN (BORN DOROTHY JOHNSON)

Dorothy Vaughn was born in 1910 and grew up in Missouri and West Virginia. She earned a scholarship to Wilberforce University, where she studied mathematics. After graduating, she taught high school mathematics, married, and raised a family. In 1943 she began work at the National Advisory Committee for Aeronautics (NACA) as a data processor, posted at the Langley Memorial Aeronautical Laboratory in Virginia. The nation was facing the challenge of the Second World War, and NACA played a crucial role in developing new aircraft technology during that conflict, as well as the Cold War that followed.

Dorothy Vaughn lived and worked in the segregated world of the United States in the mid-20th century. She spent her childhood in Southern states with active Jim Crow laws, laws that required separate facilities and schools for African Americans and restricted their access to public spaces and civic activities. In addition, the professions available to educated women at the time were largely limited to nursing and teaching. In this atmosphere of limited opportunity, Vaughn still managed to excel. On June 25, 1941, President Franklin D. Roosevelt issued Executive Order 8802, which prohibited employment discrimination in the defense industry. This presidential directive made it possible for Vaughn to bring her mathematical talents to the NACA.

In an age before modern computers, calculations were performed by men and women working with reference tables and slide rules. The aeronautical engineering work done at NACA required a great deal of mathematical figuring, and Dorothy Vaughn worked as part of a team of African American women solving these complex problems. Despite the importance of the task, they still suffered from segregation within the agency. However, Vaughn did exemplary work, and after five years she took charge of her division, West Area Computing.

In 1957 the Soviet Union launched *Sputnik*, the first artificial satellite to orbit the earth. The event added a new dimension to the Cold War that already pitted the Soviet Union against the United States; the two nations now competed to develop the technology of space flight. NACA grew into NASA, the National Aeronautics and Space Administration, in 1958. Vaughn and the other mathematicians she supervised became part of the Analysis and Computation Division at NASA. As the rivalry between the United States and the USSR became focused on putting human beings into space, she worked on the most important parts of the American effort: the Mercury Program, the Gemini Program, and the Apollo Program.

The capabilities of mechanical computing developed significantly in the 1960s, and NASA was quick to adopt the new technology. Vaughn recognized the importance of this development and worked to develop her expertise in computer programming. She became skilled in the use of FORTRAN, a computer programming language intended for scientific applications.

Dorothy Vaughn retired from NASA after twenty-eight years of government service as a mathematician, supervisor, and computer programmer. She contributed her mathematical expertise to her country during times of conflict and international tension: the Second World War, the Cold War, and the space race. She also played an important part in one of the most important engineering feats of the 20th century, the development of space flight. Despite the obstacles placed in her path by racism and sexism, she stood out in her field and embraced new technologies during a time of rapid change. Her life and career were discussed at length by Margot Lee Shetterly in her book *Hidden Figures*. In addition, Dorothy Vaughn was awarded the Congressional Gold Medal posthumously in 2019.

KATHERINE JOHNSON
(BORN CREOLA KATHERINE COLEMAN)

The effort to send human beings into outer space was a key component of the Cold War between the United States and the Soviet Union. During the 1950s and 1960s, both countries sponsored extensive programs of research and technical development aimed at achieving this goal. The testing of aircraft and spacecraft involved the mathematical evaluation of mountains of data, and the design of actual missions in space required careful calculation of flight paths, orbits, and points of rendezvous. Katherine Johnson performed crucial tasks in data analysis and mission design for NASA as it carried out some of its most important space flights and contributed substantially to the success of the American space program.

Katherine Johnson (born Creola Katherine Coleman) had great academic ability from an early age; she graduated from high school at the age of fourteen. She enrolled at West Virginia State College, where she excelled in mathematics under the guidance of Professor W.W. Schiefflin Claytor. When Johnson earned her degree in mathematics and French in 1937, she was only eighteen years old. Like many educated women of the time, she took a job as a teacher. However, in 1939 she was one of the first three African American students admitted to the graduate program at West Virginia University. She studied mathematics at the graduate level for a year, then left to start her family.

In 1953 Johnson accepted a position working for the NACA at the Langley Memorial Aeronautics Laboratory in Virginia. For five years, she performed the complex computations involved in the testing of experimental aircraft. As an American woman of color in the mid-20th century, she faced substantial discrimination. The state of Virginia still required segregated workplaces, so Johnson worked with other African American women at a separate facility, the West Area Computing site, under the leadership of Dorothy Vaughn (see earlier entry).

The Soviet Union's launch of the *Sputnik* satellite in 1957 galvanized the American space program. Johnson did mathematical analysis for the Space Task Group at the NACA, a team of engineers focused on space flight. In 1958, when NACA became NASA, workplace segregation policies were officially ended, and Johnson continued to work as a member of the engineering group. As a result, she collaborated on the design of many of NASA's earliest space missions. In 1960 Johnson wrote an agency report on the positioning of satellites in earth orbit with a colleague, Ted Skopinski—the first report on flight research ever credited to a woman at NASA.

Alan Shepherd flew the first American human spaceflight mission in the Mercury Program in May 1961, and Katherine Johnson did work on the trajectory and the launch windows for the flight. Her mathematical abilities made her an essential part of NASA's missions. In a famous story, John Glenn, the third American astronaut to fly in space and the first to orbit the earth, insisted that she check the orbital calculations for his mission. At the time, NASA was just beginning to use computers for navigation, but astronauts and pilots did not consider them to be completely dependable. Glenn placed his trust in the work of Katherine Johnson, and his mission proved successful.

On May 25, 1961, three weeks after Alan Shephard's space flight, President John F. Kennedy committed the United States to landing human beings on the moon by the end of the 1960s, and that ambitious goal dominated the activities of NASA for most of the decade. The Gemini Program was meant to develop the spaceflight techniques necessary for a trip to the moon, including the rendezvous and docking of two different craft in space. The Apollo Program included test missions for the actual spacecraft that would go to the moon as well as the lunar missions themselves. Katherine Johnson continued at NASA throughout this exciting time. She worked on the orbital paths for the rendezvous of the lunar module and the command module in America's first successful moon landing mission, *Apollo 11*. Her career at NASA encompassed the entire Apollo Program, as well as some of the early space shuttle missions. When she retired in 1986, she had played a vital part in the establishment of American space flight, and she had written twenty-six research reports for the space agency.

Katherine Johnson lived for 101 years. She overcame the racial discrimination and gender bias of the 20th century and contributed her talents to the conquest of space in the early days of that endeavor. In 2015 she was awarded the Presidential Medal of Freedom, America's highest civilian honor. However, her greatest legacy lies in her work itself.

CONCLUSION

At the University of Siegen in western Germany, the departments of Mathematics and Physics are housed in the Emmy Noether campus,

Figure 12.3 The departments of Mathematics and Physics at the University of Siegen are located on the Emmy Noether Campus. (Panther Media GmbH/Alamy Stock Photo)

named in honor of the German female mathematician (Figure 12.3). A woman who was only permitted to teach at the university through the good graces of a male sponsor is now an icon, a role model for hundreds of students. Her achievements, and those of her predecessors, will inspire the ambitions of her successors.

FURTHER READING

Astell, Mary. 1994. *A Serious Proposal to the Ladies, Part II*, in *Women Philosophers of the Early Modern Period*, edited by Margaret Atherton. Indianapolis: Hackett Publishing.

Christine de Pizan. 1982. *The Book of the City of Ladies*, translated by Earl Jeffrey Richards. New York: Persea Books.

Henrion, Claudia. 1997. *Women in Mathematics*. Bloomington: Indiana University Press.

Noble, David. 1982. *A World Without Women*. New York: Knopf.

Osen, Lynn. 1974. *Women in Mathematics*. Cambridge: MIT Press.

Schiebinger, Londa. 1989. *The Mind Has No Sex*. Cambridge: Harvard University Press.

Shetterly, Margot Lee. 2016. *Hidden Figures*. New York: Harper Collins.

Wollstonecraft, Mary. 2008. *A Vindication of the Rights of Women*. London: Folio Society.

Woolf, Virginia. 1989. *A Room of One's Own*. New York: Houghton Mifflin.

13

Numbers in the 20th Century

In the early years of the 21st century, major league baseball teams in the United States began to choose their players based on detailed statistical analyses of their performance. In a way, this trend reveals the extent to which mathematics has permeated the culture of modern America—and the modern world. We live in a data-driven society, and quantitative evidence plays a dominant role in academic discourse, in political decision making, in commerce and business, and in everyday life. It affects a wide variety of personal decisions: the brand of gum one chooses, the school one attends, the political candidates one supports. The mathematical character of modern life proceeds largely from the success of modern science and social science—from the ability of these disciplines to describe the nature of the physical universe and the fabric of human society in precise terms. The techniques of modern statistical analysis were developed in the early 20th century by Francis Galton and Karl Pearson, and they refined the way data was evaluated, the way quantitative evidence was used, to establish validity. As a result, the 20th and early 21st centuries have been an age of statistics, as scientists, social scientists, governments, businesses, and individuals gather quantitative information and seek useful patterns in that information. Some modern philosophers point out the illusory character of such efforts and argue that human subjectivity colors all forms of argument or assessment. Yet the success of the sciences and social sciences is indisputable, and the culture of numbers appears irresistible. The back of a baseball card in the 1970s was covered with statistics, the career of a human being expressed as lists of figures. Increasingly, that is true of all of us.

Technology has played a central role in this culture of data and prediction. The Industrial Revolution created the capacity to produce machines according to precise technical specifications, and these machines revolutionized the production of commodities, the nature of transportation, the character of warfare. They also made it possible to create new generations of ever-more-precise machines: mechanical, electrical, electronic. The invention of computer technology was a result of this growing wave of technical power and precision. Once introduced, such technology became a major factor in its own evolution, as computers, microchips, and software aided in the development of ever more sophisticated generations of—computers, microchips, and software. Computers are a product of a mathematical, technical culture, and they have become the primary means of managing that culture. They facilitate the storage and evaluation of the mountain of data that increasingly shapes modern life. In short, in the 21st century, computers pick baseball players.

It is not possible to adequately describe the place of mathematics and number in the culture of the 20th century in a short chapter. However, four brief treatments can help illustrate the impact of quantification and precision on the modern world.

MATHEMATICS AND SCIENCE IN THE 20TH CENTURY: DINOSAUR PALEONTOLOGY AND MASS EXTINCTION

Since the time of Galileo and Newton, measurement and the mathematical analysis of data played an ever-increasing part in scientific thinking. This trend continued throughout the Enlightenment and the 19th century and reached a crescendo in the 20th. New instruments, new ways to manage data, and new ideas resulted in a view of the natural world defined by mathematics. For example, with the advent of Einstein's general theory of relativity, theoretical physics developed as a largely mathematical discipline.

Chemistry serves as an effective illustration of the importance of mathematics in 20th-century science. In the 19th century, the categorization of atoms in terms of atomic weight led to a mathematical ordering of matter itself: the periodic table of the elements. The discovery of basic subatomic particles in the years between 1895 and 1932—electrons, protons, and neutrons—resulted in a new understanding of the structures and patterns inherent in the periodic table. The order of the elements and the nature of their chemical properties could be explained by mathematical parameters: the number of protons in the nucleus, the number and distribution of electrons moving around it, the presence of electrical charge. Further, new kinds of instrumentation yielded valuable data on the structure of molecules. Infrared spectrometry, x-ray diffraction, nuclear magnetic resonance spectrometry—all yielded measurements that contributed to the understanding of complex compounds and the ability to manipulate or synthesize them. For example, Linus Pauling derived the secondary structure of protein chains based on measurements of chemical bond angles gathered through x-ray diffraction. Clearly, the development of chemistry in the 20th century depended heavily upon the gathering and interpretation of new forms of quantitative information.

At first glance, paleontology appears different from chemistry. A science with seemingly little room for mathematics, it focuses on the search for fossils and their subsequent interpretation and classification. Yet the history of paleontology in the second half of the 20th century demonstrates the great potential of new forms of data analysis. While fossils remained at the heart of the discipline, new quantitative techniques provided paleontologists with powerful new tools and revolutionized the understanding of life in the distant past.

Dinosaurs capture the imagination of adult and child alike. Their sheer size and the diversity of their forms inspire curiosity and wonder. Yet when one is confronted with the physical evidence of their existence—their massive fossils—the question of their ultimate extinction arises. They dominated the terrestrial environments of the earth for more than 140 million years, then died out at the end of the Cretaceous Period, 66 million years ago. In short, what happened to the dinosaurs?

For much of the 20th century, paleontologists answered that question by pointing to the evolutionary superiority of mammals. Mammals were quick, agile, intelligent, adaptable, and attentive to their young. Dinosaurs, as reptiles, were slow, dull-witted, bound to a tropical or semi-tropical lifestyle, and neglectful of their offspring. Thus, as the world slowly changed at the end of the Cretaceous Period, dinosaurs could not adapt, and mammals came to replace them as the dominant group of vertebrates on the land. However, this view of a gradual mammalian triumph was called into question by a series of dramatic dinosaur discoveries in the 1960s and 1970s. New fossils suggested that some dinosaurs had highly active lifestyles and were probably endothermic, or warm-blooded. Other fossils indicated that dinosaurs exercised a high degree of parental care, rather like modern birds. John Ostrom, a paleontologist at Yale University, proposed the evolutionary link between dinosaurs and birds that has now become accepted. By the end of the 1970s, the prevailing view of dinosaurs was changing rapidly, and it seemed that they were far more adaptable than previously thought. This realization gave new force to an old question: What happened to the dinosaurs?

In June 1980 a paper was published in the journal *Science* by Luis and Walter Alvarez, Frank Asaro, and Helen Michel. The paper argued that the disappearance of the dinosaurs resulted from the impact of a large extraterrestrial body, an asteroid or a comet, that took place 66 million years ago and caused a mass extinction event. Paleontologists already knew that approximately 60 percent of the species on earth had died out at the end of the Cretaceous Period, but most had assumed that this process occurred over millions of years. The new paper claimed that a catastrophic event had drastically altered the environment of the earth (presumably, for several years) with devastating results for living organisms.

The theory originated with the geological work of Walter Alvarez. Alvarez had discovered a curious phenomenon in his studies of paleomagnetism in the rocks of central Italy. Near the city of Gubbio, he found a layer of clay at the upper boundary of rock strata from the Cretaceous Period—that is, a clay layer laid down at the end of the time of the dinosaurs. When a sample of that clay was analyzed, it was found to have an unusually high level of iridium, a heavy metallic element. The clay layer contained 100 times more iridium than the usual level found in sedimentary rock. Walter Alvarez brought this iridium anomaly to the attention of his father, the Nobel Prize–winning physicist Luis Alvarez. The two scientists knew that iridium occurred in high concentrations in certain meteorites, and with the assistance of two chemists, Frank Asaro and Helen Michel, they confirmed the level of iridium in the clay layer and formulated the impact theory. They maintained that an asteroid 6 to 10 kilometers in diameter hit the earth and blasted out an immense crater. In addition to the massive shock of the explosion and the resulting firestorm, such an impact would cast millions of tons of dust and debris into the atmosphere, material from the asteroid itself and from the crater. This dust would surround the earth and obscure the sun, interfering with photosynthesis and causing the death of plants and the animals that depended upon them for food. The effect would be especially acute in the ocean, where photosynthetic algae would be severely affected. Further, the dust would gradually settle to the earth over several years, explaining the presence of high levels of iridium in the Gubbio clay layer.

The Alvarez hypothesis caused a sensation in paleontology and in popular culture. Other scientists found high iridium levels at the upper boundary of Cretaceous rocks in many sites around the world, confirming the global character of the phenomena. Additional evidence also suggested the likelihood of an extraterrestrial impact: microscopic globules of glass and samples of quartz transformed by great pressure were found along with the iridium anomalies. In 1996 scientists identified the likely site of the impact off the coast of the Yucatan Peninsula in southern Mexico; the crater had once been 300 kilometers wide (Figure 13.1). The new theory transformed

Numbers in the 20th Century 311

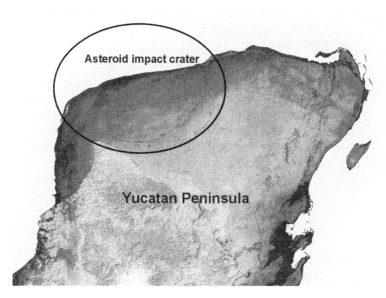

Figure 13.1 The Chicxulub Crater in the Yucatan Peninsula has been identified as the site of the asteroid impact that came at the end of the Cretaceous Period. Examination of the crater has confirmed many aspects of the impact theory. (NASA/Jet Propulsion Laboratory/Modified by ABC-CLIO)

the scientific understanding of evolutionary change, as paleontologists and biologists placed new emphasis on the notion of mass extinction. Evolution, it seemed, might occur at different rates over time, as mass extinctions radically changed the earth's biosphere. Ordinary people were fascinated (or terrified) by the idea of extraterrestrial impact, and Hollywood promptly produced feature films on the subject. Two movies were released in the year 1998, and both Robert Duvall and Bruce Willis saved the earth from dangerous asteroids.

The great mystery of the extinction of the dinosaurs had apparently been solved, one of the central questions of paleontology. The answer did not originate with the study of fossils, but rather from the mathematical analysis of the metal content in a sample of clay. The creative interpretation of quantitative evidence had inspired a genuine revolution in the way scientists thought about the biological past and led to a great deal of new scientific work. Some scientists would dispute the rate and timing of dinosaur extinction and claim that the dinosaurs were already in decline at the time of the impact. While such arguments were intriguing, no one could deny the significance of the Alvarez hypothesis. Physics, chemistry, and mathematics had been combined with paleontology in a fascinating new way.

Mass extinction is a fascinating concept. The Darwinian model for extinction asserts that as the conditions of life gradually change, some species of animals and plants cannot effectively compete in the ongoing struggle for existence. These eventually die out as their places in the ecosystem become filled with other, more successful species. The concept of mass extinction does not discard this model; rather, it suggests that events may take place (volcanic eruptions, extraterrestrial collisions) that change the conditions of life very quickly; so quickly that many species cannot adapt. Large numbers of species disappear in a short time, geologically speaking. Some species will survive, and their descendants will evolve rapidly into new species once conditions stabilize. The question is, how common are mass extinctions, and how do they show up in the fossil record?

Scientists can attempt to track the number of species that disappear over a given span of time in the fossil record. After the advent of the Alvarez hypothesis, paleontologists identified five clear mass extinctions: the end of the Cretaceous Period (66 million years ago), the end of the Triassic Period (201 million years ago), the end of the Permian Period (250 million years ago), the end of the Devonian Period (365 million years ago), and the end of the Ordovician Period (440 million years ago). The age of these events have been established through the use of radiocarbon dating methods. Different extinction events affected different numbers of species: the Cretaceous event resulted in 65 percent species loss, while the Permian event wiped out 95 percent of all the species on earth.

In 1984 two paleontologists, David Raup and Jack Sepkoski, published a statistical study of mass extinctions in the fossil record. The two scientists had created a database of known extinctions, and they performed complex statistical analyses to look for correlations and trends. They argued that the data supported a 26-million-year periodicity for mass extinctions; in other words, a mass extinction event had taken place every 26 million years. The end of the dinosaurs was part of a much larger, recurring pattern in the history of life on earth. Astronomers and astrophysicists sought to explain these findings, and their proposal was startling. If the sun had a dark companion star with a 26-million-year orbit, that star would pass close to the solar system periodically. Its gravity would disrupt the Oort Cloud, a theoretical region of bodies in orbit around the sun beyond Pluto. This disruption would lead to the inner solar system being flooded with asteroid-sized objects and vastly increase the likelihood of an extraterrestrial body colliding with the earth. Sometimes, a small collision would occur, resulting in a small extinction event. Other times, a larger body (or several bodies) would hit the earth, and this would cause a more substantial extinction. However, impacts would take place every 26 million years, accounting for the patterns in the data. A fitting name was conferred on the dark companion star: Nemesis.

In an extraordinary blend of paleontology, astrophysics, and statistics, members of the scientific community were suggesting that the course of life on earth was regularly interrupted by the movements of a dark companion star. The "Nemesis hypothesis" has met with a mixed reception since it was proposed in the 1980s. Some scientists challenged the statistical models that lay behind it and denied the periodic character of mass extinction. Others argued that it did not fit available evidence, particularly with respect to impact craters. Still others pointed out that a companion star so far away would have an extremely unstable orbit. There are members of the scientific community who continue to support it, with revised statistical studies of extinction and new ideas on astronomical causes. Yet the Nemesis hypothesis, and the Alvarez hypothesis that arguably inspired it, both illustrate the transforming effect of quantitative methods on the sciences of evolutionary biology and paleontology in the late 20th century. As the 21st century has proceeded, new quantitative techniques have allowed for exciting interpretations of fossil evidence. The creative combination of chemistry, geology, paleontology, and mathematics continues to shape our understanding of the history of life.

MATHEMATICS AND FINANCE IN THE 20TH AND 21ST CENTURIES: THE FEDERAL RESERVE

Economic activity has involved numbers since humanity first learned to count, and modern commercial and financial institutions depend upon complex mathematics as a matter of course. Investment houses employ elaborate algorithms to predict the performance of their assets; insurance companies quantify risk in order to determine appropriate premiums; banks calculate interest payments and the amount of money they need to keep on hand in order to fund daily operations. A detailed description of the role of mathematics in contemporary commerce and finance would require a textbook of 1,000 pages. In order to illustrate the intersection of mathematics, business, banking, and the everyday life of people, it is useful to consider the operations of one crucial institution: the U.S. Federal Reserve.

The Federal Reserve is the central bank of the United States. Founded by congressional action in 1913, it was intended to manage the stability of the U.S. economy. The United States did not have a central bank for much of its history; President Andrew Jackson closed the last such institution in 1833. The Federal Reserve was established to prevent banking crises, for a series of such crises had caused national turmoil in the late 19th and early 20th centuries. The main components of the Federal Reserve system are the twelve Federal Reserve banks, located in major cities in the United States, and a central board of governors. That board is drawn from the presidents of the Federal Reserve banks and presided over by the chair of the Federal Reserve. While the president and Congress act to choose these leaders, the Federal Reserve system is designed to be largely independent of political control, and it does not depend on the federal government for its operating expenses. This prevents politicians from attempting to influence the central bank's actions for their own political purposes.

The Federal Reserve (hereafter referred to as the "Fed") maintains the stability of the U.S. economy by controlling inflation, working to ensure healthy levels of employment, and supervising the banking system. The Fed supports the stability of banks by lending them money to fund their everyday operations. These short-term loans are designed to help banks maintain a steady level of cash despite the fluctuations of their own loans and investments; banks can draw on these loans through the "discount window"—the lending department of the Fed. The Fed's lending function is one of the major ways that new money is introduced into the economy: the Fed lends it to banks, and banks lend it to businesses and consumers. The Fed charges banks interest on these loans at the "discount rate," and the banks make a profit by lending those funds out at higher rates of interest.

Inflation may be defined as an increase in prices across the economy; a certain level of inflation is considered positive and shows a healthy degree of demand for products and services. The Fed gathers extensive quantitative data on the state of the economy, including inflation rates, and publishes it eight times per year in a lengthy report known as the "Beige Book." It acts on that data using a variety of tools, especially the level of the discount rate. For example, low inflation and high unemployment indicate that the U.S. economy is sluggish because of low demand for products and workers. In that situation, the board of the Federal Reserve might vote to reduce the level of the discount rate as a stimulus to the economy. This allows banks to reduce the rates of interest that they charge to their various customers while still making a profit. When banks reduce interest rates, it makes it less expensive to borrow money across the entire range of American business transactions. Rates on credit cards, automobile loans, mortgages, and corporate loans all come down, and consumers and companies respond by borrowing money and spending it. This enhances economic activity, increases employment, and raises inflation to healthy levels.

For example, imagine that the Fed reduces interest rates by one half of a percentage point, from 2 percent to 1.5 percent. Banks will respond by reducing mortgage rates, perhaps from 4 percent to 3.5 percent. This will inspire people to buy houses, because they will be paying less interest, and the increase in home sales will stimulate other sectors of the economy in a number of ways. Homebuilders will have more business and hire more workers, who will then have money to spend on different products, further increasing demand. Manufacturers of construction materials and equipment will sell more product, make more profit, and hire more people—who will spend their wages and increase demand. People who buy houses will also hire movers and buy furniture and paint and carpeting—this will employ more people and increase demand. Thus, the Fed has the power to revitalize the economy in times of economic downturn, in ways that directly affect the lives of ordinary people, as well the fortunes of companies and financial institutions.

A heightened level of economic activity can have negative effects as well. Imagine that there are low levels of unemployment—everyone is working, earning money, and buying products, so demand is high. High demand naturally leads to inflation, and if inflation increases too much, it can be destructive. As prices go up, the value of money goes down, reducing the value of people's savings and salaries. Further, high inflation can have a distorting effect on financial markets and interest rates. In such a situation, the Federal Reserve can raise the discount rate in an attempt to slow down the economy. A higher discount rate results in higher interest rates for credit cards or other kinds of loans. This reduces demand, which in turn will reduce prices and the rate of inflation. Such an action may result in an increase in unemployment, but this might be considered necessary to combat rising prices.

So, a stimulus to the economy will reduce unemployment but may raise inflation; a check on the economy might cut down on inflation but increase unemployment. The Federal Reserve manages this balancing act, constantly evaluating the economy, gathering data, and using interest rates to maintain stability. Its policies are not always effective, for sometimes inflation or unemployment are caused by factors beyond the control of a central bank; for example, in the 1970s high petroleum prices helped trigger substantial inflation in the United States. The Fed responded as well as it could, but it did not have the power to offset such a powerful economic force.

In times of economic emergency, the Federal Reserve can take extraordinary steps to prevent economic collapse. In 2008 the collapse of the American housing market triggered a crisis, as banks were forced to sell assets in an attempt to raise the cash necessary to cover their obligations. Banks around the world had billions of dollars' worth of bad loans, loans that could drive them to insolvency—the condition of owing more money than one's assets are worth. Insolvent banks would have to close, crippling the American and world economies. In this delicate and dangerous time, the Federal Reserve acted to recapitalize the banks—to provide them with enough money to operate until the economy recovered. It bought billions of dollars' worth of bad loans and other nonperforming assets from banks at face value, thus transferring cash to the banks and taking over their risks. The Fed also made billions more available to foreign banks in danger of insolvency, reaching outside America's borders in a time of international financial stress. The strategy worked; the banks recovered, and eventually most of the bad loans and assets purchased by the Fed recovered their value. In the end, the Federal Reserve made a significant profit on its emergency purchases.

The Fed also maintains a relationship with the U.S. Treasury. It can create new money and put it into circulation by buying Treasury bonds (T bills), essentially lending money to the U.S. government. The government spends the money, thus introducing it into the economy and increasing business activity. Such a policy is called "quantitative easing" and represents an increase in the money supply. While it carries a substantial risk of inflation, it can also provide great relief to a stressed economy in a very short time.

In short, the Federal Reserve may be considered as an economic watchdog, constantly scrutinizing the American economy for signs of instability. It monitors volumes of quantitative data and uses financial tools to manage the economic life of the nation in response to that data. It also serves as a potent example of mathematical intricacy of modern financial administration.

MATHEMATICS AND WARFARE IN THE 20TH CENTURY: THE SECOND WORLD WAR

Two hundred thirty-five. Not a large number, a prime number, or a perfect square. Yet 235 has a special significance in the history of science and the history of warfare, for uranium 235 is the isotope of uranium that fueled the first atomic bomb. The discovery of nuclear fission and the application of that discovery to the production of nuclear weapons would change human history forever. It is perhaps the most potent example of the role played by the physical sciences and the modern mathematical view of nature in modern warfare.

Applied mathematics and engineering have shaped warfare throughout the modern period. In the late 16th century, the military compass (or sector) had a wide range of mathematical uses, including the aiming of artillery. One of those who claimed credit for its invention, Galileo Galilei, particularly emphasized this application of the instrument. Three-quarters of a century later, the French engineer Sebastien Le Piestre de Vauban used his mathematical skill in the design of fortifications and military assault techniques. Jean Baptiste Vaquette de Gribeauval created a standardized system for the production of French artillery to exact specifications in the late 1700s. Conflict in the late 19th century was revolutionized by the introduction of weapons produced by precision machine tools in modern factories. This trend culminated in the First World War, where industrially produced weaponry—breech-loading artillery, machine guns, airplanes, submarines—resulted in horrendous casualties for the combatant nations.

The military technologies of the Second World War far surpassed those of the First. Mathematics influenced these technologies in many ways (excluding issues of inventory, logistics, and supply). The aiming of weapons often required complex calculations, and in the days before modern electronic computing technology, such calculations were often performed with mechanical devices. In addition, the design of modern industrial weaponry required a culture of precise engineering. The emergence of remote-sensing technology depended heavily on the mathematics of radio waves and sound waves. Finally, the scientific research that led to the atomic bomb proceeded from a mathematical approach to nature and matter, an approach rooted in the European Enlightenment.

During the Second World War, many weapons were aimed by eye. However, more elaborate armaments required sophisticated mathematical targeting systems, often dependent upon mechanical calculating devices. For example, by the end of the war, the large guns of American warships were aimed with mechanical analog computers based on complex systems of gears; these would determine the elevation and direction of the ship's guns. Such computers required a variety of data, particularly the range, course, and speed of the target and the course and speed of the firing ship. Before the advent of radar, range-finding instruments used the principles of trigonometry to find the distance to a target.

The problem of ballistics, the aiming of artillery, affected the army as well. Soldiers in the field did not have access to the massive mechanical computers used by the navy. They made use of prefigured tables with data on range, wind direction, air temperature, and other parameters; these tables were compiled laboriously by hand, often by female personnel working back in the United States. ENIAC, one of the earliest modern electronic computers, was originally designed in 1943 to facilitate the production of artillery tables.

The aiming of bombs from an airplane involved similar mathematics. In this case, the altitude and airspeed of the plane, its direction of flight, and its distance from the target were the essential parameters in determining when a bomb should be released. These factors determined the parabolic path the bomb would follow as it fell. Before the war, the United States developed the Norden bombsight, a mechanical computer designed to make these calculations continually as the plane approached the target. The Norden automatically adjusted for the airspeed of the plane, its altitude, and its compass heading, while the operator of the sight, the plane's bombardier, would make alterations based on wind speed and local conditions. The U.S. military had great expectations of the Norden bombsight, but under wartime conditions it did not perform as well as expected. Nevertheless, it was used until the end of the war (Figure 13.2).

Figure 13.2 The Norden bombsight was used by the U.S. Army Air Corps during World War II. It was essentially a mechanical analog computer. (U.S. Air Force)

The mechanized weapons and equipment of the Second World War required a high degree of mathematical precision to design and produce. High-performance engine parts and the components of powered machinery had to be manufactured to demanding specifications and tolerances. The turrets of tanks, the propellers of planes, the barrels of cannon—all were products of industrial processes based on careful engineering. The use of wind tunnel data in the evaluation of airplane design illustrates the use of quantitative data in the design process. For example, the American P-38 fighter often suffered from control failures during high-speed dives, and this resulted in pilot fatalities. Wind tunnel data identified the source of the problem, and the plane was altered in response to those measurements.

Remote sensing technology was used on a large scale in the Second World War, and it depended heavily upon the mathematical analysis of information. Radar (radio direction and range-finding) is based on bursts of radio waves with a specific wavelength emitted by the sensing unit. Such waves are reflected back by the objects to be sensed—in the Second World War, these were primarily planes and ships. The reflected waves are then detected by a radio receiver and displayed visually on a viewing scope or screen. By measuring the direction of the returning pulses and the time elapsed between emission and reception, radar could determine the range and direction of those objects—crucial information in a combat situation. Radar was a deciding factor in many Allied victories during the war. Notoriously, American radar operators detected the incoming Japanese planes bound for Pearl Harbor on December 7, 1941, but their warnings were ignored by army air forces supervisors.

The development of the atomic bomb required close coordination between cutting-edge scientific research, massive engineering projects, and precise design of the weapon itself. All of these activities required extremely complex mathematics. The theoretical origins of the atomic bomb lay in Einstein's special theory of relativity. By positing a relationship between mass and kinetic energy, the theory led to Einstein's famous equation $E = mc^2$, where E is energy, m is mass, and c is the speed of light. This equation demonstrated that mass and energy were interconvertible—that matter may become energy. Nuclear power and nuclear weapons proceed from this relationship.

The research done on the physics of radioactivity and subatomic particles in the early 20th century lay the groundwork for nuclear fission, the splitting of atoms. In 1938 German physicists used neutrons to induce the fission of uranium nuclei (at first, they did not recognize what they had done). When the nucleus of a uranium atom split into smaller nuclei, some of its mass—the mass defect—was converted into energy. In 1940, at the urging of Leo Szilard, Albert Einstein sent a letter to President Franklin Roosevelt informing him of the possibility of initiating a chain reaction—a self-sustaining fission reaction—in a mass of uranium, resulting in the release of a tremendous amount of energy. A bomb built on such principles would be of unbelievable destructive power, and Einstein warned that Nazi Germany was working towards such technology. In response to Einstein's and Szilard's letter, the United States initiated the Manhattan Project, one of the most extensive—and expensive—government programs in history up to that time.

The project faced a number of great challenges at the outset. Only uranium 235 could fuel a fission reaction, but because U_{235} made up less than 1 percent of natural uranium, a separation process had to be created. In addition, the critical mass of uranium, the amount necessary to sustain a fission reaction, had to be determined. An artificial element, plutonium, had been discovered in 1940, and it could also serve as fuel for a fission reaction, but it had different characteristics and would require a different bomb mechanism. The Manhattan Project thus pursued two different approaches to atomic bomb production, two different designs, and ultimately, both were successful.

Uranium separation took place at the Y-12 plant constructed in Oak Ridge Tennessee. Natural uranium is only 0.7 percent U_{235}; the rest is made up of a heavier isotope, U_{238}. The physicist Ernest Lawrence designed immense devices called calutrons that operated on the basis of the difference in mass between the two isotopes. Uranium was ionized (given an electric charge), and the charged atoms were accelerated in an electric field and then deflected by a strong magnetic field. Because U_{235} was lighter than U_{238}, it was deflected differently, and this allowed the collection of U_{235}, literally atom by atom, for the purposes of nuclear fission.

On December 2, 1942, the world's first controlled nuclear chain reaction took place in Chicago under the supervision of Enrico Fermi, and the data from the experiment allowed for an estimation of the critical mass of U_{235}. While uranium bombs were relatively easy to design once sufficient U_{235} was refined, plutonium bombs operated on an implosion principle and required elaborate engineering. The actual construction of the atomic bombs was carried out at a facility in Los Alamos, New Mexico. Trinity, the code name for the first nuclear test explosion, took place on July 16, 1945, using a plutonium-fueled device. The first uranium-fueled atomic bomb was dropped on the city of Hiroshima in Japan on August 6, and a plutonium bomb destroyed the city of Nagasaki three days later.

Technology had a tremendous effect on the outcome of the Second World War, and the development of modern technology is a process informed by mathematics. In the decades since the end of World War II, military technology has grown more powerful, more elaborate, and more dependent on scientific research and mathematical calculation. One hopes that the intelligence that lies behind the production of modern weapons will also be sufficient to prevent their use.

NUMBER, COMPUTERS, AND DIGITAL CULTURE

Consider a simple, commonplace action: watching a movie on Netflix. Millions of people do it every day. Yet that choice embodies the digital culture of the early 21st century, a culture that rests upon a foundation of computing power, digital data storage, and the rapid transfer of information. To understand the roots of this digital culture, it is necessary to examine, in a brief manner, the development of computer technology and its use of binary data.

A history of computers is beyond the scope of this book. The story of their development is a complex one involving a dizzying number of individuals and the machines they conceived of and constructed. During the 19th century, scientists and inventors sought to speed the process of performing laborious mathematical calculations and to cut down on human error. A variety of hand-operated adding machines, cash registers, and calculators were developed, based on complex physical mechanisms and clockwork. The Difference Engine of Charles Babbage, a complex machine designed for the solution of polynomials, was an elaborate example of such a mechanical calculating device (see Chapter 11). With the invention of electric motors in the late 19th century, calculating machines could be driven by electrical power.

Babbage never completed his Difference Engine; his ideas went beyond the technical capacities of the mid-19th century. He went on to design a far more complicated and capable machine, his Analytical Engine, but he never constructed it. The Analytical Engine, as he conceived of it, would have performed many of the functions of a true computer. It included an input unit, which served as the machine's interface with the user; a control unit, which governed the machine's order of operations; a "store," a mechanical form of artificial memory that could retain data; a "mill," the processing unit that carried out calculations; and finally, a printer. Most importantly, Babbage's analytical engine was not tied one specific kind of problem; it could be programmed with different algorithms using a form of punch cards. Babbage's friend, Lady Ada Lovelace, recognized that the Analytical Engine had the potential to go beyond mathematical calculations, because it could be programmed with a number-based code that would allow it to solve various kinds of problems. She envisioned a form of programming in which the machine's internal logic could be harnessed to different functions.

Babbage's machines were analog computers—devices whose internal operations mirror the problems that it is created to solve. In other words, the structure of the machine reflects its function. For example, if an adding machine has gears with ten teeth, and one of those wheels is turned by four "clicks" in the process of adding the number four, that is an act of analog computing. The most commonly known analog devices are mechanical clocks, where the movement of the second, minute, and hour hands correspond to the actual turning of geared wheels inside the mechanism, driven by a spring. However, analog computers are not necessarily mechanical. Electronic analog computers were developed in the middle of the 20th century, in which the electrical relays corresponded to specific mathematical operations.

While analog computers could serve a variety of functions, they were limited in terms of flexibility and data storage. Digital computers would come to dominate computing in the second half of the 20th century. Such computers operate on the basis of coded instructions in binary form, and in order to understand the basic architecture of digital computing, one must briefly explore the history of binary number systems.

Binary numbers are base 2, just as decimal numbers are base 10. In a binary system, only two characters exist: 0 and 1. Binary numbers employ place-values, and each place represents a power of 2. Thus, the number 3 in binary is written as "11": a "1" in the place-value representing 2 to the first power and a "1" in the place representing units or ones. Similarly, the number 4 in binary is written as "100": a "1" in the place-value representing 2 to the second power (4) and "0" in the other place-values.

Gottfried Wilhelm Leibniz, a philosopher and mathematician in the late 17th century, became interested in binary numbers for religious and philosophical reasons. He viewed the system as a link between logical and mathematical reasoning. He also believed that the yin/yang hexagrams of the *Yijing*, one of the Chinese classics, were a form of binary coding.

A century and a half later, George Boole published his *Mathematical Analysis of Logic* (1847) in which he developed a symbolic approach to logical thinking. Boole used binary symbols to represent validity. In his system, "0" and "1" did not stand for quantities, but rather for the concepts of false (0) and true (1). He constructed a nonmathematical form of algebra that represented the logical relationships between propositions. For Boole, three basic relationships may exist between statements x and y:

x **and** y   are true

x **or** y   is true

x **not** y   is true

Boole's symbolic logic would become the basic architecture of digital computing in the 20th century.

The invention of electronics would revolutionize the development of computers. Electronic circuits allow the control or modulation of a large electrical current by a small one. For example, in a cable television set, the components allow the small current that comes from the cable signal to control the larger current that comes from wall socket, switching it off and on rapidly to form video effects on the screen and audio effects in the speakers. The first electronic components, invented in 1906, were vacuum tubes. In 1937 an American graduate student at MIT named Claude Shannon recognized that the off/on character of vacuum tubes reflected the binary character of Boole's symbolic logic. In other words, the action of electronic components could reproduce the operations of Boole's logical algebra; it could evaluate the relationship between data points in terms of **and**, **or**, and **not**. Theoretically, mathematical operations and other kinds of problems could be coded or expressed in Boolean logical terms and solved electronically. Data could be treated by electronic components if it was in binary form, for the on/off switching of these components was compatible with coded instructions. These operations would become the essence of digital computing.

The Second World War would create the necessity, and the funding, for projects that developed digital computing technology on a large scale. In Great Britain, the Colossus at Bletchley Park was built to aid in the breaking of German codes. Alan Turing, the pioneering thinker in computing and artificial intelligence, served in the Government Code and Cypher

School in Bletchley Park during the war. In the United States, ENIAC—the Electronic Numerical Integrator and Computer—was designed in 1943 for the purpose of calculating artillery tables. Completed in 1945, it contained 18,000 vacuum tubes.

Computer technology continued to grow steadily in the decades after the Second World War. Alan Turing and John von Neumann did crucial work on the nature of computer programming, and Turing's work served as the basis for a stored program that the machine could access on its own (Ifrah 1994, 280). Vacuum tubes were made obsolete in the 1950s and 1960s by solid-state components—transistors. In the final decades of the 20th century, transistors would be progressively miniaturized and coordinated, first into integrated circuits, then into microchips. Large mainframe computers were joined by smaller personal computers. The basic parts of these machines corresponded closely to Charles Babbage's vision of his Analytical Engine: input/output, storage of instructions, control unit, memory, and processing unit (Ifrah 1994, 235–236).

The foundations of modern digital culture lay in the pursuit of mechanical calculation, in the structure of symbolic logic, in the design of electronic logic circuits, in the architecture of computer programs, and in the continuing development of more powerful processing and memory technology. All for Netflix.

FURTHER READING

Brusatti, Steve. 2018. *The Rise and Fall of the Dinosaurs*. New York: William Morrow.

Ifrah, George. 1994. *The Universal History of Numbers III: The Computer and the Information Revolution*. London: Harvill Press.

Marshall, Stephen. 2015. *The Story of the Computer*.

Raup, David. 1986. *The Nemesis Affair*. New York: Norton.

Rhodes, Richard. 1987. *The Making of the Atomic Bomb*. New York: Simon and Schuster.

Appendix: A Brief Look at Navigation

The evolution of stellar navigation serves as an excellent example of the transmission of mathematical knowledge between cultures. While navigation was a practical discipline, its techniques combined elements of astronomy, technology, and pure mathematics. Imagine a mariner determining the position of a wooden ship at sea in the late 19th century, well before the rise of global positioning technology. That mariner might determine the ship's heading using a magnetic compass (as opposed to a gyrocompass on a steel ship). The magnetic compass was invented in China, probably during the 10th century CE. The compass heading would be expressed in terms of a 360-degree circle with the ship at the center. Such a division of the circle finds its roots in the mathematics of the Hellenistic Greek Mediterranean; Hellenistic mathematicians were influenced in turn by the sexagesimal (that is, base 60) mathematics of ancient Mesopotamia.

Hellenistic mathematicians also envisioned coordinate systems of latitude and longitude, and their work influenced the *Geography* of Claudius Ptolemy, written in the 2nd century CE. However, the ability to actually find one's latitude and longitude at a given point on the earth's surface was developed over many centuries. Our 19th-century mariner would find latitude in the Northern Hemisphere by measuring the angle of elevation of the North Star (Polaris) above the horizon—that angle itself represents one's latitude, for Polaris is located above the earth's north pole. In the Southern Hemisphere, the same operation could be performed using different groupings of stars that rest above the south pole.

A modern navigator would measure such angular values in degrees using a sextant, but ancient or medieval navigators employed other methods to determine stellar elevation. Polynesian mariners guided their open craft across the vast reaches of the Pacific using the changing positions of constellations as their guide. The astronomers of the Hellenistic Mediterranean developed the astrolabe and the quadrant to measure stellar elevation;

Ptolemy refers to a version of the latter instrument in the *Almagest*. The medieval Arabs refined both devices and used them to find stellar positions (angle of elevation and azimuth). An Arab mariner of the period might also employ a wooden apparatus called a *kamal*, and the use of this tool spread to India and China. European sailors adopted these instruments in the Late Middle Ages. Columbus used a quadrant on his voyages, and the *kamal* evolved into the European cross-staff by the early 16th century.

Stars may be obscured for long periods of time at sea, so the sun provides an alternative reference—the altitude of the sun at noon varies with one's latitude. However, the elevation of the sun also changes over the course of a year at a given latitude, so a table of solar positions, or declinations, is needed. The data for such tables was collected by Arab and Persian astronomers during the Middle Ages. In the late 15th century, Abraham Zacuto, a Jewish professor at the University of Salamanca in Spain, constructed such a table, and he ultimately brought it to Portugal.

The problem of longitude challenged astronomers and mariners around the world for many centuries. To determine longitude, it was necessary to establish a reference meridian, that is, a line of longitude that served as the basis for measurement. The Hellenistic mathematician Eratosthenes used a meridian that ran through the city of Alexandria. Modern longitude coordinates are based on the prime meridian that runs through the cities of Greenwich and London in Britain. One could theoretically discover one's longitude by comparing the time at the reference meridian to the local time—the latter value could be found by observing the sun at the top of its arc at noon. However, mariners could not establish the time at the refence meridian as they traveled. Different astronomical methods were explored, including the use of lunar eclipses or the orbits of the moons of Jupiter as a timekeeping standard. In the mid-18th century, an approach based on varying lunar distances proved somewhat successful.

Mechanical clocks provided another alternative. The European clocks of the 16th and 17th centuries were too cumbersome and inaccurate to be of any use in maritime navigation. The pendulum clocks developed by the late 17th century represented a significant improvement, but they could not be employed on a moving ship. In 1714 the British Parliament passed the Longitude Act, establishing a prize of £20,000 for the solution to the problem of determining longitude. The prize inspired significant innovation, including the work of John Harrison. Harrison produced a series of clocks designed to withstand the difficult conditions of sea travel while remaining accurate. His H4 (1761) and H5 (1772) models proved to keep time at sea to an acceptable degree of precision. While the parliamentary prize was never actually awarded, he did receive significant compensation for his decades of effort. His designs served as the basis for marine chronometers until modern times, and our 19th-century mariner would certainly have used one to find his longitude.

Compasses, latitude calculations, and longitude calculations all proceed from the interaction of mathematics with technology. Mathematics also came to play an important role in the plotting of courses and the determination of distances at sea. These activities require the application of spherical trigonometry, the analysis of triangles formed by intersecting great circles on the surface of a sphere (in this case, the earth). Trigonometric techniques pioneered in medieval India, Arabia, and Persia lie at the heart of the calculations of modern navigators.

The global positioning system (GPS) is now available to anyone with a smart phone; computers and satellite technology dominate the contemporary process of finding one's location on the globe. Yet the calculations required for global positioning find their roots in centuries of human maritime navigation, a fascinating blend of technology, astronomy, and mathematics.

Bibliography

Adamson, Peter. 2019. *Medieval Philosophy*. Oxford: Oxford University Press.
Alder, Ken. 2002. *The Measure of All Things*. New York: The Free Press.
Alhazen. 1974. *Book of Optics*. In *A Source Book in Medieval Science*, edited by Edward Grant. Cambridge: Harvard University Press.
Al-Khwārizmī. 1974. *The Book of Restoring and Balancing*. In *A Source Book in Medieval Science*, edited by Edward Grant. Cambridge: Harvard University Press.
Anglin, W.S. and Lambech, Joachim. 1995. *The Heritage of Thales*. New York: Springer Press.
Astell, Mary. 1994. *A Serious Proposal to the Ladies, part II*, in *Women Philosophers of the Early Modern Period*, edited by Margaret Atherton. Indianapolis: Hackett Publishing.
Atherton, Margaret (ed). 1994. *Women Philosophers of the Early Modern Period*. Indianapolis: Hackett Publishing.
Aveni, Anthony. 2001. *Skywatchers*. Austin: University of Texas Press.
Belting, Hans. 2011. *Florence and Baghdad*. Cambridge: Belknap Press.
Berlinski, David. 2014. *The King of Infinite Space: Euclid and His Elements*. New York: Basic Books.
Bergren, J.L. 1986. *Episodes in the Mathematics of Medieval Islam*. New York: Springer.
Casson, Lionel. 2001. *Libraries in the Ancient World*. New Haven: Yale University Press.
Clagett, Marshall. 1999. *Ancient Egyptian Science, A Source Book, vol 3*. Philadelphia: American Philosophical Society.
Clark, Walter. 2006. *The Aryabhatiya of Aryabhata*. Whitefish: Kessinger Publishing.
Cohen, I.B. 1985. *The Birth of a New Physics*. New York: W.W. Norton.

Colish, Marcia. 1997. *Medieval Foundations of the Western Intellectual Tradition*. New Haven: Yale University Press.
Cropsey, Joseph. 1995. *Plato's World*. Chicago: University of Chicago Press.
Culbert, Patrick. 1993. *Maya Civilization*. Montreal: St Remy Press.
Dalley, Stephanie. 1989. *Myths from Mesopotamia*. Oxford: Oxford University Press.
Dauben, Joseph. 2007. *Chinese Mathematics*. In *The Mathematics of Egypt, Mesopotamia, China, India, and Islam—A Sourcebook*, edited by Victor Katz. Princeton: Princeton University Press.
Dear, Peter Robert. 1995. *Discipline and Experience: The Mathematical Way in the Scientific Revolution*. Chicago: University of Chicago Press.
Demaraest, Arthur. 2004. *Ancient Maya*. Cambridge: Cambridge University Press.
de Pizan, Christine. 1982. *The Book of the City of Ladies*, translated by Earl Jeffrey Richards. New York: Persea Books.
Drake, Stillman (ed). 1957. *Discoveries and Opinions of Galileo*. New York: Anchor Books.
Ebrey, Patricia Buckley. 1993. *Chinese Civilizations, A Sourcebook*. New York: Simon and Schuster Free Press.
Edwards, A.W.F. 2003. *Pascal's Work on Probability*. In *The Cambridge Companion to Pascal*, edited by Nicholas Hammond. Cambridge: Cambridge University Press.
Eliot, Simon and Stern, Beverly. 1979. *The Age of Enlightenment*. New York: Harper and Row.
Euclid. 1933. *Elements*. Edited by Isaac Todhunter. London: J.M. Dent and Sons, Everyman's Library.
Field, Judith Veronica. 1997. *The Invention of Infinity*. Oxford: Oxford University Press.
Galilei, Galileo. 1957. *The Assayer*. In *Discoveries and Opinions of Galileo*, edited by Stillman Drake. New York: Anchor Books.
Gay, Peter. 1967. *The Enlightenment*. New York: Alfred Knopf.
Gay, Peter (ed). 1973. *The Enlightenment, A Comprehensive Anthology*. New York: Simon and Schuster
Gillings, Richard. 1982. *Mathematics in the Time of the Pharaohs*. New York: Dover Publications.
Gingerich, Owen. 2005. *The Book Nobody Read*. New York: Penguin Books.
Goldstein, Thomas. 1980. *Dawn of Modern Science*. New York: Houghton Mifflin.
Grant, Edward (ed). 1974. *A Source Book in Medieval Science*. Cambridge: Harvard University Press.
Grant, Edward. 1977. *Physical Science in the Middle Ages*. Cambridge: Cambridge University Press.

Hammond, Nicholas (ed). 2003. *The Cambridge Companion to Pascal.* Cambridge: Cambridge University Press.
Hankins, Thomas. 1985. *Science in the Enlightenment.* Chicago: University of Chicago Press.
Heilbroner, Robert. 1999. *The Worldly Philosophers.* New York: Touchstone Books.
Henrion, Claudia. 1997. *Women in Mathematics.* Bloomington: Indiana University Press.
Herodotus. 1958. *History, Book II.* New York, Heritage Press.
Hirshfeld, Alan. 2009. *Eureka Man: The Life and Legacy of Archimedes.* New York: Walker and Company.
Hobsbawm, Eric. 1962. *The Age of Revolutions.* New York: Vintage Books.
Hourani, Albert. 1991. *A History of the Arab Peoples.* Cambridge: Belknap Press.
Ifrah, Georges. 1994. *The Universal History of Numbers II—The Modern Number-System.* London: Harvill Press.
Ifrah, Georges. 1998. *The Universal History of Numbers: The World's First Number Systems.* London: The Harvill Press.
Imhausen, Annette. 2016. *Mathematics in Ancient Egypt.* Princeton: Princeton University Press.
Isidore of Seville. 1974. *Etymologies.* In *A Source Book in Medieval Science*, edited by Edward Grant. Cambridge: Harvard University Press.
Jones, Alexander. 2017. *A Portable Cosmos.* New York: Oxford University Press.
Joseph, George Gheverghese. 2016. *Indian Mathematics.* London: World Scientific Publishers.
Katz, Victor (ed). 2007. *The Mathematics of Egypt, Mesopotamia, China, India, and Islam—A Sourcebook.* Princeton: Princeton University Press.
Katz, Victor (ed). 2016. *Sourcebook in the Mathematics of Medieval Europe and North Africa.* Princeton: Princeton University Press.
King, William. 1902. *Seven Tablets of Creation.* https://www.sacred-texts.com/ane/stc/index.htm
Knight, David. 2014. *Voyaging in Strange Seas.* New Haven: Yale University Press.
Knowlton, Timothy. 2010. *Maya Creation Myths.* Boulder: University of Colorado Press.
Leicester, Henry. 1971. *The Historical Background of Chemistry.* Mineola: Dover Publications.
Libbrecht, Ulrich. 1973. *Chinese Mathematics in the 13th Century.* Cambridge: MIT Press.
Lindberg, David. 1976. *Theories of Vision from al-Kindi to Kepler.* Chicago: University of Chicago Press.

Lindberg, David. 1978. *Science in the Middle Ages*. Chicago: University of Chicago Press.
Lloyd, G.E.R. 1974. *Early Greek Science: Thales to Aristotle*. New York: W.W. Norton and Company.
Lloyd, G.E.R. 1974. *Greek Science after Aristotle*. New York: W.W. Norton and Company.
Lloyd, G.E.R. 1970, 1973, 2010. *Greek Science*. New York: W.W. Norton & Company. London: Folio Society.
Lyons, Jonathan. 2009. *The House of Wisdom*. New York: Bloomsbury.
MacGregor, Neil. 2010. *History of the World in 100 Objects*. London: Viking Press.
Martianus Capella. 1977. *The Marriage of Philology and Mercury*. In *Martianus Capella and the Seven Liberal Arts*, translated by William Harris Stahl and Richard Johnson. New York: Columbia University Press.
Merzbach, Uta and Boyer, Carl. 2011. *A History of Mathematics*. Hoboken: John Wiley and Sons.
Milbrath, Susan. 1999. *Star Gods of the Maya*. Austin: University of Texas Press.
Montesquieu. 1977. *The Spirit of the Laws*. Berkeley: University of California Press.
Needham, Joseph. 1978. *The Shorter Science and Civilization in China*. Cambridge: Cambridge University Press.
Needham, Joseph. 1981. *Science in Traditional China*. Cambridge: Harvard University Press.
Neuenschwander, Dwight. 2017. *Emmy Noether's Wonderful Theorem*. Baltimore: Johns Hopkins University Press.
Newton, Isaac. 1953. *Mathematical Principles of Natural Philosophy*. In *Newton's Philosophy of Nature*, edited by H.S. Thayer. New York: Hafner Publishing.
Noble, David. 1982. *A World Without Women*. New York: Knopf.
North, John. 1995. *The Norton History of Astronomy and Cosmology*. New York: W.W. Norton and Co.
Oates, Joan. 1986. *Babylon*. London: Thames and Hudson.
O'Leary, Delacy. 2001. *How Greek Science Passed to the Arabs*. Abingdon: Routledge Press.
Osen, Lynn. 1974. *Women in Mathematics*. Cambridge: MIT Press, 1974.
Pascal, Blaise. 1941. *Pensees*. New York: Random House.
Peterson, Mark. 2011. *Galileo's Muse*. Cambridge: Harvard University Press.
Plato. 1961. *The Collected Dialogues*, edited by Edith Hamilton and Huntington Cairns. Princeton: Princeton University Press.

Plato. 1961. *Meno*. In *The Collected Dialogues*, edited by Edith Hamilton and Huntington Cairns. Princeton: Princeton University Press.

Plato. 1961. *Theaetetus*. In *The Collected Dialogues*, edited by Edith Hamilton and Huntington Cairns. Princeton: Princeton University Press.

Plato. 1965. *Timaeus*. New York: Everyman's Library.

Plofker, Kim. 2007. *Mathematics in India*. In *The Mathematics of Egypt, Mesopotamia, China, India, and Islam*, edited by Victor Katz. Princeton: Princeton University Press.

Plofker, Kim. 2009. *Mathematics in India*. Princeton, Princeton University Press.

Pollard, Justin and Reid, Howard. 2006. *The Rise and Fall of Alexandria*. New York: Viking Penguin.

Rao, S. Balachandra. 1994. *Indian Mathematics and Astronomy: Some Landmarks*. Bangalore: Jnana Deep Publications.

Robson, Eleanor. 2008. *Mathematics in Ancient Iraq*. Princeton: Princeton University Press.

Romer, John. 2016. *A History of Ancient Egypt*. New York: St. Martin's Press.

Schiebinger, Londa. 1989. *The Mind Has No Sex*. Cambridge: Harvard University Press.

Schiele, Linda and Freidel, David. 1990. *A Forest of Kings*. New York: Harper Collins.

Schiele, Linda and Freidel, David. 1993. *Maya Cosmos*. New York: Harper Collins.

Schneider, Thomas. 2013. *Ancient Egypt in 101 Questions and Answers*. Ithaca: Cornell University Press.

Schwartz, Benjamin. 1985. *The World of Thought in Ancient China*. Cambridge: Harvard University Press.

Scriba, Christopher. 2015. *5000 Years of Geometry*. London: Springer Basel.

Shaw, Ian. 2004. *Oxford History of Ancient Egypt*. Oxford: Oxford University Press.

Shekoury, Raymond. 2010. *Mesopotamians: Pioneers of Mathematics*. Createspace.

Shetterly, Margot Lee. 2016. *Hidden Figures*. New York: Harper Collins.

Smith, Adam. 1973. *The Wealth of Nations*. In *The Enlightenment, A Comprehensive Anthology*, edited by Peter Gay. New York: Simon and Schuster.

Smith, Tim Mackintosh. 2020. *Arabs: A 3000-Year History of Peoples, Tribes, and Empires*. New Haven, Yale University Press.

Snow, John. 1855. *On the Mode of Communication of Cholera*. London: John Churchill. at UCLA Department of Epidemiology, Fielding School of Public Health, https://www.ph.ucla.edu/epi/snow/snowbook3.html

Staeger, Rob. 2007. *Ancient Mathematicians*. Greensboro: Morgan Reynolds.

Thayer, H.S. (ed). 1953. *Newton's Philosophy of Nature*. New York: Hafner Publishing.

Tuplin, Christopher and Rihll, Tracey. 2002. *Science and Mathematics in Ancient Greek Culture*. Oxford: Oxford University Press.

Unknown. 1972. *The Epic of Gilgamesh*, translated by N.K. Sandars. London: Penguin Books.

Wagner, David (ed). 1983. *The Seven Liberal Arts in the Middle Ages*. Bloomington: Indiana University Press.

Watts, Edward. 2017. *Hypatia, The Life and Legend of an Ancient Philosopher*. New York: Oxford University Press.

Westfall, Richard. 1978. *The Construction of Modern Science*. Cambridge: Cambridge University Press.

Wheelwright, Philip. 1966. *The PreSocratics*. New York: Macmillan Publishing Company.

Wollstonecraft, Mary. 2008. *A Vindication of the Rights of Women*. London: Folio Society.

Woolf, Virginia. 1989. *A Room of One's Own*. New York: Houghton Mifflin.

Index

Abacus, 10, 18, 116, 199
  in China, 116
  in Medieval Europe, 199
  in Mesopotamia, 10, 18
Abbasid Dynasty, 167, 171, 173
  decline, 176
  Mongol conquest of, 176
Abū al-Wafā, 160
  work in trigonometry, 160
Adelard of Bath, 193, 195
  translation of Euclid's *Elements*, 195
African number systems, x
Akbar the Great, 161
  Indian mathematics and, 161
Akkadians, 1, 4, 7, 9
Al-Andalus. *See* Islamic Spain
Al-Biruni, 160, 175, 187
  correspondence with al-Wafā, 175, 187
  correspondence with ibn Sina, 175, 187
  views on Indian culture, 160
Al-Haytham (Alhazen), 182–184, 186–187, 273
  Doubts on Ptolemy, 183–184, 186–187
  Euclid's fifth postulate and, 273
  Greek influences on, 182–183
  study of optics, 182–183
  use of experiment, 183–184

Al-Khwārizmī, 160, 174, 175, 179–182, 186
  Book of Restoring and Balancing (treatise on algebra), 160, 174, 180–181; text, 191
  commentary on the *Sindhind*, 181, 186
  Greek and Indian influences, 180–181
  influence on medieval Europe, 179, 200
  relationship with al-Ma'mun, 174, 179, 181
  treatise on the astrolabe, 168
  treatise on Indian numbers, 160, 174, 179–180
Al-Kindi, 174
  Greek philosophy and, 174
  Indian numbers and, 174
Al-Ma'mun, 167, 173–174
  astrology and, 174
  interest in Aristotle, 174, 186
  patronage of learning, 174
  pursuit of Ptolemy's *Almagest*, 174, 186
Al-Mansur, 167
Al-Tusi, 176, 187–188
  Euclid's fifth postulate and, 273
  leadership of Maragha observatory, 176, 187–188
  Mongol patronage of, 176, 187
  responses to Ptolemy, 187

336 Index

Al-Wafā, Abū, 175, 178, 187
  correspondence with al-Biruni, 175, 187
  study of trigonometric functions, 175
Alberti, Leon Battista, 216
  influence of al-Haytham (Alhazen) on, 216
  influence of Euclid on, 216
  On Painting, 216
Alexander the Great, 79, 158
Alexandria, 76–78, 80–85, 94
  Christianity in, 85, 95–96
  library of. *See* Great Library of Alexandria
  scholarship in. *See* Mouseion
Algebra, 121, 122, 123, 148, 160, 162–163, 174, 180, 181, 185, 208–209, 230
  Al-Khwārizmī and, 160, 174, 180–181
  Bhāskara II, and, 162–163
  Descartes and, 230
  in early modern Europe, 230
  in Imperial China, 121, 122, 123
  in Indian Sanskrit tradition, 148, 162
  techniques of Leonardo of Pisa, 208–209
  Omar Khayyam and, 185
Algorism, 208, 210, 213, 217
  adoption in medieval Europe, 217
  role in medieval Italian commerce, 213
Almagest. *See* Ptolemy, Claudius
Altitude (celestial), 169–170
  measured with astrolabe, 169
Alvarez, Luis, 310
Alvarez, Walter, 310
Alvarez hypothesis, 310
Amate, 128
Analog computer, 321
Analysis (mathematical). *See* Infinitesimal analysis
Analytical geometry, 230, 239, 241
  Descartes' *Geometry* and, 241
Antikythera Mechanism, 89–91, 96
  influences on, 90
  Koine inscriptions on, 90

Apollonius of Perga, 82, 91, 93
  astronomy and, 91
  Conics of, 82
  planetary epicycles and, 93
Arab Commerce, 172
  Saharan routes, 190
  Seaborne routes, 172
  Silk Road, 172
Arab Empire, 159, 170–171
Arabic culture, 170, 172
Archimedes of Syracuse, 82, 87–89, 200, 226–227
  availability in early modern Europe, 226–227
  correspondence with Eratosthenes, 82
  medieval translations of, 200
  principle of buoyancy, 87
  Sand-Reckoner of, 88
  semiregular solids and, 87–88
Area of a circle, Egyptian rule for, 25–28
Aristarchus' cosmology, 89, 91, 94
Aristotelian logic, 51, 54–55, 160, 182, 200, 207, 210, 226
  al-Biruni and, 160
  al-Haytham (Alhazen) and, 182
  Boethius and, 198, 199, 204
  Gerard de Cremona and, 196
  ibn Rushd and, 187, 189
  in medieval universities, 200, 207, 210, 226
Aristotle, 47, 50–52, 54–55, 56, 58, 67, 79, 94, 174, 175, 182, 186, 187, 196, 200–201, 207, 210–211, 215, 217, 226–227, 233, 236
  al-Ma'mun and, 174, 186
  Analytics of, 51
  cosmology of, 47–50, 94, 186, 233
  challenges to, 202, 226–227
  Hellenistic reaction to, 79
  ibn Sina and, 175, 187
  influence on medieval university culture, 210–211, 217, 226
  medieval translations of, 196, 200, 201

Index 337

Organon of, 54–55
Pythagoras and, 58
in Raphael's *School of Athens*, 215
relation to geometry, 55
seven liberal arts and, 207
Thales and, 56
views on motion, 236
views on vision, 182
Arithmetical triangle, 123, 151, 185, 231, 242–243
in Indian mathematics, 123
Omar Khayyam and, 185
Pascal and, 231, 242–243
Pingala and, 151
Yang Hui and, 123
Zhu Shijie and, 123
Āryabhaṭa, 145, 152, 154, 155–157
Āryabhatiya, 152, 155–157
calculation of pi and, 157
circular astronomy and, 156
ganita and, 156
treatment of Pythagorean theorem, 157
use of sine trigonometry, 156, 157
Assyrians, 1, 4, 5, 7, 9, 16, 20
Astell, Mary, 288–289
Astrolabe, 167–170, 199
in Arabic culture, 167–170
Gerbert of Aurillac and, 199
Hellenistic origins of, 168, 325
Astrology, 20, 131, 174, 224
Al-Ma'mun and, 174
Babylonian, 20
Kepler and, 224
Maya, 131
Astronomy, 9, 19–21, 32, 45, 89, 91, 93, 96, 127–129, 131–132, 136, 138, 139, 140, 141–143, 148, 149, 151–152, 153–155, 155–157, 157–159, 159–161, 161, 163, 167–170, 173, 174, 175, 176, 177–178, 186–188, 222, 233
Arabic, 167–170, 173, 174, 175, 176, 177, 178, 186–188
Babylonian, 9, 19–21, 96

circular geometry and, 91, 93, 156, 158, 186, 187, 222, 233; al-Tusi and, 187; Apollonius of Perga and, 93; Āryabhaṭa and, 156; Copernicus and, 233; Hellenistic culture and, 158; influence on Arabic scholars, 186; Kepler's rejection of, 222; Ptolemy and, 93
Egyptian, 32, 45
heliocentric. *See* Heliocentrism
Hellenistic, 75–78, 83, 89–91, 157–159; influence on Indian astronomy, 152, 155, 157–159
Indian, 148, 149, 151–152, 153–155, 155–157, 157–159, 161, 163, 173, 177–178; influence on Arab culture, 159–161, 173, 177–178; *See also Siddhānta* texts
Maya, 127–129, 131–132, 136, 138, 139, 140, 141–143
Ptolemaic. *See* Ptolemy, Claudius
Atomic bomb, 316, 319–320
Atomic number, 280, 308
Atomic weight, 280, 281, 308
Atomism (Greek), 280
Averroes. *See* Ibn Rush'd
Avicenna. *See* Ibn Sina
Azimuth, 169–170
measured with astrolabe, 169

Babbage, Charles, 253–255, 320, 321
Analytical Engine, 253, 254, 321
Difference Engine #1, 253–255, 320, 321
Babylonian gods, 6–7, 20
Babylonians, 1, 2, 3, 4, 5, 6, 7, 9, 16, 17, 19–21
Bacon, Roger, 193
text, 217–218
Baghdad, 167, 173
construction of, 173
Mongol conquest of, 167, 176
Bakhshali Manuscript, 145–147
disputes over date, 147

338   Index

Ballistics, 317
Bassi, Laura, 289
　Newtonianism and, 289
Bayt al-Hikma. *See* House of
　Wisdom
Bentham, Jeremy, 257
Bernoulli, Daniel, 256, 261–262
　Hydrodynamica, 261–262
Bernoulli, Jacques, 261
　analytical work, 261
　correspondence with Leibniz, 261
Bhagavad Gītā, 150
Bhāskara I, 145, 152, 153, 156–157
　commentary on *Āryabhatiya*,
　　156–157
　Indian numbers and, 153
Bhāskara II, 145, 161–163
　algebra and, 162–163
　Bīja-ganita, 163
　Līlāvatī, 161–162
　Siddhānta tradition and, 163
Binary numbers, 321–322
Bletchley Park, 322–323
Boethius, 193, 198, 199, 204–204
　influence on European logic, 204
　translations of Aristotle and Euclid,
　　198, 199, 204
Book of the Dead, 28–29
Boole, George, 322
　development of symbolic logic,
　　322
Boyle, Robert, 280–281
　predictive chemical laws and, 280
　notion of atoms characterized by
　　shape, 281
Bradwardine, Thomas, 193, 201–202
Brahe, Tycho, 222
Brahmagupta, 145, 152
　Indian numbers and, 152, 153
　use of zero, 152
Brâhmî, 153
Bretuil, Émilie de, Marquise du
　Châtelet, 263–265, 290,
　　294–296
　Institutions de Physique, 264, 295
　Newtonianism and, 264, 295
　Voltaire and, 264, 294–295

Brunelleschi, Filippo, 215–215
　contributions to perspective in
　　painting, 215–216
Buffon (Georges-Louis Leclerc, the
　　Count de Buffon), 265–266
　Epochs of Nature, 266
　Natural History, 265–266
　study of the age of the earth, 266
Bullae, 10
Byzantine Empire, 197

Calculation methods, 37–39, 128, 162,
　179, 180–181, 191, 200
　Al-Khwārizmī and, 180–181, 191;
　　influence on medieval Europe,
　　200
　Arab algorism. *See* Algorism
　Bhāskara II and, 162
　in Egypt, 37–39
　in India, 179; influence on Arabic
　　culture, 179
　in Maya culture, 128
Calculus, 231–233, 245, 260,
　261–262
　and analysis, 261–262
　dispute over discovery of, 232
　Leibniz and, 231, 260, 261
　Newton's work, 231, 245, 260,
　　261
Calendars, 21, 32, 44–46, 96, 129, 131,
　151, 158, 185
　Arabic *Jalali* Calendar, 185
　Babylonian, 32
　Egyptian, 32, 44–46, 96
　Greek, 32
　Indian *Jyotiṣa* tradition, 151, 158
　Lunar, 32
　Maya. *See* Maya calendars
　Metonic, 21, 44
　Sidereal, 45
Capella, Martianus, 193, 205–207
　Marriage of Mercury and Philology,
　　206–207; Cassiodorus and, 206;
　　treatment of geometry in, 206;
　　influence on monastic education,
　　206
Capital (financial), 248

Carnot (Lazare Nicolas Marguerite), 270
Cassiodorus, 205, 206
Cathedral schools, 199
Cavendish, Margaret, 288–289
  Royal Society and, 289
Celsius scale, xi
Chandayoga Upanishad, 149
Chichen Itza, 132, 139, 141, 143
Chicxulub Crater, 311
Chinese Classics, 103
Chinese counting board, 114–116
Chinese numeration, 112–116
Cholera, 277
Christianity, 132, 197, 200, 224, 226, 270
  challenged in French Revolution, 270
  growth in 4th century, 197
  influence on Kepler's astronomy, 224
  Maya culture and, 132
  reconciled with Aristotelian ideas, 200
  role in medieval universities, 226
  prominence in early medieval intellectual life, 197
Christine de Pizan, 288
Colossus computer. See Bletchley Park
Compass (magnetic), 325
Computer technology, 308
Condorcet (Marie Jean Antoine Nicolas de Caritat), 270
Confucianism, 106, 108, 111, 112
Confucius. See Kong fuzi
Copernicus, Nicholas, 221, 223, 227–228, 233–235
  De Revolutionibus, 234–235
  degree of accuracy, 223
  heliocentrism, 233, 235
  parallax and, 235
  relationship with Ptolemaic astronomy, 227–228, 235
Cuneiform writing, 2, 4, 7, 8, 9, 13–16
  numeration and, 8, 9, 13–16
  surviving tablets, 7
Cylinder seals, 10

Dalton, John, 281
  atomic weight and, 281
  quantitative conception of matter, 281

Dao De Jing, 107–108; text, 107
Daoism, 107–108, 111, 118–119
  and wuxing, 118–119
De Landa, Diego, 132
  destruction of Maya manuscripts and, 132
Decimal number systems, 9, 16, 18, 33, 104, 112–115, 146–147, 148, 177–179, 199, 200, 207–208
  in Arabic world, 177–179
  in China, 104, 112–115
  in Egypt, 33
  in Europe, 199, 200, 207–208
  in India, 146–147, 148
  in Mesopotamia, 9, 16, 18
Delambre, Jean Baptiste, xii, 271
Demotic script, 77–78
Descartes, Rene, 221, 228, 230, 239–241, 242, 245
  analytical geometry and, 230, 239, 241
  cosmology of, 245
  Discourse on Method, 239
  Geometry, 241
  influence of, 241
  laws of matter and motion, 240
  method of doubt, 239–240
  views of matter, 240–241
Diagonal of a square, 2–3
Digital computers, 321–323
Digital culture, 320, 323
Dinosaur paleontology, 308–313
Diophantus, 84, 95, 160, 180, 209
  Arithmetica of, 84, 180
  Hypatia of Alexandria and, 95
  influence on Arabic culture, 160, 180
  influence on Leonardo of Pisa (Fibonacci), 209
Double-entry accounting, 213–214
Dresden Codex, 127–129
Dutch Capitalism, 248–250

Earth-centered cosmology. See Geocentricism
Eastern Roman Empire. See Byzantine Empire

Eccentric orbits, 92, 154, 158
  in Hellenistic astronomy, 92, 158
  in Indian astronomy, 154, 158
Eclipses, 21, 128–129, 154, 326
  in Babylonian astronomy, 21
  in Indian astronomy, 154
  in Maya astronomy, 128–129
  lunar eclipses and longitude, 326
Economic crisis of 2008, 315
Egyptian Numeration, 32–37
Egyptian weighing of the heart, 28–29
Einstein, Albert, 308, 319
El Caracol, 132, 141–143
El Castillo, 139–140
Electronic computing, 322
  Claude Shannon and, 322
  use of vacuum tubes, 322, 323
  development of transistors, 323
Elements (of matter), 62–63, 236, 279–282, 308
  Chinese. *See* Wuxing (Chinese five element theory)
  Greek, 62–63, 236
  modern chemistry, 308
  periodic table, 279–282
Empedocles, 62
ENIAC, 291–292, 317, 322
  female programmers and, 292
Enlightenment, 255
Ennuma Elish, 6–7; text, 22–23
Enuma Anu Enlil, 20
Epicycles, 93, 154, 158
  in Hellenistic astronomy, 93, 158
  in Indian astronomy, 154, 158
Eratosthenes of Cyrene, 75–76, 82, 91, 326
  astronomy and, 91
  calculation of earth's circumference, 75–78, 82
  correspondence with Archimedes, 82
  longitude and, 326
Error in measurement, xii
Experiment, 183–184, 227, 236–237, 239, 243–244
  al-Haytham (Alhazen) and, 183–184
  Descartes' rejection of, 239
  Galileo and, 237–238
  Newton's method of, 246
  Pascal's view of, 243–244
  Scientific Revolution and, 227
Euclid, 55, 67, 79, 82, 85–87, 182, 216
  influence of Aristotle on, 86
  influence of Eudoxus on, 86
  influence on Alberti, 216
  Optics, 86, 182
  Ptolemy I and, 81, 85
Euclid's *Elements*, 55, 67, 85–87, 95
  al-Ma'mun and, 174
  al-Tusi and, 176
  cultural impact of, 272
  Hypatia of Alexandria and, 95
  influence on Spinoza, 228, 271
  in late medieval thought, 201
  medieval Arabic translations of, 173, 174
  medieval European translations of, 200, 201, 207
  text, 96–99, 173, 174, 176, 200, 201, 207, 228, 272
  Theon of Alexandria and, 95
  universities and, 200, 217
Euclid's Fifth Postulate (Parallel Postulate), 184–185, 272–273
  al-Haytham (Alhazen) and, 273
  al-Tusi and, 273
  Lambert and, 273
  Lobachevsky and, 273
  Omar Khayyam and, 184–185
  Saccheri and, 273
Eudoxus of Cnidus, 47–49, 50, 54, 56, 64–66, 67, 79, 86, 93, 94
  circular geometry and, 48–49
  cosmology, 47–50, 94
  Hellenistic response to, 79
  influence on Euclid, 65, 86
  influence on Ptolemy, 93
  irrational numbers and, 64
  law of proportionality, 64–65
  method of exhaustion, 65
  Plato's Academy and, 64
  travels in Egypt, 56, 64

Euler, Leonard, 262
  differential equations and, 262
  mathematical notation of, 262
  techniques of analysis and, 262

Fall of Roman Empire. *See* Rome
Faraday, Michael, 282–283
  discovery of electromagnetic induction, 283
  relationship with Maxwell, 283
  study of electromagnetic fields, 282–283
Federal Reserve, 313–316
  discount rate of, 314
Fermat, Pierre, 221, 230, 231, 245
  Analytical geometry and, 230
  correspondence with Pascal, 242
  work on infinitesimal analysis, 231
Fibonacci. *See* Leonardo of Pisa
Fibonacci sequence, 208
Field's Medal, 285
Fivefold view of universe (Chinese), 111, 116, 119
  five elements and, 119
Fractions, 36–37, 40, 41, 103, 162, 208
  Bhāskara II and, 162
  in Egyptian mathematics, 36–37, 40, 41
  in Chinese mathematics, 103
  Leonardo of Pisa and, 208
French Revolution, xi–xii, 269–271
Fresnel, Augustin, 282
Fibonacci. *See* Leonardo of Pisa

Galilei, Galileo, 221, 228, 236–239, 242, 245, 246, 247, 308, 316
  Assayer, 237–239
  experimental method and, 228, 236–237
  heliocentrism, 236, 238
  mathematical studies of motion, 228, 237–238
  Newton and, 247
  skepticism of, 237
  trial of, 236, 238
  Two New Sciences, 238
  views of matter, 238–239

Galton, Francis, 307
Ganita, 154, 155, 156, 161
Gauss, Carl Freidrich, 263, 290, 297
  correspondence with Sophie Germain, 290, 297
  differential geometry and, 263
  precursor of non-Euclidean geometry, 263
Geocentrism, 47–50, 91–94, 154–155, 155–157, 158, 160
  in Arabic culture, 160
  Aristotle and, 47, 49–50
  Eudoxus and, 47–50
  in Greece, 47–50
  in Hellenistic culture, 91–94;
    possible influence on India, 158
  in India, 154–155, 155–157, 158, 160; influence on Arabic culture, 160
Geological Time, 260, 265–267
Geometry, 3, 19, 25–28, 42–44, 47–50, 52–53, 55, 56–58, 60–62, 62–63, 64–65, 96, 103, 120–121, 148, 150, 152, 153–155, 155–157, 158, 162–174, 176, 177, 181, 215–217, 230, 239–240, 263
  Descartes and, 230, 239–240
  Fermat and, 230
  Gauss and, 263
  in ancient Egypt, 25–28, 42–44, 57–58
  in ancient Greece, 47–50, 52–53, 55, 56–58, 60–62, 62–63, 64–65, 96
  in Arabic culture, 174, 176, 181
  in imperial China, 103, 120–121
  in Hellenistic culture, 158
  in Islamic religious art, 177
  in Indian Sanskrit tradition, 148, 150, 152, 153–155, 155–157, 162
  in Mesopotamia, 3, 19
  non-Euclidean. *See* Non-Euclidean geometry
  Renaissance perspective and, 215–217
  Thales and, 56–58

Gerard de Cremona, 193, 194–196, 200, 233
  mathematical translations, 194, 200
  translation of *Almagest*, 194, 195, 196, 200
Gerbert de Aurillac Sylvester II, 160, 193, 199
  Indo-Arabic numbers and, 160, 199
Germain, Sophie, 290, 296–297
  correspondence with Gauss, 290, 297
  use of male name, 290, 296–297
  work, 297
Gilgamesh, Epic of, 5–6, 7
Gola, 154, 155
Great Library of Alexandria, 75, 80–82, 83–84
  destruction of, 83–84
  founding of, 80
Great Pyramid of Khufu, 28, 29–30

Han Dynasty, 102, 111, 116, 119
  Chinese numeration and, 112
  Nine Chapters and, 119
  wuxing and, 116
Harrison, John, 326
  H4 and H5 clocks, 326
Heliacal rising, 45
Heliocentrism, 89, 156, 222–224, 227–228, 233, 235
  Aristarchus and, 89
  Āryabhaṭa and, 156
  Copernicus and, 227–228, 233, 235
  Galileo and, 236, 238
  Kepler and, 222–224, 228
Hellenistic culture, 79–81, 84
  decline of, 84
Hellenistic Greek language, 78–82
Herodotus, 31–32, 42–43, 44, 56, 58
  Pythagoras and, 58
  Thales and, 56
Herschel, William, 260
Hieratic numeration, 35–36, 38
Hieroglyphics, 33–37, 78–79
  numeration and, 33–37
  Rosetta Stone and, 78–79

Hindu cosmology, 154
Hinduism, 149–150
  Hindu *pramāṇas*, 150
  links to mathematics, 150
Hipparchus of Nicaea, 82, 83, 91, 158
  astronomical work of, 83, 91
  influences on, 83
  trigonometry of, 83, 158
House of Wisdom, 173, 174
  al-Ma'mun's patronage of, 173
  library of, 174
Hundred schools of thought, 105, 111
Hutton, James, 266–267
  stress on geological processes, 266
  Theory of the Earth, 266
Hypatia of Alexandria, 94–96, 215, 292–294
  death of, 95–96, 293–294
  Neoplatonism and, 94, 95, 292, 293
  in Raphael's *School of Athens*, 215
  work, 95, 293

Ibn-Qurra, Thābit, 174
  translations of Greek mathematics, 174
Ibn Rush'd, 189
Ibn Sina, 175, 187
  correspondence with al-Biruni, 175
  discussion of Aristotle's cosmology, 187
Indian Numeration, xi, 146, 148, 152, 153, 173, 174, 177–179
  influence on Arabic culture, 159, 168, 173, 174, 177–179
  links to *Brâhmî*, 153
  possible Chinese influence on, 153
Indo-Arabic Numbers, 133, 134, 177–179, 199, 200, 208, 217
  influence on medieval Europe, 199, 200, 208, 217
Indo-Greek states, 158
Industrial Revolution, 257–259, 308
  cloth production and, 258
  culture of precision, 259
  steam engine and, 258

Infinitesimal analysis, 231, 260–262
  links to Eudoxus' method of exhaustion, 231, 261
  role in modern mathematics, 261–262
Inflation, 314–316
Insurance, 249–250
Iridium anomaly, 310
Iron Law of Wages, 259, 269
Irrational numbers, 61–62, 64, 66–67
Isidore of Seville, 193, 205
Islam, 170–171, 177, 178
  geometry and Islamic art, 177
  number and Islamic mysticism, 178–179
Islamic Spain (al-Andalus), 188–189, 194–195, 198
  books available in, 194–195
  libraries in, 188–189, 195
  Umayyad control of, 188
Italian Humanism, 202–203, 226
  cause of skepticism, 226
  challenges to Aristotle, 202, 226
  recovery of Greek philosophy and mathematics, 202, 226
Italian Renaissance, 196, 202–203

Jai Singh II, 160–161
Johnson, Katherine, 302–304
  John Glenn and, 303
  Presidential Medal of Freedom awarded to, 304
  work at NASA, 303–304
Jordanus of Nemore, 201

Kamal, 326
Kepler, Johannes, 221–225, 228, 242, 245, 247
  acceptance of heliocentrism, 222–223, 228
  elliptical orbits and, 222, 245
  influences on, 224, 228
  laws of planetary motion, 221–225
  Newton and, 247
  perfect solids and, 223–224

Kerala School, 148, 149, 163–165
  ideas of Mādhava. *See* Mādhava
  Siddhānta tradition and, 163
  study of infinite series, 163
Khayyam, Omar, 184–185, 242
  arithmetical triangle and, 185, 242
  study of Euclid's fifth postulate, 184–185
Koine. *See* Hellenistic Greek language
Kong fuzi, 106
Kovalevsky, Sonya Corvin-Krukovsky, 291, 298–299
  awarded Prix Bordin, 299
Kukulkan, 129, 132, 139–140

Labor movements, 259, 276–277
Labor Theory of Value, 267–269, 277
  Adam Smith and, 267–268
  David Ricardo and, 269
  Karl Marx and, 277
Lagrange, Joseph-Louis, 256, 262
  calculus-based mechanics, 262
  work on gravitation, 262
Lambert, Heinrich. *See* Euclid's Fifth Postulate
Land measurement, 26–27, 209
  in Egypt, 26–27
  Leonardo of Pisa and, 209
Laozi, 107
Laplace, Pierre-Simon, 256, 262–263
  conversation with Napoleon, 263
  Mécanique celeste, 263
Latitude, 169, 325
  measured with astrolabe, 169
Lavoisier, Antoine, 256, 280
Lawrence, Ernest, 319
  design of calutron, 319–320
Leibniz, Gottfried Wilhelm, 221, 232, 260, 261, 322
  binary numbers and, 322
  invention of calculus, 232, 260, 261
  mathematical notation of, 260
Leonardo da Vinci, 193, 203, 216
  influence on Luca Pacioli, 216

344   Index

Leonardo of Pisa (Fibonacci), 193, 201, 207–209, 213
   approaches to algebra, 208–209
   Arab influences on, 207–209
   Fibonacci sequence and, 208
   Liber Abaci, 208
Libraries
   Alexandria. *See* Great Library of Alexandria
   Assyrian, 5
   Baghdad. *See* House of Wisdom
   Cordoba, 188–189, 199
   European monastic, 198
   Pergamum, 82
   Timbuktu, 190
Li Zhi, 122
Liu Hui. *See Nine Chapters on the Mathematical Art*
Lobachevsky, Nikolai, 273–274
   New Foundations of Geometry, 274
   non-Euclidean geometry and, 273–274
   notion of curved space, 274
Locke, John, 255
Longitude, 326–327
Los Alamos nuclear facility, 320
Lovelace, Ada, 254–255, 321
   computer programming and, 254, 321

Machine tools, 258–259
   John Wilkinson and, 259
   role in industry, 259
   steam engine and, 258–259
Mādhava, 164
   followers in Kerala school, 164
   founding of Kerala school, 164
   trigonometry, 164
   work on infinite series, 164
Mali, 190
Malthus, Thomas, 268
   Essay on the Principle of Population, 268
   study of population growth, 268
Manhattan Project, 319–320

Manuscripts, 127–129, 132–133, 173, 174, 186, 189–190, 194–196, 198–201, 202
   Arab translation of Greek manuscripts, 173, 174, 186
   European translation of Arabic manuscripts, 194–196, 198–201
   Italian Renaissance recovery of Greek works, 202
   Maya manuscripts, 127–129, 132–133; destruction of, 132
   Timbuktu manuscripts, 189–190
Maragha observatory, 176, 187–188, 234
   leadership of al-Tusi, 187–188
   Mongol patronage of, 187
   responses to Ptolemy, 187–188, 234
Marxism, 275–277
   Adam Smith and, 275
   concept of surplus value, 277
   Das Kapital, 276
   labor theory of value and, 277
   links to Enlightenment natural law, 275
   notion of historical materialism, 276
Mass extinction, 310–312
Mathematical Classic of Master Sun, 112
   text, 124
Maxwell, James Clerk, 260, 282–283
   electromagnetic fields described by differential equations, 283
   scientific influence of, 283
   Treatise on Electricity and Magnetism, 283
Maya calendars, 129, 131, 133, 136–137, 137–139
   calendar round, 138, 140
   haab', 131, 135, 137–138, 140
   Long Count, 131, 136–137
   tzolk'in, 131, 137–138
Maya Civilization, 127, 130, 132
   city-states in, 130
   decline of, 132
   geographical extent of, 130
   periods of, 127, 130
   population, 130
Maya kingship, 130–132, 143

Maya Numeration, x, 133–136, 136–137
Maya view of time, x, 131, 136, 137–139, 140
Mean speed theorem. *See* Merton College (Oxford)
Méchain, Pierre François, xii, 271
Mendeleev, Dmitri, 281–282
Mercator, Gerardus, 229
Merton College (Oxford), 201–202, 210
   studies of uniform acceleration, 210
   mean speed theorem, 210
Metonic cycle, 21
Metric system, xi–xii, 271
   nature and measurement, xi, 271
   French revolution and, xi, 271
   role of Delambre, xii, 271
   role of Méchain, xii, 271
Meyer, Lothar, 281
Middle class culture, 260
Miletus, 56
Mirzakhani, Miriam, 285–286
Mohism, 106, 108, 111
   Mo jing, 106
   Mozi and, 106
Monasteries (European), 198
   libraries in, 198
   role in early medieval intellectual life, 198
Monge, Gaspard, 270
Montesquieu (Charles-Louis de Secondat), 256
   Spirit of the Laws, 256
Mouseion, 75, 80–82
   Great Library of Alexandria and, 75, 80–82
Moscow Mathematical Papyrus, 43
Mozarabs, 195–196, 200
Mughal Empire, 161
Muhammad, 170–171

National Advisory Committee for Aeronautics (NACA), 292
   women's role in, 292
National Aeronautics and Space Administration (NASA), 292
   women's role in, 292
Natural law, 237–238, 240, 243–244, 246–248, 256
   Descartes and, 240
   Enlightenment views of, 255–257
   Galileo and, 237–238
   Newton and, 246–248
   Pascal and, 243–244
Needham, Joseph, 106, 116, 118
   Chinese chemistry and, 118
Nemesis hypothesis, 312–313
Neoplatonism, 94, 95, 223, 292, 293
   Hypatia of Alexandria and, 94, 95, 292, 293
   influence on Kepler, 223
Neptune, discovery of, 260
Newton, Isaac, 221, 229, 245–248, 253, 260–261, 265, 308
   approach to experimental method, 229, 245, 246, 247
   discovery of calculus, 245, 260–261
   laws of motion, 226, 229, 246–247
   law of universal gravitation, 247–248
   Mathematical Principles of Natural Philosophy (Principia), 245, 255, 256; Rules of Reasoning, 246; text, 251–252
   method of divided differences, 253
   stress on quantitative measurement, 246
   views of religion and creation, 247–248, 265
Newtonianism, 227, 255, 256, 257, 282
   Voltaire and, 255–256, 257
Nilakantha, 164
   astronomical innovations, 164
   work on infinite series, 164
Nile River, 28, 30–32, 45
   flooding of the Nile, 30–32, 45
Nine Chapters on the Mathematical Art, 104, 111, 119–121
   algebraic problem solving and, 121
   geometry and, 120–121

346 Index

Liu Hui and, 111, 119–121; Chinese cosmology and, 119–120; value of pi and, 120–121
use of Pythagorean theorem, 121
Noether, Emmy, 299–301
  invariant analysis and, 300
  emigration from Nazi Germany, 300–301
Non-Euclidean geometry, 185, 263, 271–275
  Gauss and, 263
  ideas of Riemann, 274–275
  impact of, 275
  Lambert and, 273
  link to Euclid's Fifth Postulate, 272–273
  Lobachevsky and, 273–274
  relationship to Omar Khayyam, 185
Norden bombsight, 317
Nuclear fission, 316
Numerology, 59–60, 111, 129, 131, 136, 137, 138, 140
  in Maya culture, 129, 131, 136, 137, 138, 140
  in Pythagoreanism, 59–60
  in *wuxing*, 111

Oak Ridge, Tennessee, 319
  Y12 isotope separation plant in, 319
Olmecs, 130
Oresme, Nicole, 193, 201–202, 210–212
  graphic representation of motion, 210–211
Orthogonal lines, 216–217

Pacioli, Luca, 203, 214, 216
  Divina Proportione and Renaissance perspective, 216
  treatment of double-entry accounting, 214
Palermo Stone, 28, 30–32
Pāṇini, 145, 149, 151
  relationship with Indian numbers, 151

Papermaking, 172
  origins in China, 172
  importance to Arabic intellectual culture, 172
Parallax, 91, 235
  effect on Ptolemy's *Almagest*, 91
  influence on Copernicus, 235
Pascal, Blaise, 221, 231, 229, 242–244, 246
  arithmetical triangle and, 231, 242–243
  correspondence with Fermat, 242
  experimental method and, 243
  mystical experience of, 244
  "problem of points" and, 242
  rejection of Cartesian certainty, 244
  treatment of conics, 242
  work on probability theory, 231, 242
Pascal's triangle. *See* Arithmetical triangle
Pauling, Linus, 308
Pearson, Karl, 307
Perfect Solids, 62–63, 66, 223–225
  Kepler and, 223–225
  Plato and, 62–63
  Theaetetus and, 66
Periodic Table of the Elements, 260, 279–282, 308
  concept of atomic weight, 280, 281
  role of atomic number, 280
  work of Meyer and Mendeleev, 281–282
Persian culture in the Middle Ages, 175
Perspective in painting, 215–217
Pi (π), 87, 120–121, 157, 164
  Archimedes' value, 87
  Āryabhaṭa's value, 157
  Liu Hui's value, 120–121
  Mādhava's work, 164
Piscopia, Elena, 289
Place value, 12–13, 16–18, 19, 21, 104, 114–116, 133–136, 147, 148, 153, 160, 174, 177–180, 199, 200, 208
  in Arabic culture, 177–179

in Chinese numbers, 104, 114–116, 153
in Indian numbers, 147, 148, 153, 177; influence on al-Khwārizmī, 160, 174, 179–180; links to Sanskrit, 153
in Maya numbers, 133–136
in medieval Europe, 199, 200, 208
in Mesopotamian numbers, 16–18, 19, 21
Plato, 47, 48–49, 50–51, 52–53, 54, 58, 62–63, 66–67, 67–74, 203, 215, 226–227
  Academy of, 54, 64
  availability in early modern Europe, 226–227
  Circular cosmology and, 48–49, 91
  Empedocles and, 62
  four elements and, 62–63
  geometry and, 49, 52–53, 54
  Hellenistic reaction to, 79
  Italian Humanist art and, 203, 215
  Meno of, 52–53; text, 67–74
  perfect solids and, 62–63
  Pythagoreans and, 53–54, 58, 62
  Republic of, 53
  Timaeus of, 53, 62–63
  Theaetetus of, 66–67
Plimpton 322, 9, 10, 17
Plutonium, 319–320
Polynesian navigation, 325
Polytheism, 130, 131, 139–140, 141–143, 149
  Babylonian, 130
  Egyptian, 130
  Greek, 130
  Indian, 149
  Maya, 130, 131, 139–140, 141–143
Popol Vuh, 137
Printing, 123, 226
  in China, 123
  in early modern Europe, 226
Probability theory, 231
  Pascal and, 231

Ptolemaic dynasty, 75, 76, 78–81, 83
Ptolemy, Claudius, 91–94, 158, 223, 233–234, 325
  Almagest of, 91–94, 95, 158, 160, 174, 186–187, 194–196, 223, 227–228, 233–234, 293, 326; al-Ma'mun and, 186; basis of al-Tusi's work, 187; challenged by al-Haytham (Alhazen), 186–187; circular astronomy and, 91, 92, 186, 233–234; Copernicus and, 227–228, 233–234; degree of accuracy, 223; effect on medieval universities, 207; Hypatia of Alexandria and, 95, 293; influence on Arabic culture, 160, 174, 186; translated by Gerard de Cremona, 194–196; Theon of Alexandria and, 95, 293
  Geography of, 325
  influence on Renaissance painting, 215
  influences on, 91, 93, 94
  legacy of, 94, 158, 186
  use of eccentric orbits, 92, 234
  use of planetary epicycles, 93, 233
  views on vision, 182
Ptolemy I Soter, 80
Puranas, 154
Pyramids, 43–44
  Egyptian calculation and, 43–44
  See also El Castillo; Great Pyramid of Khufu
Pythagoras of Samos, 3, 47, 56, 58–62
  travels in Egypt, 56
Pythagorean theorem, 3, 9–10, 21, 60–61, 86, 121, 157, 162
  in China, 121
  in Euclid's *Elements*, 86
  in Greek culture, 60–61
  in India, 157, 162; Bhāskara II and, 162
  in Mesopotamia, 3, 9–10, 21

Pythagoreans, 53–54, 58–62, 67, 223, 226
  availability in early modern Europe, 226
  influence on Kepler, 223
  influence on Plato, 53–54, 58
  irrational numbers and, 61–62
  mysticism and, 59
  reincarnation and, 58
  views of harmony and harmonics, 58–61

Qin Dynasty, 101, 111
Qin Jiushao, 122
  Shushu jiuzhang, 122
Qin Shi Huang, 111
  Book burning and, 111, 120
Quadrant, 325, 326
Quadrivium, 206–207

Radar, 318–319
Raphael, 50–51, 193, 214–217
  School of Athens, 50–51, 214–217
Rational numbers. See Fractions
Raup, David, 312
Retrograde motion of planets, 92
Rhind Mathematical Papyrus, 25–28, 32, 40–42, 43, 44, 103
  area of a circle and, 25–28
  geometry and, 25–28, 42, 43
  mathematical tables in, 40, 41
  pefsu problems in, 41
  problem types in, 32, 40–44
  seqed problems in , 43–44
  Suàn shù shū and, 103
Ricardo, David, 257, 259, 268–269
  Iron Law of Wages, 259, 269
  labor theory of value and, 269
  Principles of Political Economy and Taxation, 257, 268
  stress on free trade, 269
Riemann, G.F.B., 274–275, 285
  notion of multiple geometries, 274
Rig-Veda, 145, 149–150
  decimal numbers and, 150
Robert of Chester, 196

Rome, 83–84, 196–197
  conquest of Hellenistic kingdoms, 83
  early medieval population decrease, 196
  fall of Roman Empire, 196;
    intellectual consequences of, 196, 197
  patronage of Alexandrian scholarship, 84
Rosetta Stone, 78–79
Rousseau, 256
  Discourse on the Origin of Inequality, 256

Saccheri, Girolamo. See Euclid's Fifth Postulate
Sanskrit, 145, 147, 148–149
  Classical, 149
  mathematical texts, 145, 147, 148
  poetry, x
  Vedic Sanskrit, 149
  Vedic texts, 149
Scientific Revolution, 221, 226, 227, 250
Sepkoski, Jack, 312
Serapeum, 81, 84, 85
  closure, 85
  library of, 81, 84
Seven Liberal Arts, 199, 205–207
  Cassiodorus and, 205
  in cathedral schools, 199, 207
  enhanced by transmission of ancient and Arabic texts, 207
  Martianus Capella and, 205–206
  in medieval European universities, 200, 207
  in monastic schools, 206
Sexagesimal numbers, x, 2, 8, 10, 12–16, 18, 90, 209, 325
  Sumerian System, x, 2, 8, 10, 12–16, 18
  influence of, 21, 90, 325
  Leonardo of Pisa and, 209
Shujing, 103
  text, 105

Siddhānta texts, 149, 151–152, 154–155, 155–157, 157–159, 161, 163, 173, 177, 186
    Āryabhaṭa and, 152
    Bhāskara I and, 152
    Bhāskara II and, 161, 163
    Brahmagupta and, 152
    calendars and, 151
    circular geometry and, 152, 154, 156, 158–159
    Hellenistic influence on, 152, 157–159
    Indian numbers and, 151
    positional astronomy and, 151
    Sanskrit poetry and, 151–152
    transmission to Arab world, 152, 159–161, 173, 177, 186
    trigonometry of sines and 152, 154–155, 158–159
Sindhind, 159–160, 173, 177, 181, 186
    commentary by al-Khwārizmī, 181, 186
Sine function, 148, 154–155, 156, 158–159, 160
    Āryabhaṭa and, 156
    Indian origins of, 155
    influence on Arabic mathematics, 160
    use in Indian astronomy, 148, 154–155, 156, 158–159
Sirius
    in Egyptian calendar, 32, 45
Skepticism in early modern Europe, 226, 237, 239
    challenges to Aristotelian thought, 226
    Descartes and, 239
    exploration and, 226
    Galileo and, 237
    Humanist confidence and, 226
    printing and, 226
Smith, Adam, 257, 267–269
    division of labor and, 267
    labor theory of value, 267
    natural law in economics, 268
    The Wealth of Nations, 257, 267–268

Snow, John, 277–279
    data on cholera incidence, 278
    epidemiology and, 279
    statistical map of disease, 278
Social sciences, 256
Socrates, 52–53, 54, 67–74
Solar declination table, 326
Song Dynasty, 111, 121–123
Spinoza, Baruch, 221, 228, 271
    emphasis on certainty, 228
    influence of Euclid on, 228, 271
Statistical analysis, 307
Steam engine, 258
    James Watt's design, 258–259
    operation of, 258–259
Stock markets, 248–249
Suàn shù shū, 102–103, 104
    comparison with Rhind Papyrus, 103
Śulba-sūtras, 150
    geometry in, 150
Sumerian Counting Tokens, 8, 10–12, 18
Sumerians, 1, 2, 4, 5, 7, 8, 12–16
Sun-centered cosmology. *See* Heliocentricism
Supply and demand curves, 269
Symbolic logic. *See* Boole, George
Szilard, Leo, 319

Tang Dynasty, 101, 111
    Civil service exams and, 111
Tao Te Ching. *See* Dao De Jing
Thales of Miletus, 47, 56–58
    astronomy of, 56
    law of triangles, 56–57
    measurement of pyramids and, 57–58
    travels in Egypt, 56–58
Theaetetus, 54, 66
Theaetetus (of Plato), 66–67
    irrational numbers and, 66–67
Third Dynasty of Ur, 7
Timaeus (of Plato), 62–64
    perfect solids and, 62–64

Timbuktu, 189–190
  commerce with Arabs, 190
  manuscripts of, 189–190
  Sankoré Madrasah, 190
Toledo, 189, 196, 200
  Christian conquest of, 189, 196, 200
  Gerard de Cremona and, 196
  site of translation activity, 189, 200
Trigonometry, 83, 122, 148, 154–155, 156, 158–159, 160, 164, 174, 175, 176, 262
  al-Tusi and, 176
  in Arabic culture, 174
  al-Wafā and, 175
  Āryabhaṭa and, 156
  Euler and, 262
  Hipparchus of Nicaea and, 83, 158
  in Indian mathematics, 148, 154–155, 156, 158–159, 164
  Indian transmission to Arabs, 160
  Li Zhi and, 122
  Mādhava's work, 164
Turing, Alan, 322–323

Umayyad Dynasty, 167, 171
Universities in medieval Europe, 200, 201–202, 210
Upanishads, 145, 149
Uranium, 316, 319–320
Utilitarianism, 257

Vaughn, Dorothy, 301–302
  Congressional Gold Medal awarded to, 302
  work at NASA, 302
Venus, 20, 129, 131–132, 142–143
  in Babylonian astronomy, 20
  in Maya astronomy, 129, 131–132, 142–143
  synodic period of Venus, 129, 131, 142
Vigesimal numbers
  in Maya calendars, 136–137
  in Maya culture, x, 133–136
Vitruvian Man, 203
Volta, Alessandro, 282
Voltaire, 255–256, 264, 294–295
Voyager spacecraft, ix

Watt, James. *See* Steam engine
Wilkinson, John. *See* Machine tools
Wollstonecraft, Mary, 290–291
  female education and, 290
  Vindication of the Rights of Women, 290
Women's colleges, 290
Women's voting rights, 291
World wars, effect on women's roles, 291
Woodblock printing, 124
Woolf, Virginia, 286–287, 292
Wren, Christopher, 229
Wuxing (Chinese five element theory), 111, 116–119
  Daoism and, 118–119
  origins of, 118
  transformations in, 117
  Yijing and, 118–119
  Yin Yang and, 118–119

Yajur Veda, 149, 150
  devotions to number, 150
Yale Tablet, 2–4, 9, 17, 19
Yang Hui, 123
Yellow Emperor, 119–120
Yijing, 104, 108–110, 111, 119, 120
  hexagrams, 108–110, 111, 119, 120
  King Wen sequence, 109
  trigrams, 108–109, 111, 119, 120
Yin-Yang duality, 108–110, 111
Yuan (Mongol) Dynasty, 121–123

Zero, 17–18, 114, 122, 133–136, 146, 148, 152, 153, 163, 177, 178
  in Arabic mathematics, 177, 178
  in Chinese mathematics, 114, 122
  in Indian mathematics, 146, 148, 153, 177; use by Brahmagupta, 152; Bhāskara II and, 163
  in Maya mathematics, 133–136
  in Mesopotamian mathematics, 17–18
Zhu Shijie, 123
  Jade Mirror of the Four Origins, 123
Zou Yan, 111, 118–119

About the Author

Robert Kiely, PhD, teaches the history of ideas in the Liberal Arts Department of the School of the Art Institute of Chicago. He earned his doctorate in the history of ideas and the history of science at Northwestern University.